Hugo Hens

Applied Building Physics

Boundary Conditions,
Building Performance and
Material Properties

Hugo Hens

Applied Building Physics

Boundary Conditions,
Building Performance and
Material Properties

Prof. Hugo S. L. C. Hens
K. U. Leuven
Department of Civil Engineering
Building Physics
Kasteelpark Arenberg 40
3001 Leuven
Belgium

Coverphoto: Capricornhaus, Düsseldorf, Germany; Capricorn Development GmbH & Co. KG, Mönchengladbach, Germany.
Architects: Germann + Schossig, Köln, Germany, © Schüco International KG, Bielefeld, Germany

Bibliographic information published by the Deutsche Nationalbibliothek
The Deutsche Nationalbibliothek lists this publication in the Deutsche Nationalbibliographie; detailed bibliographic data are available on the Internet at http://dnb.d-nb.de.

Library of Congress Card No.: applied for

British Library Cataloguing-in-Publication Data
A catalogue record for this book is available from the British Library.

ISBN: 978-3-433-02962-6

© 2011 Wilhelm Ernst & Sohn
Verlag für Architektur und technische Wissenschaften GmbH & Co. KG, Rotherstr. 21, 10245 Berlin, Germany

All rights reserved (including this of translation into other languages). No part of this book may be reproduced in any form – by photoprint, microfilm, or any other means – nor transmitted or translated into a machine language without written permission from the publisher.
Registered names, trademarks, etc. used in this book, even when not specifically marked as such, are not to be considered unprotected by law.

Cover: Design pur, Berlin
Typesetting: Manuela Treindl, Fürth/Bayern
Printing and Binding: betz-druck GmbH, Darmstadt

Printed in the Federal Republic of Germany.
Printed on acid-free paper.

To my wife, children and grandchildren

*In remembrance of Professor A. de Grave
who introduced Building Physics as a new discipline
at the University of Louvain, K. U. Leuven, Belgium in 1952*

Contents

	Preface	XIII
0	**Introduction**	1
0.1	Subject of the book	1
0.2	Building Physics and Applied Building Physics	1
0.3	Units and symbols	2
0.4	References	5
1	**Outdoor and indoor conditions**	7
1.1	Overview	7
1.2	Outdoor conditions	7
1.2.1	Dry bulb (or air) temperature	8
1.2.2	Solar radiation	12
1.2.2.1	Beam insolation	14
1.2.2.2	Diffuse insolation	16
1.2.2.3	Reflected insolation	17
1.2.2.4	Total insolation	18
1.2.3	Long wave radiation	19
1.2.4	Relative humidity and (partial water) vapour pressure	22
1.2.5	Wind	24
1.2.5.1	Wind speed	24
1.2.5.2	Wind pressure	26
1.2.6	Precipitation and wind-driven rain	27
1.2.6.1	Precipitation	27
1.2.6.2	Wind driven rain	30
1.2.7	Standardized outside climate values	32
1.2.7.1	Design temperature	32
1.2.7.2	Thermal reference year	32
1.2.7.3	Moisture reference year	34
1.2.7.4	Equivalent outside temperature for condensation and drying	36
1.2.7.5	Very hot summer day, very cold winter day	39
1.3	Indoor conditions	39
1.3.1	Dry bulb (or air) temperature	39
1.3.1.1	In general	39
1.3.1.2	Measured data	40
1.3.2	Relative humidity and (partial water) vapour pressure	42
1.3.2.1	In general	42
1.3.2.2	Measured data	42
1.3.3	Water vapour release indoors	46
1.3.4	Indoor climate classes	47
1.3.5	Inside/outside air pressure differentials	49

Applied Building Physics: Boundary Conditions, Building Performance and Material Properties. Hugo Hens
Copyright © 2011 Wilhelm Ernst & Sohn, Berlin
ISBN: 978-3-433-02962-6

1.4	References.	49
2	**Performance metrics and arrays**	**51**
2.1	Definitions	51
2.2	Functional demands	51
2.3	Performance requirements	51
2.4	Some history	52
2.5	Performance arrays	53
2.5.1	Overview	53
2.5.2	In detail	55
2.5.2.1	Functionality	55
2.5.2.2	Structural adequacy	55
2.5.2.3	Building physics related quality	55
2.5.2.4	Fire safety	55
2.5.2.5	Durability	57
2.5.2.6	Maintenance	57
2.5.2.7	Costs	57
2.6	References	57
3	**Functional requirements and performances at the building level**	**59**
3.1	Thermal comfort	59
3.1.1	In general	59
3.1.2	Physiological basis	59
3.1.3	Global steady state thermal comfort	62
3.1.3.1	Clothing	62
3.1.3.2	Heat flow between the body and the environment	63
3.1.3.3	Comfort equations	64
3.1.3.4	Comfort parameters	65
3.1.3.5	Equivalent environments and comfort temperatures	65
3.1.3.6	Comfort appreciation	67
3.1.4	Local discomfort	71
3.1.4.1	Draft	71
3.1.4.2	Vertical air temperature difference	71
3.1.4.3	Radiant temperature asymmetry	72
3.1.4.4	Floor temperature	73
3.1.5	Transient conditions	75
3.1.6	Comfort-related enclosure performance	75
3.2	Health and indoor environmental quality	76
3.2.1	In general	76
3.2.2	Health	77
3.2.3	Definitions	77
3.2.4	Relation between pollution outdoors and indoors	78
3.2.5	Physical, chemical and biological contaminants	79
3.2.5.1	Process related	79
3.2.5.2	Material related	80

3.2.5.3	Ground related	82
3.2.5.4	Combustion related	83
3.2.6	Bio-germs	84
3.2.6.1	Bacteria	84
3.2.6.2	Mould	84
3.2.6.3	Dust mites	86
3.2.6.4	Insects	87
3.2.6.5	Rodents	88
3.2.6.6	Pets	88
3.2.7	Human related contaminants	88
3.2.7.1	Carbon dioxide (CO_2)	88
3.2.7.2	Water vapour	89
3.2.7.3	Bio-odours	89
3.2.7.4	Environmental tobacco smoke	89
3.2.8	Perceived indoor air quality	91
3.2.8.1	Odour	91
3.2.8.2	Inside air enthalpy	92
3.2.9	Sick building syndrome	93
3.2.10	Contaminant control by ventilation	94
3.2.10.1	Ventilation effectiveness	94
3.2.10.2	Ventilation needs	94
3.2.11	Ventilation requirements	99
3.3	Energy efficiency	100
3.3.1	In general	100
3.3.2	Some statistics	101
3.3.3	End energy use in buildings	102
3.3.3.1	Lighting and appliances (function)	103
3.3.3.2	Domestic hot water	104
3.3.3.3	Space heating, space cooling and air conditioning	105
3.3.4	Space heating	105
3.3.4.1	Terminology	105
3.3.4.2	Steady state energy balance for heating at zone level	106
3.3.4.3	Steady state energy balance for heating at building level	112
3.3.5	Parameters defining the net energy demand for heating in residential buildings	113
3.3.5.1	Overview	113
3.3.5.2	In detail	114
3.3.6	Parameters defining the net energy demand for cooling in residential buildings	129
3.3.7	Gross energy demand, end energy use	131
3.3.8	Residential buildings ranked in terms of energy efficiency	131
3.3.8.1	Insulated buildings	131
3.3.8.2	Energy efficient buildings	132
3.3.8.3	Low energy buildings	132
3.3.8.4	Passive buildings	132
3.3.8.5	Zero energy buildings	133
3.3.8.6	Energy plus buildings	133
3.3.8.7	Energy autarkic buildings	133
3.4	Durability	133
3.4.1	In general	133

3.4.2	Loads.	134
3.4.3	Damage patterns	135
3.4.3.1	Decrease in thermal quality	135
3.4.3.2	Decrease in strength and stiffness.	136
3.4.3.3	Stress, strain, deformation and cracking.	137
3.4.3.4	Biological degradation	139
3.4.3.5	Frost damage.	142
3.4.3.6	Salt attack	145
3.4.3.7	Chemical attack	149
3.4.3.8	Corrosion	151
3.5	Life cycle costs	152
3.5.1	In general	152
3.5.2	Total and net life cycle cost	152
3.5.3	Examples.	153
3.5.3.1	Optimal insulation thickness.	153
3.5.3.2	Whole building optimum from an energy perspective	155
3.6	Sustainability	161
3.6.1	In general	161
3.6.2	Life cycle analysis	162
3.6.2.1	Definition	162
3.6.2.2	Impacts	163
3.6.2.3	Total energy use	163
3.6.2.4	Total environmental load	164
3.6.2.5	Recycling	164
3.6.3	High performance buildings	165
3.7	References.	168
4	**Heat-air-moisture performances at the envelope level**	**175**
4.1	Introduction.	175
4.2	Air-tightness	175
4.2.1	Flow patterns	175
4.2.2	Performance requirements	177
4.2.2.1	Air in and outflow.	177
4.2.2.2	Inside air washing, wind washing and air looping	177
4.3	Thermal transmittance (U)	179
4.3.1	Definitions	179
4.3.1.1	Envelope parts	179
4.3.1.2	Envelope.	179
4.3.2	Basis for performance requirements.	180
4.3.2.1	Envelope parts	180
4.3.2.2	Envelope.	180
4.3.3	Examples of performance requirements.	180
4.3.3.1	Envelope parts	180
4.3.3.2	Envelope.	180
4.4	Transient thermal response.	184

4.4.1	Properties of importance	184
4.4.1.1	Opaque envelope parts	184
4.4.1.2	Transparent envelope parts	185
4.4.2	Performance requirements	185
4.4.2.1	Opaque envelope parts	185
4.4.2.2	Transparent envelope parts	185
4.4.3	Consequences for the building fabric	186
4.4.3.1	Opaque envelope parts	186
4.4.3.2	Transparent envelope parts	187
4.5	Moisture tolerance	188
4.5.1	In general	188
4.5.2	Built-in moisture	188
4.5.2.1	Definition	188
4.5.2.2	Requirements	188
4.5.2.3	Consequences for the building fabric	189
4.5.3	Rain and rain penetration	191
4.5.3.1	Definition	191
4.5.3.2	Requirements	192
4.5.3.3	Modelling	192
4.5.3.4	Consequences for the envelope	194
4.5.4	Rising damp	197
4.5.4.1	Definition	197
4.5.4.2	Requirements	197
4.5.4.3	Modelling	197
4.5.4.4	Consequences for the building fabric	200
4.5.5	Water heads	201
4.5.5.1	Definition	201
4.5.5.2	Requirements	202
4.5.5.3	Modelling	202
4.5.5.4	Consequences for the building fabric	202
4.5.6	Hygroscopic moisture	203
4.5.6.1	Definition	203
4.5.6.2	Requirements	204
4.5.6.3	Modelling	204
4.5.6.4	Consequences for the building fabric	204
4.5.7	Surface condensation	204
4.5.7.1	Definition	204
4.5.7.2	Requirements	204
4.5.7.3	Modelling	205
4.5.7.4	Consequences for the building fabric	206
4.5.8	Interstitial condensation	206
4.5.8.1	Definition	206
4.5.8.2	Modelling	207
4.5.8.3	Requirements	211
4.5.8.4	Consequences for the building fabric	211
4.5.8.5	Remark	212
4.5.9	All sources combined	212
4.5.9.1	Modelling	212

4.5.9.2	Requirements	213
4.5.9.3	Where actual models fail	215
4.6	Thermal bridges	220
4.6.1	Definition	220
4.6.2	Requirement	220
4.6.3	Consequences for the envelope	220
4.7	Contact coefficients	221
4.8	Hygrothermal stress and strain	221
4.9	Example of performance control: timber-framed walls	222
4.9.1	Assembly (from inside to outside)	222
4.9.2	Air-tightness	222
4.9.3	Thermal transmittance	223
4.9.4	Transient response	225
4.9.5	Moisture tolerance	226
4.9.5.1	Built-in moisture	226
4.9.5.2	Rain	226
4.9.5.3	Rising damp	226
4.9.5.4	Hygroscopic moisture and surface condensation	226
4.9.5.5	Interstitial condensation	226
4.10	References	233

5 Heat-air-moisture material properties 237

5.1	Introduction	237
5.2	Dry air and water	238
5.3	Building and isolating materials	239
5.3.1	Thermal properties	239
5.3.1.1	Definitions	239
5.3.1.2	List of design values, non-certified materials (ISO 10456)	239
5.3.1.3	Design values according to the standard NBN B 62-002 (edition 2001)	244
5.3.1.4	Measured data	253
5.3.2	Air properties	261
5.3.2.1	Design values	261
5.3.2.2	Measured data	261
5.3.3	Moisture properties	271
5.3.3.1	Design values (ISO 10456)	271
5.3.3.2	Measured data	276
5.3.4	Radiant properties	295
5.4	References	296

Appendix 297

Preface

Overview

Until the first energy crisis of 1973, building physics existed as a shadow field in building engineering, with seemingly limited applicability. While soil mechanics, structural mechanics, building materials, building construction and HVAC were seen as essential, designers only sought advice on room acoustics, moisture tolerance, summer comfort or lighting when really needed or when, after construction, problems arose. Energy was not even a concern, while indoor environmental quality was presumably guaranteed thanks to ever present infiltration, window ventilation and the heating system. The energy crises of the seventies, persisting moisture problems, complaints about sick buildings, thermal, visual and olfactory discomfort, and the move towards more sustainability changed that all. The societal pressure to diminish energy consumptions in buildings without degrading usability acted as a trigger that activated the whole notion of performance based design and construction. As a result, building physics and its potential to quantify performances was suddenly pushed to the forefront of building innovation.

As all engineering sciences, building physics is oriented towards application, which is why, after a first book on fundamentals, this second volume examines the performance rationale and performance requirements as a basis for sound building engineering. Choices have been made, among others to limit the text to a thorough discussion of the heat-air-moisture performances only. The subjects treated are: the climate outdoors and conditions indoors, the performance concept, performances at the building level, performances at the building enclosure level and heat-air-moisture material properties. The book incorporates thirty five years of teaching architectural, building and civil engineers, bolstered by forty years of experience, research and consultancy. Where needed information and literature from international sources has been used, which is indicated by an extensive list with references at the end of each chapter.

The book is written in SI-units. It should be usable for undergraduate and graduate students in architectural and building engineering, although mechanical engineers studying HVAC, and practising building engineers who want to refresh their knowledge, will also benefit from it. The level of presentation assumes that the reader has a sound knowledge of the basics treated in the first book on fundamentals, along with a background in building materials and building construction.

Acknowledgments

A book of this magnitude reflects the work of many, not only that of the author. Therefore, I want first of all to thank my thousands of students. They gave us the opportunity to test the content and helped us in upgrading with the corrections they proposed and the experience they offered in pointing out which parts should be better explained.

The book could not have been written in this form without standing on the shoulders of those who preceded me. Although I started my career as a structural engineer, my predecessor, Professor Antoine de Grave planted the seeds that slowly nurtured my interest in building physics and its applications. The late Bob Vos of TNO, the Netherlands, and Helmut Künzel of the Fraunhofer Institut für Bauphysik, Germany, showed the importance of experimental work and field testing for understanding building performance, while Lars Erik Nevander of Lund University, Sweden, showed that application does not always correspond to extended modelling, mainly because reality in building construction is much more complex than any model may be.

During the four decades at the Laboratory of Building Physics, several researchers and PhD-students got involved. I am very grateful to Gerrit Vermeir, Staf Roels and Dirk Saelens who became colleagues at the university; to Jan Carmeliet, professor at the ETH-Zürich; to Piet Standaert, a principal at Physibel Engineering; to Jan Lecompte at Bekaert NV; Filip Descamps, a principal at Daidalos Engineering and part-time professor at the Free University Brussels (VUB); Arnold Janssens, professor at the University of Ghent (UG); Hans Janssen, associate professor at the Technical University Denmark (TUD); Rongjin Zheng, associate professor at Zhejiang University, China, and Bert Blocken, associate professor at the Technical University Eindhoven (TU/e), who all contributed by their work. The experiences gained by operating four Annexes of the IEA, Executive Committee on Energy Conservation in Buildings and Community Systems, also forced me to rethink the performance approach. The many ideas I exchanged in Canada and the USA with Kumar Kumaran, Paul Fazio, Bill Brown, William B. Rose, Joe Lstiburek and Anton Ten Wolde, were also of great help. A number of reviewers took time to examine the book. Although we do not know their names, we thank them here.

Finally, I want to acknowledge my family. My loving mother who died too early. My late father, who reached a respectable age. My wife, Lieve and my three children who managed living together with a busy engineering professor, and my many grandchildren.

Leuven, May 2010 *Hugo S. L. C. Hens*

0 Introduction

0.1 Subject of the book

This is the second volume in a series of books on Building Physics and Applied Building Physics:

- Building Physics: Heat, Air and Moisture
- **Applied Building Physics: Boundary Conditions, Building Performance and Material Properties**
- Applied Building Physics and Performance Based Design 1
- Applied Building Physics and Performance Based Design 2

In this volume the subjects are: indoor and outdoor climate, the performance concept, performances at the building and building enclosure level and heat-air-moisture material properties. The book thereby functions as a hinge between 'Building Physics: Heat, Air and Moisture' and 'Applied Building Physics and Performance Based Design 1 and 2'. Although acoustics and lighting are not treated in detail, they form an integral part of the performance array and are mentioned where and when needed.

In Chapter 1 outdoor and indoor conditions are described and design and calculation values discussed. Chapter 2 specifies the performance concept and its hierarchical structure, from the urban environment over the building level down to the material's level. Aspects typical for performances are definability in an engineering way, their predictability at the design stage and controllability during decommissioning. In Chapter 3, the main heat, air, moisture related performances at the building level are discussed. Chapter 4 analyzes the hygrothermal performance requirements of importance for a good building enclosure design, while Chapter 5 treats the material properties needed for predicting the heat, air, moisture response of buildings and building enclosures.

A performance approach helps designers, consulting engineers and contractors to guarantee building quality. However physical requirements are not the only track that adds value. Although functionality, spatial quality and aesthetics, i.e. aspects belonging to the architect's responsibility, are of equal importance, they should not become an argument for neglecting the importance of a highly performing building and building services design.

0.2 Building Physics and Applied Building Physics

For the readers who like to know more about the engineering field of 'Building Physics', its importance and history, we refer to 'Building Physics: Heat, Air and Moisture'. Honestly, the term 'applied' may be perceived to be a pleonasm. Building Physics is by definition applied. But by inserting the word, the focus is unequivocally directed towards usage of the knowledge building physics generates, in building and building services design plus building construction.

0.3 Units and symbols

The book uses the SI-system (internationally mandated since 1977). The base units are the meter (m), the kilogram (kg), the second (s), the Kelvin (K), the ampere (A) and the candela. Derived units, which are important when studying building physics and applied building physics, are:

Unit of force: Newton (N); $1\ N = 1\ kg \cdot m \cdot s^{-2}$
Unit of pressure: Pascal (Pa); $1\ Pa = 1\ N/m^2 = 1\ kg \cdot m^{-1} \cdot s^{-2}$
Unit of energy: Joule (J); $1\ J = 1\ N \cdot m = 1\ kg \cdot m^2 \cdot s^{-2}$
Unit of power: Watt (W); $1\ W = 1\ J \cdot s^{-1} = 1\ kg \cdot m^2 \cdot s^{-3}$

For the symbols, the ISO-standards (International Standardization Organization) are followed. If a quantity is not included in these standards, the CIB-W40 recommendations (International Council for Building Research, Studies and Documentation, Working Group 'Heat and Moisture Transfer in Buildings') and the list edited by Annex 24 of the IEA, ECBCS (International Energy Agency, Executive Committee on Energy Conservation in Buildings and Community Systems) are applied.

Table 0.1. List with symbols and quantities.

Symbol	Meaning	Units
a	Acceleration	m/s^2
a	Thermal diffusivity	m^2/s
b	Thermal effusivity	W/(m$^2 \cdot$ K $\cdot$ s$^{0.5}$)
c	Specific heat capacity	J/(kg $\cdot$ K)
c	Concentration	kg/m^3, g/m^3
e	Emissivity	–
f	Specific free energy	J/kg
	Temperature ratio	–
g	Specific free enthalpy	J/kg
g	Acceleration by gravity	m/s^2
g	Mass flow rate, mass flux	kg/(m$^2 \cdot$ s)
h	Height	m
h	Specific enthalpy	J/kg
h	Surface film coefficient for heat transfer	W/(m$^2 \cdot$ K)
k	Mass related permeability (mass may be moisture, air, salt.)	s
l	Length	m
l	Specific enthalpy of evaporation or melting	J/kg
m	Mass	kg
n	Ventilation rate	s^{-1}, h^{-1}
p	Partial pressure	Pa
q	Heat flow rate, heat flux	W/m^2
r	Radius	m
s	Specific entropy	J/(kg $\cdot$ K)

0.3 Units and symbols

Symbol	Meaning	Units
t	Time	s
u	Specific latent energy	J/kg
v	Velocity	m/s
w	Moisture content	kg/m^3
x, y, z	Cartesian co-ordinates	m
A	Water sorption coefficient	kg/(m$^2 \cdot$ s$^{0.5}$)
A	Area	m^2
B	Water penetration coefficient	m/s$^{0.5}$
D	Diffusion coefficient	m^2/s
D	Moisture diffusivity	m^2/s
E	Irradiation	W/m^2
F	Free energy	J
G	Free enthalpy	J
G	Mass flow (mass = vapour, water, air, salt)	kg/s
H	Enthalpy	J
I	Radiation intensity	J/rad
K	Thermal moisture diffusion coefficient	kg/(m $\cdot$ s $\cdot$ K)
K	Mass permeance	s/m
K	Force	N
L	Luminosity	W/m^2
M	Emittance	W/m^2
P	Power	W
P	Thermal permeance	W/(m$^2 \cdot$ K)
P	Total pressure	Pa
Q	Heat	J
R	Thermal resistance	m$^2 \cdot$ K/W
R	Gas constant	J/(kg $\cdot$ K)
S	Entropy, saturation degree	J/K, –
T	Absolute temperature	K
T	Period (of a vibration or a wave)	s, days, etc.
U	Latent energy	J
U	Thermal transmittance	W/(m$^2 \cdot$ K)
V	Volume	m^3
W	Air resistance	m/s
X	Moisture ratio	kg/kg
Z	Diffusion resistance	m/s
α	Thermal expansion coefficient	K^{-1}
α	Absorptivity	–
β	Surface film coefficient for diffusion	s/m
β	Volumetric thermal expansion coefficient	K^{-1}

Symbol	Meaning	Units
η	Dynamic viscosity	$N \cdot s/m^2$
θ	Temperature	°C
λ	Thermal conductivity	$W/(m \cdot K)$
μ	Vapour resistance factor	–
ν	Kinematic viscosity	m^2/s
ρ	Density	kg/m^3
ρ	Reflectivity	–
σ	Surface tension	N/m
τ	Transmissivity	–
ϕ	Relative humidity	–
α, ϕ, Θ	Angle	rad
ξ	Specific moisture capacity	kg/kg per unit of moisture potential
Ψ	Porosity	–
Ψ	Volumetric moisture ratio	m^3/m^3
Φ	Heat flow	W

Table 0.2. List with suffixes and notations.

Symbol	Meaning
Indices	
A	Air
c	Capillary, convection
e	Outside, outdoors
h	Hygroscopic
i	Inside, indoors
cr	Critical
CO_2, SO_2	Chemical symbol for gasses
m	Moisture, maximal
r	Radiant, radiation
sat	Saturation
s	Surface, area, suction
rs	Resulting
v	Water vapour
w	Water
ϕ	Relative humidity
Notation	
[ ], bold	Matrix, array, value of a complex number
dash	Vector (ex.: $\bar{a}$)

0.4 References

[0.1] CIB-W40 (1975). Quantities, Symbols and Units for the description of heat and moisture transfer in Buildings: Conversion factors. IBBC-TNP, Report No. BI-75-59/03.8.12, Rijswijk.

[0.2] ISO-BIN (1985). Standards series X02-101 – X023-113.

[0.3] Kumaran, K. (1996). *Task 3: Material Properties.* Final Report IEA EXCO ECBCS Annex 24. ACCO, Louvain, pp. 135.

1 Outdoor and indoor conditions

1.1 Overview

In building physics, the in- and outdoor conditions have a role comparable to the loads in structural mechanics, reason why the term 'environmental loads' is often used. A good knowledge of that load is essential, if one aims making environmentally correct design decisions.

The different parameters that make up the load are:

Outside		Inside	
Air temperature	θ_e	Air temperature	θ_i
		Radiant temperature	θ_R
Relative humidity	ϕ_e	Relative humidity	ϕ_i
(Partial water) vapour pressure	p_e	(Partial water) vapour pressure	p_i
Solar radiation	E_S		
Under-cooling	q_{rL}		
Wind	v_w	Air speed	v
Rain and snow	g_r		
Air pressure	$P_{a,e}$	Air pressure	$P_{a,i}$

In the next paragraphs, each parameter is discussed separately. One should however keep in mind that in- and outside conditions are coupled. The more decoupling is demanded, the more severe the envelope and HVAC performance requirements become and the more energy will be needed. Predicting future outdoor conditions is also not possible. Not only are most parameters measured in a few locations only, but the future never is an exact copy of the past. The weather in fact does not obey the paradigm 'the more data available, the better the forecast'. Moreover, glooming global warming may disturb any such trial. A typical way of by-passing that problem is using reference values and reference years for each performance evaluation needing outside climate data, such as quantifying the heating and cooling load, predicting end energy consumption, evaluating overheating, judging moisture performance, looking to durability issues others than moisture induced, etc.

In the paragraphs that follow, data from the weatherstation of Uccle, Belgium, 50° 51' North, 4° 21' East, are used to illustrate statements and trends. Reason for that is we disposed of a rich documentation analyzing the measurements done there during the last century.

1.2 Outdoor conditions

Weather patterns are to a large extent defined by the geographical location – northern or southern latitude –, proximity to the sea and height above sea level. But micro-climatic factors also intervene. Thanks to the urban heat island effect, the air temperature in city centres is higher than at the country side whereas air pollution makes solar radiation less intense and relative humidity is lower. Table 1.1 illustrates the differences by listing the monthly mean air temperatures measured under thermometer hut in Uccle and Sint Joost for the period 1901–1930.

Table 1.1. Monthly mean dry bulb temperature in Uccle and Sint-Joost, Brussels, for the period 1901–1930 (°C).

Location	Month											
	J	F	M	A	M	J	J	A	S	O	N	D
Uccle	2.7	3.1	5.5	8.2	12.8	14.9	16.8	16.4	14.0	10.0	5.2	3.7
Sint Joost	3.8	4.2	6.8	9.4	14.6	16.7	18.7	18.0	15.4	11.2	6.4	4.7

Both stations are situated in Brussels but Uccle faces a green area, while 'Sint Joost' lays straight in the centre of the city.

The outside climate varies periodically with as main cycles:

- The year with its succession of winter, springtime, summer and autumn in moderate regions and the wet and dry season in the equatorial band. The annual cycle is governed by the earth's elliptic orbit around the sun
- The sequence of low and high pressure fronts. In moderate and cold climates, high pressure stands for warm weather in summer and cold weather in winter. Low pressure instead gives cool, wet weather in summer and fresh, wet weather in winter
- Day and night. The daily cycle is a consequence of the earth's autorotation.

Reference years and standardised quantities are focusing on annual and daily cycles and daily values. Meteorological reference data instead consider 30 year averages, for the 20–21 century: 1901–1930, 1931–1960, 1961–1990, 1991–2020. Per period the averages vary due to long term climate changes such as induced by global warming, relocation of meteorological stations, more accurate measuring apparatus and the way averages are calculated. Between 1901 en 1930, the daily mean temperature was the average between the daily minimum and the daily maximum, the one logged with a minimum and the other logged with a maximum mercury thermometer. Today, the air temperature in many weather stations is logged each 10′ and the daily average is calculated as the mean of these 144 values.

1.2.1 Dry bulb (or air) temperature

The air temperature plays an important role in the heating and cooling load and the annual end energy use for heating and cooling. The load fixes the HVAC-investment, while the energy consumed defines part of the annual costs. The air temperature is also linked to overheating and participates in the heat, air, and moisture load the envelope experiences. Measurement proceeds in open field under thermometer hut, 1.5 m above ground level. The accuracy as imposed by the WMO (World Meteorological Organization) is $\leq \pm 0.5$ °C. Table 1.2 gives monthly mean values for several weather stations in Europe. All are quite well represented by an annual mean and a first harmonic, although two harmonics give better results:

1.2 Outdoor conditions

Table 1.2. Monthly mean dry bulb temperatures for several locations in Europe (°C).

Location	Month											
	J	F	M	A	M	J	J	A	S	O	N	D
Uccle (B)	2.7	3.1	5.5	8.2	12.8	14.9	16.8	16.4	14.0	10.0	5.2	3.7
Den Bilt (NL)	1.3	2.4	4.3	8.1	12.1	15.3	16.1	16.1	14.2	10.7	5.5	1.2
Aberdeen (UK)	2.5	2.7	4.5	6.8	9.0	12.1	13.7	13.3	11.9	9.3	5.3	3.7
Eskdalemuir (UK)	1.8	1.9	3.9	5.8	8.9	11.8	13.1	12.9	10.9	8.5	4.3	2.7
Kew (UK)	4.7	4.8	6.8	9.0	12.6	15.6	17.5	17.1	14.8	11.6	7.5	5.6
Kiruna (S)	−12.2	−12.4	−8.9	−3.5	2.7	9.2	12.9	10.5	5.1	−1.5	−6.8	−10.1
Malmö (S)	−0.5	−0.7	1.4	6.0	11.0	15.0	17.2	16.7	13.5	8.9	4.9	2.0
Västerås (S)	−4.1	−4.1	−1.4	4.1	10.1	14.6	17.2	15.8	11.3	6.3	1.9	−1.0
Lulea (S)	−11.4	−10.0	−5.6	−0.1	6.1	12.8	15.3	13.6	8.2	2.9	−4.0	−8.9
Oslo (N)	−4.2	−4.1	−0.2	4.6	10.8	15.0	16.5	15.2	10.8	6.1	0.8	−2.6
München (D)	−1.5	−0.4	3.4	8.1	11.9	15.6	17.5	16.7	13.9	8.8	3.6	−0.2
Potsdam (D)	−0.7	−0.3	3.5	8.0	13.1	16.6	18.1	17.5	13.8	9.2	4.1	0.9
Roma (I)	7.6	9.0	11.3	13.9	18.0	22.3	25.2	24.7	21.5	16.8	12.1	8.9
Catania (I)	10.0	10.4	12.0	14.0	18.0	22.0	25.2	25.6	23.2	18.4	15.2	11.6
Torino (I)	1.6	3.5	7.6	10.8	15.4	19.0	22.3	21.6	17.9	12.3	6.2	2.4
Bratislava (Sk)	−2.0	0.0	4.3	9.6	14.2	17.8	19.3	18.9	15.3	10.0	4.2	0.1
Copenhagen (Dk)	−0.7	−0.8	1.8	5.7	11.1	15.1	16.2	16.0	12.7	9.0	4.7	1.1

1 harmonic

$$\theta_e = \overline{\theta}_e + A_1 \sin\left(\frac{2\pi t}{365.25}\right) + B_1 \cos\left(\frac{2\pi t}{365.25}\right) \tag{1.1}$$

2 harmonics

$$\theta_e = \overline{\theta}_e + A_1 \sin\left(\frac{2\pi t}{365.25}\right) + B_1 \cos\left(\frac{2\pi t}{365.25}\right) \\ + A_2 \sin\left(\frac{4\pi t}{365.25}\right) + B_2 \cos\left(\frac{4\pi t}{365.25}\right) \tag{1.2}$$

where $\overline{\theta}_e$ is the annual mean dry bulb temperature. t is time and d the phase shift in days. For three of the locations listed in Table 1.2 one gets (see also Figure 1.1):

	$\overline{\theta}_e$ °C	A_1 °C	B_1 °C	A_2 °C	B_2 °C
Uccle	9.8	−2.3	−6.9	0.45	−0.1
Kiruna	−1.2	−4.2	−11.6	1.2	0.5
Catania	17.2	−4.1	−6.6	0.8	0.2

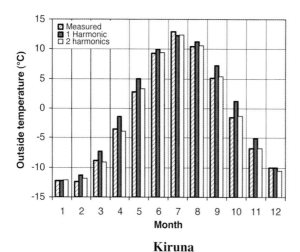

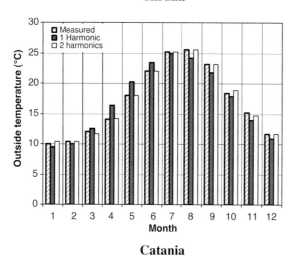

Figure 1.1. Outside air temperature: annual course, one and two harmonics.

For Uccle the average difference between the monthly mean daily minimum and maximum ($\theta_{e,max,day} - \theta_{e,min,day}$) during the period 1931–1960 looked like:

J	F	M	A	M	J	J	A	S	O	N	D
5.6	6.6	7.9	9.3	10.7	10.8	10.6	10.1	9.8	8.0	6.2	5.2

Combination with the annual course gives (time in hours):

$$\theta_e = \bar{\theta}_e + \hat{\theta}_e \cos\left[\frac{2\pi(t-h_1)}{8766}\right]$$
$$+ \frac{1}{2}\left\{\Delta\bar{\theta}_{e,dag} + \Delta\hat{\theta}_{e,dag} \cos\left[\frac{2\pi(t-h_2)}{8766}\right]\right\} \sin\left[\frac{2\pi(t-h_3)}{24}\right] \quad (1.3)$$

1.2 Outdoor conditions

with:

| $\Delta\bar{\theta}_{e,dag}$ | $\Delta\hat{\theta}_{e,dag}$ | h_1 | h_2 | H_2 |
°C	°C	h	h	h
8.4	2.8	456	−42	8

Equation (1.3) assumes the daily values fluctuate harmonically. This is untrue. The difference between the minimum and maximum daily value swings considerably without even a glimpse of a harmonic course. To give an example, the difference for January and July 1973 in Leuven, Belgium, was purely stochastic, with average values of 4.0 °C and 8.9 °C respectively and a standard deviation mounting to 60% of the average in January and 39% of the average in July.

A question of particular importance is if the dry bulb temperature recorded during the last decades reflects global warming. For that, we turned to the data recorded by a weather station in Leuven, Belgium, between 1997 and 2007. The annual means are shown in Figure 1.2.

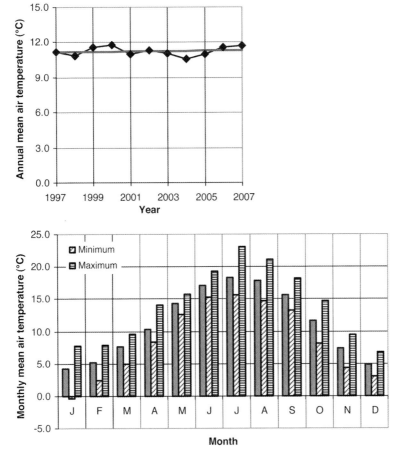

Figure 1.2. Weather station, Leuven, Belgium: annual mean dry bulb temperature (left), average, minimum and maximum monthly mean recorded between 1996 and 2007 (right).

With an average of 11.2 °C and the slope of the least square line close to 0.02 °C, a recorded increase seems hardly present. However, in Uccle, which is 30 km South West of Leuven, the average annual mean between 1901 and 1930 touched 9.8 °C, i.e. 1.4 °C lower than the value measured in Leuven between 1997 and 2007. Between 1952 and 1971, a same 9.8 °C was recorded. Since then, however, the moving 20 years average increased slowly, with the highest values noted between 1988 and 2007.

1.2.2 Solar radiation

Solar radiation means free heat gains. These decrease the end energy needed for heating but increase the end energy needed for cooling. Too much gain may also cause overheating. The sun further increases the outside surface temperatures the irradiated envelope parts experience. Although enhancing drying, these higher temperatures activate solar driven diffusion to the inside of moisture stored in the exterior rain buffering layers. In air-dry outer layers in turn, the higher temperature and accompanying drop in relative humidity aggravates hygrothermal stress and strain.

The sun acts as a 5762 K hot black body at a distance of 150 000 000 km from the earth. That large distance allows considering solar radiation as being emitted by a flat body with the rays in parallel. Above the atmosphere the solar spectrum looks as drawn in Figure 1.3 with a mean irradiation equal to:

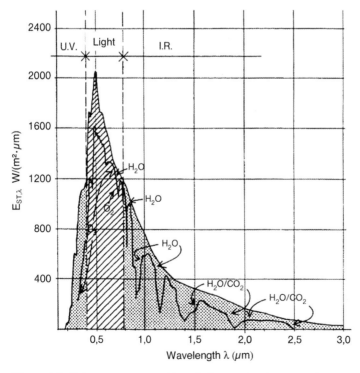

Figure 1.3. Solar spectrums outside and after passing the atmosphere.

1.2 Outdoor conditions

$$E_{ST} = 5.67 \left(\frac{T_S}{100}\right)^4 \left(\frac{r_S}{D_{SE}}\right)^2 = 5.67 \cdot (57.64)^4 \cdot \left(\frac{0.695 \cdot 10^6}{1.496 \cdot 10^8}\right)^2 = 1332 \text{ W/m}^2 \quad (1.4)$$

In that formula r_S is the radius of the sun and D_{SE} the distance between earth and sun, both in km. The value 1332 W/m² is called the average solar constant (E_{ST_0}). It reflects the mean solar radiation per square meter the earth should receive upright the beam if no atmosphere existed. That intensity is quite thin. Burning one litre of fuel for example gives $4.4 \cdot 10^7$ J. To collect that amount of energy a square meter large terrestrial surface upright the beam should need constant solar irradiation during nine hours. That thinness explains why large collecting surfaces are needed when the aim is using solar energy for heat or power production.

A more exact description of the solar constant accounts for the annual variation in distance between earth and sun and the annual cycle in solar activity (with d the number of days beyond December 31/Januari 1, at midnight):

$$E = 1373 \left\{ 1 + 0.03344 \cos\left[\frac{2\pi}{365.25}(d - 2.75)\right] \right\} \text{ (W/m}^2) \quad (1.5)$$

The azimuth (a_s) and solar height (h_s) or the time angle (ω) and declination (δ) which is the angle between the sun's highest position and the equator plane, fix the solar position in the sky, see Figure 1.4. The first two describe the sun's movement as seen locally whereas the second two relate that movement to the equator. The time angle varies from 180° at 0 p.m. over 0° at noon to –180° at 12 a.m. One hour thus corresponds to 15°. The declination in radians is given by:

$$\delta = b\ g \sin\left\{-\sin\left(\frac{\pi}{180} 23.45\right) \cos\left[\frac{2\pi}{365.25}(d + 10)\right]\right\} \quad (1.6)$$

where 23.45 is the latitude in degrees between the equator and the tropics. The maximum solar height in radians follows from:

$$h_s = \frac{\pi}{2} - \varphi + \delta \quad (1.7)$$

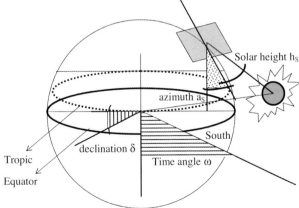

Figure 1.4. Solar angles.

φ being the latitude in radians (positive in the northern, negative in the southern hemisphere).

1.2.2.1 Beam insolation

During its trek through the atmosphere, selective absorption by the ozone, oxygen, hydrogen, CO_2, CH_4 present tempers the radiation and changes its spectrum, while scatter transforms the beams into beam and diffuse radiation. The longer the distance the solar beam has to traverse through the atmosphere, the larger that tempering. She intervenes in the equation for beam insolation through the air factor m: the ratio between the real distance traversed and the distance to sea level if the sun stood in zenith position (Figure 1.5). As an equation (solar height h_s in radians):

$$m = \frac{1 - 0.1 z}{\sin(h_s) + 0.15 \left(\frac{180}{\pi} h_s + 3.885\right)^{-1.253}} \quad (1.8)$$

where z is the height above sea level in km.

Beam insolation on a surface upright the solar rays is then given by:

$$E_{SD,n} = E_{STo} \exp(-m \, d_R \, T_{Atm}) \quad (1.9)$$

with d_R the optic factor (a measure for the scatter of the beam per unit of distance traversed through the atmosphere) en T_{Atm} the turbidity of the atmosphere. The optic factor follows from (solar height h_s in radians):

$$\begin{aligned} d_R = &\, 1.4899 - 2.1099 \cos(h_s) + 0.6322 \cos(2 h_s) + 0.0253 \cos(3 h_s) \\ &- 1.0022 \sin(h_s) + 1.0077 \sin(2 h_s) - 0.2606 \sin(3 h_s) \end{aligned} \quad (1.10)$$

Atmospheric turbidity on a clear day is given by (solar height h_s in radians):

Mean air pollution $T_{Atm} = 3.372 + 3.037 \, h_s - 0.296 \cos(0.5236 \, mo)$
Minimal air pollution $T_{Atm} = 2.730 + 1.549 \, h_s - 0.198 \cos(0.5236 \, mo)$

where mo is the month's ranking (1 for January and 12 for December).

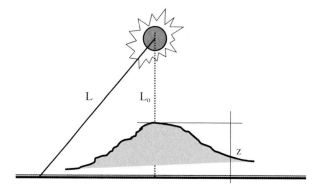

Figure 1.5. L is the atmospheric distance traversed, z the height in km and m the air factor equal to L/L_o.

1.2 Outdoor conditions

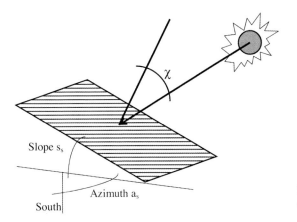

Figure 1.6. Direct radiation on a surface with slope s_s.

Beam insolation on a tilted surface, whose normal forms an angle χ with the solar rays, is calculated as (Figure 1.6):

$$E_{SD,s} = \max\left(0, E_{SD,n} \cos\chi\right) \tag{1.11}$$

with:

$$\begin{aligned}\cos\chi &= \sin\delta \sin\varphi \cos s_s - \sin\delta \cos\varphi \sin s_s \cos a_s \\ &+ \cos\delta \cos\varphi \cos s_s \cos\omega + \cos\delta \sin\varphi \sin s_s \cos a_s \cos\omega \\ &+ \cos\delta \sin s_s \sin a_s \sin\omega\end{aligned} \tag{1.12}$$

where s_s is the slope of the surface (0° for a horizontal surface, 90° ($\pi/2$) for a vertical surface, between 0 and 90° for a tilted surface seeing the sun, between 0 and –90° (–$\pi/2$) for a tilted surface away from the sun), a_s the surface's azimuth (south 0°, east 90°, north 180°, west –90°) and ω the time angle. The zero in Formula (1.11) holds as long as the solar height stays below zero, the height being given by:

$$h_s = a\sin\left(\cos\delta \cos\varphi \cos\omega + \sin\varphi \sin\delta\right) \tag{1.13}$$

For a horizontal surface Equation (1.12) simplifies to:

$$\cos\chi_h = \sin\delta \sin\varphi + \cos\delta \cos\varphi \cos\omega$$

For vertical surfaces one has:

South $\quad \cos\chi_{v,south} = -\sin\delta \cos\varphi + \cos\delta \sin\varphi \cos\omega$
West $\quad \cos\chi_{v,west} = -\cos\delta \sin\omega$
North $\quad \cos\chi_{v,north} = -\sin\delta \cos\varphi - \cos\delta \sin\varphi \cos\omega$
East $\quad \cos\chi_{v,east} = \cos\delta \sin\omega$

In case beam insolation on a horizontal surface is known ($E_{SD,h}$), the value on a tilted surface ($E_{SD,s}$) follows from:

$$E_{SD,s} = \max\left(0, E_{SD,h} \frac{\cos\chi_s}{\cos\chi_h}\right)$$

Beam insolation seems well predictable. However, unknown in most cases is the atmospheric turbidity (T_{Atm}). Cloudiness, air pollution and relative humidity define its value but their real impact is difficult to guess.

1.2.2.2 Diffuse insolation

Scattering of the solar beam results in a diffuse component. Diffuse insolation touches the globe from sunrise to sunset, independent of the type of sky, blue or cloudy. On the earth it is as if radiation comes from all directions. The simplest model considers the sky as a uniformly radiating hemisphere. Any tilted surface, whose slope fixes the angle with the mid-plane, sees a part of that hemisphere. As black body on constant temperature each point on the hemisphere has equal luminosity. That turns the view factor between the tilted surface and part of the hemisphere seen into:

$$F_{s,sk} = \frac{1 + \cos s_s}{2} \tag{1.14}$$

If $E_{Sd,h}$ stands for the diffuse insolation on a horizontal surface, then diffuse insolation on a tilted surface becomes:

$$E_{Sd,s} = E_{Sd,h} \frac{1 + \cos s_s}{2} \tag{1.15}$$

A hemisphere with largest luminosity close to the solar disk, called zenith luminosity, and lowest at the horizon, nears reality more closely. If the position of every point P on the hemisphere is characterized by an azimuth a_P and a height h_P, its luminosity may be written as $L(a_P, h_P)$. The angle Γ between the normal on the surface with slope s_s and the line from the centre of that surface to the point considered is then written as:

$$\cos \Gamma = \sin s_s \cos h_P \cos(a_s - a_P) + \cos s_s \sin h_P$$

Diffuse insolation on a tilted surface so becomes:

$$E_{Sd,s} = K_D \iint_{a_P, h_P} \left(L_{a_P, h_P} \cos h_P \cos \Gamma\right) dh_P \, da_P$$

where $K_D = 1 + 0.03344 \cos[0.017202(d - 2.75)]$, $0 \leq a_P \leq 2\pi$, $0 \leq h_P \leq \pi/2$ and $\cos \Gamma \geq 0$. Luminosity L_{a_P, h_P} in the formula looks like:

$$L_{a_P, h_P} = L_{zenith} \underbrace{\frac{\left[0.91 + 10 \exp(-3\varepsilon) + 0.45 \cos^2 \varepsilon\right]\left\{1 - \exp\left[-0.32 \csc(h_P)\right]\right\}}{0.27385 \left\{0.91 + 10 \exp\left[-3\left(\frac{\pi}{2} - h_S\right)\right] + 0.45 \sin^2 h_S\right\}}}_{f} \tag{1.16}$$

with L_{zenith} the zenith value and ε the angle between the line from the centre of the surface to the point P and, the solar beam ($\cos \varepsilon = \cos h_S \cos h_P \cos(a_S - a_P) + \sin h_S \cos h_P$). Entering that luminosity in the equation for diffuse insolation results in (with the zenith luminosity multiplier from Equation (1.16) written as f):

$$E_{Sd,s} = K_D \, L_{zenit} \iint_{a_P, h_P} \left[f \cos h_P \cos \Gamma\right] dh_P \, da_P$$

1.2 Outdoor conditions

Table 1.3. Uccle, Belgium, multiplier f_{mo} for total monthly diffuse radiation.

Slope	Azimuth								
	0 S	22.5	45	67.5	90 E, W	112.5	135	157.5	180 N
0	1.00	1.00	1.00	1.00	1.00	1.00	1.00	1.00	1.00
22.5	1.03	1.03	1.02	1.01	1.00	0.99	0.98	0.97	0.96
45	1.05	1.04	1.03	1.01	0.99	0.96	0.94	0.92	0.92
67.5	1.06	1.05	1.03	0.99	0.94	0.90	0.86	0.84	0.83
90	1.06	1.04	1.00	0.94	0.87	0.81	0.76	0.73	0.71
112.5	0.98	0.97	0.92	0.85	0.76	0.68	0.63	0.60	0.60
135	0.80	0.78	0.74	0.67	0.59	0.53	0.49	0.47	0.47
157.5	0.58	0.56	0.51	0.48	0.46	0.43	0.41	0.40	0.34
180	0.00	0.00	0.00	0.00	0.00	0.00	0.00	0.00	0.00

Zenith luminosity equals $0.8785\,h_s - 0.01322\,h_s^2 + 0.003434\,h_s^3 + 0.44347 + 0.0364\,T_{Atm}$ with T_{Atm} atmospheric turbidity and h_s solar height in degrees.

On a monthly basis, the complex formulas above are typically simplified to:

$$E_{Sd,s} = E_{Sd,h}\, f_{mo}\, \frac{1+\cos s_s}{2} \tag{1.17}$$

For Uccle, f_{mo} takes the values listed in Table 1.3.

1.2.2.3 Reflected insolation

Each surface on earth reflects part of the beam and diffuse insolation it receives. How much, depends on the short wave reflectivity. A simple model replaces the real terrestrial environment by a horizontal plane with average short wave reflectivity a_S, called the albedo, equal to 0.2. Apart of beam and diffuse insolation, every surface receives reflected insolation from that fictitious horizontal plane proportional to the view factor (F_{se}):

$$F_{se} = \frac{1-\cos s_s}{2}$$

Reflected insolation on a surface so becomes:

$$E_{Sr,s} = 0.2\left(E_{SD,h} + E_{Sd,h}\right)\left(\frac{1-\cos s_s}{2}\right) \tag{1.18}$$

That value turns zero for a receiving horizontal surface (which is not always true. A flat roof may receive reflected insolation from surrounding higher buildings).

1.2.2.4 Total insolation

Total insolation on a surface is found by summing up the beam, diffuse and reflected component. As an example, appendix 1 gives detailed information on total insolation for Uccle. Table 1.4 summarizes the mean and extreme monthly values measured there on a horizontal surface, together with the monthly mean cloudiness data, calculated as one minus the ratio between measured and clear sky insolation.

Table 1.4. Monthly total insolation on a horizontal surface and average cloudiness for Uccle (MJ/(m^2 · month)). Mean, maximum and minimum between 1958–1975.

	J	F	M	A	M	J	J	A	S	O	N	D
Solar gains												
Mean	72	129	247	356	500	538	510	439	327	197	85	56
Min.	61	104	177	263	406	431	408	366	279	145	63	41
Max.	93	188	311	485	589	640	651	497	444	274	112	78
Average cloudiness												
Mean	0.47	0.44	0.42	0.42	0.36	0.35	0.38	0.38	0.34	0.39	0.49	0.50
Min.	0.55	0.55	0.58	0.57	0.48	0.48	0.51	0.48	0.44	0.55	0.62	0.63
Max.	0.31	0.19	0.27	0.20	0.25	0.22	0.21	0.30	0.11	0.15	0.33	0.30

Table 1.5 lists the average monthly gains on a horizontal surface for different locations in Europe whereas Figure 1.7 shows the annual totals. The ratio between the least and most sunny location listed approaches a value 2.

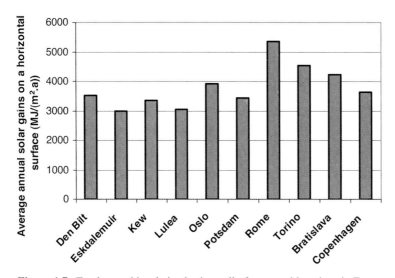

Figure 1.7. Total annual insolation horizontally for several locations in Europe.

1.2 Outdoor conditions

Table 1.5. Total monthly insolation horizontally for several locations in Europe (MJ/(m² · month)).

Location	Month											
	J	F	M	A	M	J	J	A	S	O	N	D
Den Bilt (NL)	72	132	249	381	522	555	509	458	316	193	86	56
Eskdalemuir (UK)	55	112	209	345	458	490	445	370	244	143	70	39
Kew (UK)	67	115	244	355	496	516	501	434	311	182	88	54
Lulea (S)	6	52	182	358	528	612	589	418	211	80	14	1
Oslo (N)	44	110	268	441	616	689	624	490	391	153	57	27
Potsdam (D)	104	137	238	332	498	557	562	412	267	174	88	70
Roma (I)	182	247	404	521	670	700	750	654	498	343	205	166
Torino (I)	171	212	343	474	538	573	621	579	422	281	181	148
Bratislava (Sk)	94	159	300	464	597	635	624	544	389	233	101	72
Copenhagen (Dk)	54	114	244	407	579	622	576	479	308	159	67	38

1.2.3 Long wave radiation

Long wave radiation, which is currently called under-cooling, invokes extra heat losses. It also causes the outside surface temperature and temperature in the layers between the thermal insulation and the outside to drop below the outside dry bulb temperature, even below the outside dew point, turning the outside air into a moisture source rather than a drying medium with as most visible consequence condensation on exterior surfaces of highly insulating glazing systems, EIFS-finishes and tiled or slated well insulated pitched roofs

Under-cooling is a result of the long wave radiant balance between the atmosphere (the celestial vault), the terrestrial environment and the surface considered, with the atmosphere acting as selective radiant body, absorbing all incoming terrestrial radiation but emitting only a fraction of it:

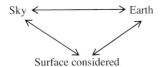

Celestial radiation is thus calculated, assuming a sky at air temperature but with emissivity below 1, absorptivity 1 and reflectivity 0. Tens of formulas have been proposed to quantify that emissivity. Some are listed below (p_e in Pa, θ_e in °C):

Clear sky

(1) $\varepsilon_{L,sky,o} = 0.75 - 0.32 \cdot 10^{-0.051\, p_e/1000}$

(2) $\varepsilon_{L,sky,o} = 0.52 + 0.065 \sqrt{p_e/1000}$

(3) $\varepsilon_{L,sky,o} = 1.24 \left(\dfrac{p_e/1000}{273.15 + \theta_e} \right)^{1/7}$

Cloudy sky

(4) $\varepsilon_{L,sky} = \varepsilon_{L,sky,o}(1 - 0.84\, c) + 0.84\, c$

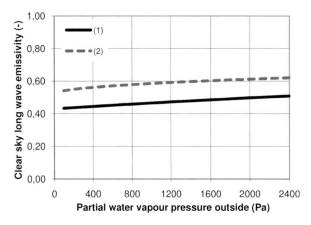

Figure 1.8. Sky emissivity: clear sky formulas (1) and (2).

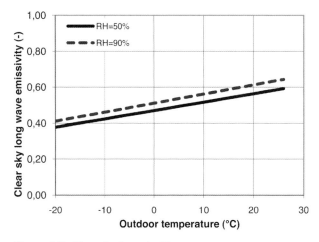

Figure 1.9. Clear sky formula (3).

For a clear sky, the value typically decreases with air temperature and increases with partial water vapour pressure, as water vapour acts as a strong greenhouse gas. In the cloudy sky formula, c stands for the cloudiness factor (0 for a clear sky, 1 for a covered sky and values in between in steps of 0.125). As Figure 1.8 shows, clear sky formulas (1) and (2) give quite different results.

Figure 1.9 underlines that formula (1) convenes with lower and formula (2) with somewhat higher outside temperatures.

The celestial/surface/earth radiant system produces the following two radiation balances:

$$M_{b,s} = \left[1 + \frac{\rho_{L,s}}{\varepsilon_{L,s}}\left(F_{s,env} + F_{s,sky}\right)\right] M'_s - \frac{\rho_{L,s}}{\varepsilon_{L,s}}\left(F_{s,env}\, M'_{env} + F_{s,sky}\, M'_{sky}\right)$$

1.2 Outdoor conditions

$$M_{b,env} = \left[1 + \frac{\rho_{L,env}}{\varepsilon_{L,env}}\left(F_{env,s} + F_{env,sky}\right)\right] M'_{env} - \frac{\rho_{L,env}}{\varepsilon_{L,env}}\left(F_{env,s}\, M'_s + F_{env,sky}\, M'_{sky}\right)$$

with $M_{b,x}$ black body emittance (x = s stands for the surface considered and x = env for the terrestrial environment). M'_{sky} is the radiosity of the sky. $F_{x,y}$ gives the view factors with x the one (f.e: the surface considered) and y the other radiant body (f.e: the terrestrial environment). As any surface is completely surrounded by the terrestrial and celestial environment, the sum $F_{s,env} + F_{s,sky}$ is 1. The same holds for the view factors between the terrestrial environment and the sky. In fact, the view factor of the terrestrial environment with the surface is so small that setting zero does not falsify the second balance. Celestial radiosity writes as:

$$M'_{sky} = \varepsilon_{L,sky}\, M_{b,sky,\theta_e}$$

turning the two balances into:

$$M_{b,s} = \frac{1}{\varepsilon_{L,s}} M'_s - \frac{\rho_{L,s}}{\varepsilon_{L,s}}\left(F_{s,env}\, M'_{env} + F_{s,sky}\, \varepsilon_{L,sky}\, M_{b,sky,\theta_e}\right) \quad (1.19)$$

$$M_{b,env} = \frac{1}{\varepsilon_{L,env}}\left(M'_{env} - \rho_{L,env}\, \varepsilon_{L,sky}\, M_{b,sky,\theta_e}\right) \quad (1.20)$$

The result is a radiant heat flow rate at the surface, expressed as:

$$q_R = q_{Rs,env} + q_{Rs,sky} = \frac{\varepsilon_{L,s}}{\rho_{L,s}}\left(M_{b,s} - M'_s\right) = \varepsilon_{L,s}\, F_{s,env}\left(M_{b,s} - \varepsilon_{L,env}\, M_{b,env}\right)$$
$$+ \varepsilon_{L,s}\, F_{s,sky}\left[M_{b,s} - \left(\rho_{L,env}\, \frac{F_{s,env}}{F_{s,sky}} + 1\right)\varepsilon_{L,sky}\, M_{b,sky,\theta_e}\right] \quad (1.21)$$

This formula is further simplified by assuming the terrestrial environment to be a black body at outside dry bulb temperature:

$$q_R = \varepsilon_{L,s}\, C_b\left[\left(F_{s,env} + F_{s,sky}\right)\left(\frac{T_{se}}{100}\right)^4 - \left(F_{s,env} + F_{s,sky}\, \varepsilon_{L,sky}\right)\left(\frac{T_e}{100}\right)^4\right] \quad (1.22)$$

Linearization of Equation (1.22) by replacing the athmosphere by a black body at temperature $\theta_{sk,e} = \theta_e - (23.8 - 0.2025\, \theta_e)(1 - 0.87\, c)$ instead of keeping it selective at outside dry bulb temperature, gives:

$$q_R = \varepsilon_{L,s} C_b\left[\left(F_{s,env}\, F_{Ts,env} + F_{s,sky}\, F_{Ts,sky}\right)\left(\theta_{se} - \theta_e\right)\right.$$
$$\left. + \left(23.8 - 0.2025\, \theta_e\right)\left(1 - 0.84\, c\right) F_{s,sky}\, F_{Ts,sky}\right] \quad (1.23)$$

That formula fits with expression (1.22) for a sky emissivity somewhat higher than given in Figure 1.9, which seems correct for real sky conditions. Combining (1.23) with the convective heat exchanged allows calculating under-cooling:

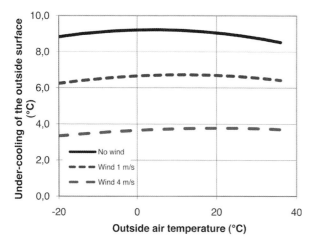

Figure 1.10. Under-cooling of the outside surface with emissivity 0.9 of a low-sloped roof with thermal transmittance 0.2 W/(m² · K) as function of the dry bulb temperature outdoors and wind speed.

$$q_{ce} + q_{Re} = \left[h_{ce} + \varepsilon_{L,s} C_b \left(F_{s,env} F_{Ts,env} + F_{s,sky} F_{Ts,sky} \right) \right] (\theta_{se} - \theta_e) \\ + \varepsilon_{L,s} C_b F_{s,sky} F_{Ts,sky} (23.8 - 0.2025 \theta_e)(1 - 0.84 c) \quad (1.24)$$

$$q_T + q_{ce} + q_{Re} = 0$$

where q_T is the heat flow rate by conduction to or from the exterior surface. Figure 1.10 shows the calculated result for a lightweight low-sloped roof with dry bulb temperature outdoors and wind speed as parameters. The roof has a thermal transmittance 0.2 W/(m² · K), a long wave emissivity 0.9 and hardly any thermal inertia. As the figure underlines, surface temperature may drop substantially below the outside air temperature.

1.2.4 Relative humidity and (partial water) vapour pressure

Relative humidity (ϕ_e) and (partial water) vapour pressure (p_e) impact the moisture response of building enclosures and buildings in a straight forward way. Table 1.6 summarizes monthly mean values for different locations all over Europe.

On the average, relative humidity remains quite constant between winter and summer. Vapour pressure instead differs a lot. In mild climates the inverse may be true between day and night: large difference in relative humidity and quite constant vapour pressure. A sudden temperature rise lowers relative humidity whereas a sudden temperature drop may push relative humidity to 100% with mist as result. During rainy weather the outside wet bulb temperature closely follows rain drop temperature. When as warm as the air, 100% relative humidity will be measured. Relative humidity and vapour pressure are also strongly influenced by the environment, with higher values in forests and alluviums than in cities.

1.2 Outdoor conditions

Table 1.6. Monthly mean relative humidity (%) and vapour pressure (Pa, bold) for several locations all over Europe.

Location	Month											
	J	F	M	A	M	J	J	A	S	O	N	D
Uccle (B)	89.6	89.0	84.0	78.5	77.8	78.9	79.9	79.8	84.2	88.3	91.2	92.7
	663	**681**	**757**	**854**	**1151**	**1334**	**1529**	**1489**	**1346**	**1084**	**806**	**780**
Aberdeen (UK)	81.5	80.3	74.9	72.3	75.0	78.5	74.1	79.1	80.6	81.6	78.1	77.1
	596	**596**	**631**	**714**	**860**	**1107**	**1161**	**1207**	**1122**	**955**	**695**	**614**
Catania (I)	66.5	72.4	68.8	69.8	71.2	70.0	62.0	68.6	69.4	69.6	68.5	65.9
	816	**912**	**964**	**1114**	**1469**	**1849**	**1985**	**2252**	**1972**	**1471**	**1183**	**900**
Den Bilt (NL)	86.1	82.1	76.0	75.9	72.9	72.7	77.1	78.9	80.7	84.4	85.5	86.8
	578	**596**	**631**	**811**	**1028**	**1263**	**1411**	**1443**	**1306**	**1086**	**772**	**587**
Kiruna (S)	83.0	82.0	77.0	71.0	64.0	61.0	68.0	72.0	77.0	81.0	85.0	85.0
	177	**171**	**221**	**324**	**476**	**710**	**1011**	**914**	**676**	**436**	**292**	**219**
Malmö (S)	87.0	86.0	83.0	76.0	73.0	74.0	78.0	77.0	82.0	85.0	87.0	89.0
	510	**496**	**561**	**711**	**958**	**1262**	**1531**	**1464**	**1269**	**969**	**753**	**627**
München (G)	83.7	81.9	76.8	72.3	74.9	76.8	74.2	76.1	79.2	82.9	83.5	85.7
	451	**484**	**598**	**780**	**1043**	**1361**	**1483**	**1446**	**1267**	**938**	**660**	**515**
Roma (I)	76.1	71.3	66.5	69.1	71.4	71.3	61.6	68.5	72.3	74.1	78.1	79.4
	794	**764**	**816**	**1048**	**1400**	**1795**	**1881**	**2042**	**1776**	**1346**	**1124**	**874**
Västerås (S)	84.0	82.0	74.0	66.0	62.0	65.0	69.0	74.0	81.0	83.0	86.0	86.0
	364	**355**	**402**	**540**	**766**	**1081**	**1354**	**1328**	**1085**	**793**	**602**	**483**

Again, the annual course may be written in terms of a Fourier series with one harmonic:

$$\phi_e = \overline{\phi}_e + \hat{\phi}_e \cos\left[\frac{2\pi(t-d)}{365.25}\right] \tag{1.25}$$

$$p_e = \overline{p}_e + \hat{p}_e \cos\left[\frac{2\pi(t-d)}{365.25}\right] \tag{1.26}$$

with:

	Relative humidity (RH), %			Vapour pressure (p), Pa			Quality of fit
	Mean	Amplitude	D Days	Mean	Amplitude	d Days	
Kiruna	75.5	11.1	346	469	380	209	$RH\pm, p-$
Roma	71.6	5.3	342	1305	627	214	$RH-, p\pm$

However, as Figure 1.11 shows, results deviate from the monthly averages.

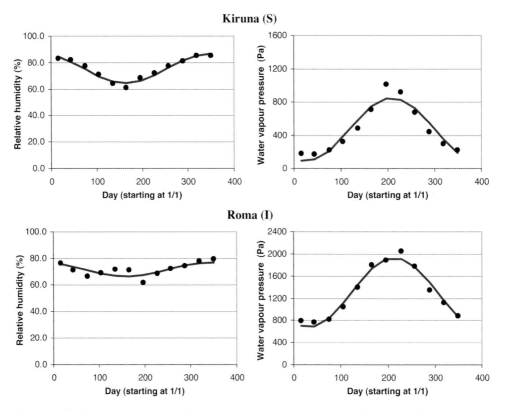

Figure 1.11. Monthly mean relative humidity and vapour pressure in two European cities, a one harmonic fit.

1.2.5 Wind

Wind impacts the hygrothermal response of building enclosures. A higher speed fastens the first drying phase, lowers the sol-air temperature and diminishes under-cooling. Especially single glass sees its thermal transmittance increased as wind augments. At the same time, infiltration, exfiltration and wind washing gain importance. Wind combined with precipitation gives wind-driven rain. Wind also governs the comfort conditions outdoors. Too much wind exerts unpleasant dynamic forces on individuals and chills people. High speeds make it also strenuous to open and close doors.

1.2.5.1 Wind speed

In meteorology, wind speed is considered to be a horizontal vector whose amplitude and direction are measured in the open field at a height of 10 m. The mean per three seconds is called the instantaneous wind, the mean per 10 minutes the average wind. Speed as a function of time looks periodically with many harmonics. The speed vector also changes direction continuously. At local level, speed and direction get shaped by the built environment. Venturi effects increase the speed in small passages between buildings. Alongside buildings, eddies

1.2 Outdoor conditions

develop, while the windward and leeward side of a building see wind still zones created. In the 500 and 2000 m thick atmospheric boundary layer, wind speed increases with height. At the same time friction induces a decrease with augmenting terrain roughness. Following logarithmic law quantifies both effects:

$$v_h = v_{10}\, K \ln\left(\frac{h}{n}\right) \tag{1.27}$$

where v_{10} is the average wind speed 10 m above ground level in flat open terrain, v_h is the average wind speed h meters above ground level, n is terrain roughness and K a number which depends on the friction velocity:

Terrain upwind	K	n
Open see	0.128	0.0002
Coastal plain	0.166	0.005
Flat grass land, runaway area at airports	0.190	0.03
Farmland with low crops, spread farms	0.209	0.10
Farmland with high crops, vineyards	0.225	0.25
Open landscape with larger obstacles, forests	0.237	0.50
Old forests, homogeneous villages and cities	0.251	1.00
City centres with high-rises, industrial developments	≥ 0.265	≥ 2

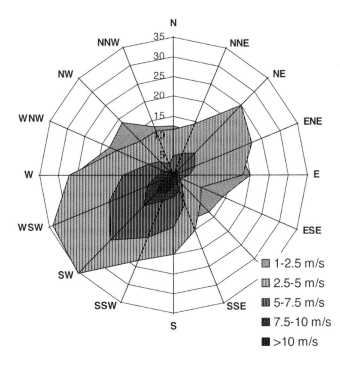

Figure 1.12. Average windrose for Uccle, Belgium.

Table 1.7. Uccle, Belgium, pro-thousands of time the wind blows from a given direction (means per month between 1931 and 1960).

Orientation	Month											
	J	F	M	A	M	J	J	A	S	O	N	D
N	18	22	43	54	59	65	57	37	54	22	15	14
NNE	20	32	48	59	62	60	57	38	47	33	30	13
NE	41	65	79	103	85	72	64	44	67	65	58	47
ENE	54	65	68	68	75	56	40	38	62	63	56	56
E	56	61	48	53	72	45	34	41	63	63	64	41
ESE	28	32	30	28	39	25	26	25	34	38	32	29
SE	36	38	34	33	31	25	25	28	45	50	39	35
SSE	60	53	57	40	38	29	26	36	55	70	60	65
S	90	98	81	49	52	44	44	53	62	100	92	109
SSW	120	109	86	62	70	55	64	88	71	118	105	132
SW	163	130	113	105	92	103	131	152	104	124	142	147
WSW	125	115	100	97	92	100	120	145	103	85	119	125
W	92	73	79	78	71	95	109	121	94	77	89	91
WNW	50	49	54	59	49	74	68	64	55	46	51	46
NW	28	35	73	62	59	84	76	50	44	26	29	32
NNW	19	23	37	50	54	68	59	40	40	20	19	18

Table 1.7 and Figure 1.12 show the annual distribution of the wind vector over the different directions for Uccle, Belgium. Wind there comes from west to south nearly half of the year. Such distributions of course change between locations.

1.2.5.2 Wind pressure

That wind exerts a pressure against an obstacle follows from Bernoulli's law. Without friction the sum of the kinetic and potential energy in a moving fluid must be constant. In a horizontal flow, potential energy is linked to pressure. When the flow stops suddenly, all kinetic energy transforms into potential energy, giving as a pressure in Pa (p_w):

$$p_w = \rho_a \frac{v_w^2}{2} \tag{1.28}$$

where ρ_a is the density of air in kg/m³ and v_w is wind speed in m/s.

As no obstacle stretches to infinity, real flow is impeded but also redirected. That makes wind pressure different from the theoretical value, a fact which is accounted for by inserting a pressure coefficient (C_p) in Formula (1.28):

$$p_w = \rho_a C_p \frac{v_w^2}{2} \tag{1.29}$$

1.2 Outdoor conditions

Table 1.8. Wind pressure coefficients: exposed low rise building with rectangular floor plan, up to three stories high and a length to width ratio of 2 : 1.

Wind speed reference: local situation, at building height

Location		0	45	90	135	180	225	270	315
					Wind Angle				
Face 1		0.5	0.25	−0.5	−0.8	−0.7	−0.8	−0.5	0.25
Face 2		−0.7	−0.8	−0.5	0.25	0.5	0.25	−0.5	−0.8
Face 3		−0.9	0.2	0.6	0.2	−0.9	−0.6	−0.35	−0.6
Face 4		−0.9	−0.6	−0.35	−0.6	−0.9	0.2	0.6	0.2
Roof pitch < 10°	Front	−0.7	−0.7	−0.8	−0.7	−0.7	−0.7	−0.8	−0.7
	Rear	−0.7	−0.7	−0.8	−0.7	−0.7	−0.7	−0.8	−0.7
Average		−0.7	−0.7	−0.8	−0.7	−0.7	−0.7	−0.8	−0.7
pitch 10–30°	Front	−0.7	−0.7	−0.7	−0.6	−0.5	−0.6	−0.7	−0.7
	Rear	−0.5	−0.6	−0.7	−0.7	−0.7	−0.7	−0.7	−0.6
Average		−0.6	−0.65	−0.7	−0.65	−0.6	−0.65	−0.7	−0.65
pitch > 30°	Front	0.25	0	−0.6	−0.9	−0.8	−0.9	−0.6	0
	Rear	−0.8	−0.9	−0.6	0	0.25	0	−0.6	−0.9
Average		−0.28	−0.45	−0.6	−0.45	−0.28	−0.45	−0.6	−0.45

v_w is the reference wind speed (for example, the value measured at the nearest weather station). The pressure coefficient C_p is quantified through field measurements, wind tunnel experiments or CFD. The value changes depending on the reference taken and the upwind and downwind environment. On a building, it varies from point to point, with highest value at the edges and upper corners and lowest value down in the middle. At the frontside, the value is positive, while at the rearside and at surfaces parallel or nearly parallel to the wind vector it is negative. Pressure coefficients can be found in literature, an example being given in Table 1.8.

1.2.6 Precipitation and wind-driven rain

In humid climates rain is the largest moisture load buildings have to cope with, causing most of the moisture deficiencies encountered in practice. In countries where rain-tightness is achieved by buffering, wind driven rain also impacts the end energy used for heating and cooling.

The term wind driven rain now applies to the horizontal component of the rain vector whereas precipitation stands for the vertical component. In windless weather, that horizontal component is zero. It gains importance with increasing wind speed. Horizontal and inclined surfaces collect rain under all circumstances while vertical surfaces need wind.

1.2.6.1 Precipitation

Precipitation is as unsteady as the wind. Table 1.9 and Figure 1.13 compile mean rain duration (hours) and mean amount data per month for Uccle, Belgium. As with wind, no annual cycle appears though more rain falls in summer than in winter. Figure 1.14 gives the absolute maxima noted during 1′, 10′ and 1 hour from 1956 to 1970 with July totalling 4 l/m² in 1′, 15 l/m² in 10′ and 42.8 l/m² in 1 hour.

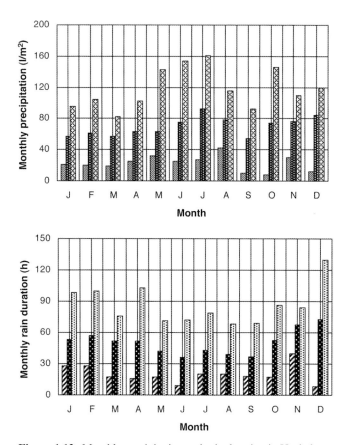

Figure 1.13. Monthly precipitation and rain duration in Uccle between 1961 and 1970 (averages, maxima and minima).

Table 1.9. Precipitation: duration in hours and amounts in l/m² per month, means for Uccle between 1961 and 1970.

Precipitation	Month											
	J	F	M	A	M	J	J	A	S	O	N	D
Duration (h)	53.6	57.3	51.9	51.5	41.9	36.0	42.9	38.8	36.5	52.5	67.3	72.7
Amount	57.4	61.4	56.8	63.6	63.1	74.9	92.3	78.6	54.6	74.5	76.5	84.5

Rain showers are characterized by a droplet size distribution. According to Best, the correlation with rain intensity looks like:

$$F_{d,precip} = 1 - \exp\left[-\left(\frac{d}{1.3\, g_{r,h}^{0.232}}\right)^{2.25}\right] \tag{1.30}$$

1.2 Outdoor conditions

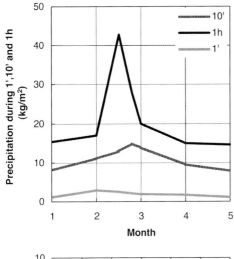

Figure 1.14. Maximum precipitation during 1′, 10′ and 1 h in Uccle (1961–1970).

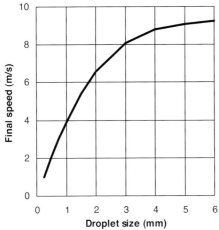

Figure 1.15. Final vertical speed of a raindrop in windless weather.

with $F_{d,precip}$ the distribution function, d droplet size in m and $g_{r,h}$ precipitation in kg/(m² · h). In windless weather, each droplet falls vertically with as final speed (d droplet size in mm):

$$v_\infty = -0.166033 + 4.91844\, d - 0.888016\, d^2 + 0.054888\, d^3 \qquad (1.31)$$

Also see Figure 1.15. Depending on the mean droplet size, following names are used to describe precipitation:

Mean droplet size d_m mm	Type of precipitation
0.25	Drizzle
0.50	Normal
0.75	Strong
1.00	Heavy
1.50	Downpour

1.2.6.2 Wind driven rain

The drag force wind exerts drives rain droplets into inclined trajectories. How steep these are, depend on the droplet size, the direction of the wind vector compared to the horizontal and the wind speed change with height. If constant and horizontal, the trajectory curve becomes a straight line. In such case wind driven rain intensity can be written as:

$$g_{r,v} = 0.222\, v_w\, g_{r,h}^{0.88} \qquad (1.32)$$

where-in $g_{r,h}$ is precipitation intensity and $g_{r,v}$ the wind driven rain intensity, both in kg/(m² · s). That equation is typically simplified to:

$$g_{r,v} = 0.2\, v_w\, g_{r,h} \qquad (1.33)$$

a relation, which fits quite well in the open field. For wind speeds beyond 5 m/s, wind driven rain surpasses precipitation.

In the built environment, Equation (1.33) does not apply. Intensity on a façade in fact depends on the horizontal rain fall intensity, raindrop size distribution, the building volumes upwind and downwind, building geometry, building orientation compared to wind direction, position on the façade and local detailing. Buildings in an open neighbourhood may catch forty times more wind-driven rain than buildings in a dense neighbourhood. Local amounts on a facade are directly bound to the wind flow patterns around the building. For low-rises, highest values are noted at the upper corners. Especially when it drizzles, high-rises catch the most at the highest floors, the higher edges and the upper corners. Differences over a facade diminish with increasing rain intensity and augmenting wind speed. As an example, Table 1.10 gives wind-driven rain deposits measured by the Fraunhofer Institut für Bauphysik on an eighteen stories high condominium, showing the trends mentioned. The formula used to describe wind driven rain impinging locally on a building enclosure is:

$$g_{r,v} = \left(0.2\, C_r\, v_w\, \cos\vartheta\right) g_{r,h} \qquad (1.34)$$

with ϑ the angle between wind direction and the normal on the surface and C_r the wind driven rain factor, a function of the type of precipitation, the environment surrounding the building, the façade spot looked for, local detailing, etc.

Table 1.10. Wind driven rain on a 18 stories condominium in Munich as measured between May and November 1972 by the Fraunhofer Institut für Bauphysik.

Measuring spot	Wind driven rain kg/m²
Façade West, middle 3th floor	29
Façade West, middle 9the floor	55
Façade West, middle 16the floor	65
Roof edge north	115
Roof edge south	130

On the average, C_r moves between 0.25 and 2. The product $0.2\, C_r\, v_w\, \cos\vartheta$ is called the catch ratio, a number which gives the amount of wind driven rain hitting a spot on a façade as fraction of the precipitation measured in open field during the same rain event. Formula (1.34) is based on experimental evidence. The alternative is using CFD to calculate the wind field

1.2 Outdoor conditions

and applying droplet tracing to get the rain trajectories. Catch ratio patterns calculated with CFD for the south-west façade of a test builing were compared with measured data using rain gauges. Figure 1.16 gives the results.
Measured values fit reasonable well in the calculated pattern.

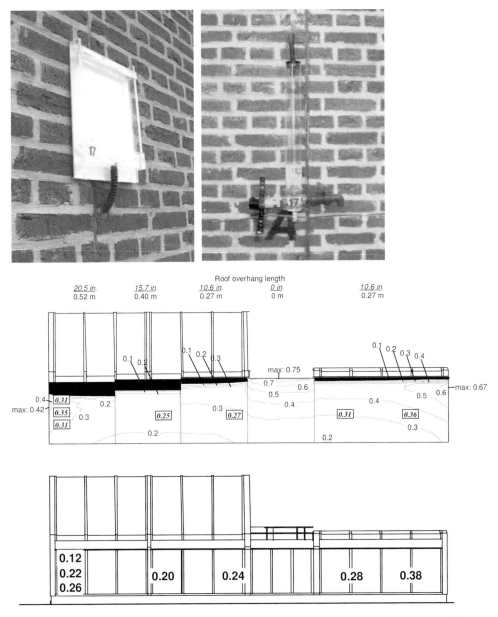

Figure 1.16. South-west façade of a test building: calculated lines of equal catch ratio using CFD and droplet tracing (top) compared with measured values (below).

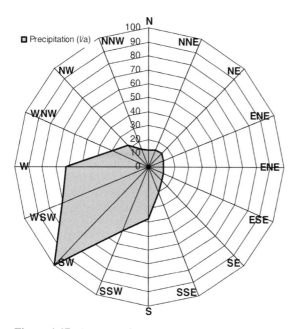

Figure 1.17. Average rainrose for Uccle.

In North-Western Europe wind-driven rain comes mainly from the South West, as Figure 1.17 shows for Uccle.

Happily, strong winds and heavy precipitation only rarely concur. Between 1956 and 1970 the most negative hourly means measured in Uccle were:

Precipitation l/h	Mean wind speed m/s
15	5.8
42	2.2

1.2.7 Standardized outside climate values

1.2.7.1 Design temperature

The design temperature (θ_{ed}, °C) applies when heat loads have to be calculated. A definition could be the lowest running two-day mean temperature, measured ten times over twenty-years (probability of 0.14%). Maps with such 0.14% design temperatures exist for most countries in Europe. Also values with other probability are used. In the USA and Canada, the designer has a choice between two probabilities: 1% and 0.4%.

1.2.7.2 Thermal reference year

Four outdoor climate components have a main part in fixing the net energy demand for heating and cooling: air temperature, solar radiation, long wave radiation and wind.

1.2 Outdoor conditions

Table 1.11. Monthly mean TRY's for Belgium.

Month	Temp.	$E_{ST,hor}$	Wind	Rel. hum.	Vapour pressure	J
	°C	W/m²	m/s	%	Pa	–
Average						
January	3.8	26.6	3.8	91	730	0.52
February	3.2	55.2	5.1	87	668	0.58
March	7.1	88.5	3.9	85	857	0.56
April	9.1	131.4	4.7	81	935	0.56
May	11.9	172.6	3.4	77	1072	0.59
June	16.7	216.9	3.2	80	1520	0.68
July	16.1	162.5	3.0	79	1445	0.53
August	17.1	158.0	3.5	83	1618	0.60
September	14.3	128.7	3.4	83	1352	0.67
October	11.0	64.9	3.6	91	1194	0.54
November	6.1	29.6	3.8	91	856	0.45
December	3.1	19.3	3.6	91	694	0.46
Cold						
January	–3.9	33.2	4.1	86	379	0.65
February	–1.4	63.1	3.3	85	462	0.66
March	3.1	94.3	3.6	81	618	0.59
April	6.7	101.5	4.0	87	853	0.43
May	10.8	151.5	3.7	84	1087	0.51
June	14.2	178.7	3.1	79	1278	0.56
July	16.1	171.4	3.7	80	1463	0.55
August	15.9	136.7	3.5	86	1552	0.52
September	12.8	123.0	2.9	82	1211	0.64
October	7.5	53.9	3.3	90	932	0.44
November	4.2	40.2	4.6	88	725	0.62
December	0.2	25.3	3.2	90	558	0.61
Warm						
January	6.8	25.3	5.2	87	859	0.50
February	7.6	55.4	4.4	89	928	0.58
March	8.5	109.4	3.5	85	942	0.69
April	12.3	130.7	3.0	86	1229	0.55
May	15.4	206.8	2.8	72	1258	0.70
June	18.8	235.2	3.1	74	1605	0.74
July	20.0	242.9	3.5	73	1706	0.79
August	19.9	181.7	2.9	74	1718	0.69
September	16.5	171.0	3.5	70	1313	0.89
October	13.8	86.2	2.6	86	1356	0.71
November	9.4	35.1	5.1	88	1036	0.54
December	7.4	15.4	5.3	87	895	0.37

Table 1.12. Energy performance legislation, reference year for Belgium.

Month	Temp. °C	Total solar radiation horizontal surface $E_{ST,hor}$ MJ/m²	Diffuse solar radiation, horizontal surface $E_{sd,hor}$ MJ/m²
January	3.2	71.4	51.3
February	3.9	127.0	82.7
March	5.9	245.5	155.1
April	9.2	371.5	219.2
May	13.3	510.0	293.5
June	16.3	532.4	298.1
July	17.6	517.8	305.8
August	17.6	456.4	266.7
September	15.2	326.2	183.6
October	11.2	194.2	118.3
November	6.3	89.6	60.5
December	3.5	54.7	40.2

To allow a calculation using Building Energy Simulation tools (BES), thermal reference years, called TRY's, were constructed. They list the hourly mean values for each of the four parameters over a whole year. They are constructed artificially and give the same net demand as should be found when averaging the annual demands over a period of X successive years. TRY's have been published for different locations in Europe and North America.

For steady state net energy demand calculations based on one month time-steps, monthly mean TRY's are used. For Belgium three TRY's were once created based on a statistical analysis of data gained during a 30 year period: average, cold and warm, see Table 1.11. The cloudiness factor J in that table stands for the ratio between the real insolation on a horizontal surface and the value one should register under clear sky conditions. The Belgian energy performance legislation reconsidered the average year based on a more recent thirty year period (1961–1990) with a split between beam and diffuse solar radiation and relative humidity, vapour pressure and cloudiness no longer considered, see Table 1.12.

1.2.7.3 Moisture reference year

Since the nineteen fifties, calculation tools have been developed for predicting the moisture response of envelope assemblies. For that to be possible the boundary conditions, the geometry, the contact conditions between layers and the heat-air-moisture properties of the material used have to be known. Looking to the boundary condition outdoors, as for energy, the concept of a moisture reference year took shape. Agreed was to take a year of highest moisture built-up in an assembly with a ten years return rate. The problem was that each moisture event demanded a different year. Rain leakage for example is best served with a wet and windy year. Solar driven vapour ingress instead scores highest in sunny summers with regular wind-driven rain spells. Interstitial condensation in turn favours from cold, humid years with little sun in summer. The exercise was therefore done for interstitial condensation only, with as a result for four European cities, the years listed in Table 1.13.

1.2 Outdoor conditions

Table 1.13. Durability reference years for interstitial condensation for four European cities.

Month	Uccle			Rome			Copenhagen			Luleå		
	θ_e °C	p_e Pa	$E_{ST,h}$ W/m²	θ_e °C	p_e Pa	$E_{ST,h}$ W/m²	θ_e °C	p_e Pa	$E_{ST,h}$ W/m²	θ_e °C	p_e Pa	$E_{ST,h}$ W/m²
J	2.7	675	24.0	8.0	747	73.3	−0.1	569	18.2	−12.6	190	2.4
F	1.7	587	50.4	10.1	877	205.3	−0.4	507	53.1	−13.2	183	22.9
M	5.9	724	99.0	10.2	811	257.8	−0.3	476	83.4	−5.2	357	54.8
A	8.5	932	115.5	13.2	997	289.2	7.4	817	159.6	−2.4	415	109.8
M	12.6	1123	180.0	18.3	1323	243.6	12.0	1080	227.5	3.6	538	179.6
J	15.5	1374	215.0	22.3	1733	257.0	14.7	1355	229.4	10.8	894	234.4
J	14.9	1423	148.2	24.5	1976	271.1	15.3	1474	186.3	15.4	1365	203.9
A	16.2	1492	172.8	22.3	1747	224.0	15.1	1435	178.3	12.3	1144	122.6
S	13.5	1300	120.6	19.4	1524	187.8	12.6	1275	125.4	7.7	883	71.2
O	11.1	1097	102.1	17.1	1446	125.5	7.7	961	54.9	−2.4	430	28.4
N	4.0	716	40.1	11.5	1029	74.9	5.0	793	26.6	−4.1	401	4.3
D	4.8	783	15.6	9.6	907	51.2	1.1	614	15.9	−17.1	123	0.5

An alternative is using a moisture index for classification. Knowing that more wetting and less drying makes a year more severe from a moisture tolerance point of view; such index must combine these two. In the index adopted by Canada the wetting (WI) and drying (DI) functions are both normalized according to:

$$\text{WI}_{norm} = \frac{\text{WI} - \text{WI}_{min}}{\text{WI}_{max} - \text{WI}_{min}} \qquad \text{DI}_{norm} = \frac{\text{DI} - \text{DI}_{min}}{\text{DI}_{max} - \text{DI}_{min}} \qquad (1.35)$$

where the wetting function equals the annual amount of wind driven rain touching a façade with given orientation while the drying function is given by:

$$\text{DI} = \sum_{h=1}^{k} \bar{x}_{sat} \left(1 - \bar{\phi}\right)$$

with $\bar{x}_{sat}$ the hourly mean vapour saturation ratio outdoors, $\bar{\phi}$ the hourly mean relative humidity and Σ the sum of all hourly means over a whole year. The moisture index itself is defined as:

$$\text{MI} = \sqrt{\text{WI}_{norm}^2 - \left(1 - \text{DI}_{norm}\right)^2} \qquad (1.36)$$

Different methodologies for quantifying MI are now possible. The min's and maxi's may be linked to one location, for example the most severe and most moderate year out of a file of thirty for the capital city. Then, in those thirty years long set of hourly meteorological data for any location, the third in severity as used for WI and DI. A variant is to take the year which has a moisture index closest to the mean of the thirty annual values as the reference for WI and DI.

Drawbacks of a moisture index as defined are twofold. Only problems caused by wind-driven rain are considered, overlooking problematic moisture deposit due to interstitial condensation by diffusion and convection. Solar gains are also not included, although drying is primarily influenced by the difference between the saturation water vapour ratio at the surface and the water vapour ratio in the air rather than by the drying function as defined above.

1.2.7.4 Equivalent outside temperature for condensation and drying

Let us consider an interface in an envelope assembly where condensate is deposited. Saturation pressure there will fluctuate together with temperature. As the relationship between both is exponential, higher temperatures have a greater weight in the mean saturation pressure than lower ones. Thus, the average saturation pressure over a longer period will differ from the one at mean temperature. The correct average instead is found by calculating the mean of the ever changing saturation pressures. Qualified for condensation and drying is the temperature belonging to that mean. Its value is easily transposed into a converted outside temperature, called 'equivalent temperature for condensation and drying'. In practice, quantification goes as follows:

- First, calculate for the month considered, using hourly data for temperature, solar radiation and long wave radiation, per azimuth and slope the sol-air temperature
- Then, fix the hourly temperature in the interface where condensate is deposited. As that interface is normally close to the outside, a steady state calculation suffices, or:

$$\theta_j = \theta_e^* + \frac{R_e^j}{R_a}\left(\theta_i - \theta_e^*\right) \quad (1.37)$$

where θ_i is the inside temperature, θ_e^* the sol-air temperature, R_e^j the thermal resistance between the interface (j) and the outside and R_a the thermal resistance environment to environment of the whole assembly

- Link saturation pressure to the hourly interface temperature
- Calculate the mean saturation pressure for the month considered:

$$\bar{p}_{sat,j} = \frac{1}{T}\int_0^T p_{sat,j}\, dt \quad (1.38)$$

- Define the temperature belonging to that mean
- Transpose that temperature into a monthly mean equivalent outside temperature for condensation and drying using as a formula:

$$\bar{\theta}_{ce}^* = \frac{\bar{\theta}_{\bar{p}_{sat,j}} + \bar{\theta}_i\left(R_e^j/R_a\right)}{1 - R_e^j/R_a} \quad (1.39)$$

- Once the value known per month, then calculate the annual average and the first and second harmonic using a Fourier transform

A calculation for Uccle considering a surface with shortwave absorptivity 1, long wave emissivity 0.9, different azimuths and several slopes, gave as result the temperatures of Table 1.14.

1.2 Outdoor conditions

Table 1.14. Uccle, equivalent temperature for condensation and drying ($a_K = 1$, $e_L = 0.9$).

Slope	Azimuth									
	Annual mean $\overline{\theta}_{ce}^*$ (°C)					Annual amplitude $\hat{\theta}_{ce}^*$ (°C)				
	N	NW NE	W E	SW SE	S	N	NW NE	W E	SW SE	S
0	14.4	14.4	14.4	14.4	14.4	12.6	12.6	12.6	12.6	12.6
15	13.7	13.9	14.4	14.7	14.8	12.4	12.5	12.6	12.7	12.7
30	12.9	13.4	14.2	14.9	16.0	11.7	12.0	12.3	12.4	12.4
45	12.2	12.8	13.9	14.7	14.9	10.9	11.3	11.6	12.0	11.8
60	11.9	12.6	13.7	14.6	14.7	9.8	10.6	11.2	11.3	11.1
75	11.8	12.4	13.5	14.3	14.6	9.3	10.0	10.7	10.6	10.3
90	12.1	12.6	13.6	14.2	14.4	9.1	9.5	9.9	9.7	9.5

Table 1.15. Uccle, month-based time function.

Month	C(t)	Month	C(t)
January	–0.98	July	+1.00
February	–0.85	August	+0.85
March	–0.50	September	+0.55
April	–0.10	October	–0.10
May	+0.55	November	–0.55
June	+0.90	December	–0.90

These values have to be corrected for surfaces with different shortwave absorptivity. That is done by:

- Combining the annual mean and amplitude into a relation of the form:

$$\theta_{ce}^* = \overline{\theta}_{ce}^* + \hat{\theta}_{ce}^* \, C(t) \tag{1.40}$$

where $C(t)$ is a month-based time function derived from a two harmonics analysis, see Table 1.15

- Correct the annual mean and amplitude for the actual shortwave absorptivity of the surface:

$$\left[\overline{\theta}_{ce}^*\right] = \alpha_K \left(\overline{\theta}_{ce}^* - \overline{\theta}_e'\right) + \overline{\theta}_e' \qquad \left[\hat{\theta}_{ce}^*\right] = \alpha_K \left(\hat{\theta}_{ce}^* - \hat{\theta}_e'\right) + \hat{\theta}_e' \tag{1.41}$$

In the Equations (1.41) $\overline{\theta}_e'$ and $\hat{\theta}_e'$ equal the annual mean equivalent temperature for condensation and drying and its amplitude for a north oriented surface with slope 45°. Their value for Uccle is 8.5 °C and 7.1 °C respectively for a shortwave absorptivity 0 and long wave emissivity 0.9. If the long wave emissivity equals 0, the mean becomes 9.8 °C and the amplitude 6.9 °C. For long wave emissivities above 0 but below 0.9, a linear interpolation between 8.5/9.8 and 7.1/6.9 applies:

Table 1.16. Uccle: very cold winter day (declination −17.7°) and hot summer day (declination 17.7°). Probability 0.012%.

Hour	Cold winterday			Hot summerday		
	Temp. °C	Direct sun, ⊥ W/m^2	Diffuse sun, hor. surface W/m^2	Temp. °C	Direct sun, ⊥ W/m^2	Diffuse sun, hor. surface W/m^2
0	−13.3	0	0	20.2	0	0
1	−13.6	0	0	19.8	0	0
2	−13.8	0	0	19.5	0	0
3	−14.2	0	0	19.0	0	0
4	−14.6	0	0	18.5	0	0
5	−15.0	0	0	18.0	136	6
6	−15.4	0	0	17.8	301	50
7	−15.5	0	0	17.8	440	87
8	−15.5	42	8	20.0	555	121
9	−14.9	336	47	22.0	645	146
10	−14.3	553	72	25.5	711	165
11	−13.1	636	89	28.0	752	180
12	−11.8	669	100	30.0	768	187
13	−11.7	656	105	30.8	759	188
14	−11.5	611	94	30.5	726	179
15	−11.9	564	72	30.0	669	165
16	−12.3	375	39	29.0	587	147
17	−12.8	47	11	27.8	480	121
18	−13.3	0	0	26.5	348	90
19	−13.7	0	0	25.5	192	49
20	−14.0	0	0	24.2	11	7
21	−14.5	0	0	23.0	0	0
22	−15.0	0	0	22.0	0	0
23	−15.1	0	0	20.5	0	0
24	−15.2	0	0	20.2	0	0

1.3 Indoor conditions

$$\bar{\theta}'_e = 8.5 + 1.3\left(\frac{0.9 - e_L}{0.9}\right) \qquad \hat{\theta}'_e = 7.1 - 0.2\left(\frac{0.9 - e_L}{0.9}\right) \tag{1.42}$$

The effective monthly mean equivalent temperature for condensation and drying so becomes:

$$\left[\theta^*_{ce}\right] = \left[\bar{\theta}^*_{ce}\right] + \left[\hat{\theta}^*_{ce}\right] C(t) \tag{1.43}$$

For permanently shadowed surfaces the air temperature is used, for Uccle $[\bar{\theta}^*_{ce}] = \bar{\theta}_e = 9.8\,°C$, $[\hat{\theta}^*_{ce}] = \hat{\theta}_e = 6.9\,°C$.

The monthly mean vapour pressure is calculated using a same time function $C(t)$:

$$p_e = \bar{p}_e + \hat{p}_e\, C(t) \tag{1.44}$$

with $\bar{p}_e$ de annual mean and $\hat{p}_e$ the amplitude. For Uccle: $\bar{p}_e = 1042$ Pa, $\hat{p}_e = 430$ Pa.

Quantifying the monthly equivalent temperature for condensation and drying for any location worldwide may proceed as explained. One could have expected that the European standard EN ISO 13788 'Hygrothermal performance of Building Components and Building Elements' applies the concept. It doesn't. Apparently, the standard considers shadowed surfaces only and uses monthly mean dry bulb temperatures.

1.2.7.5 Very hot summer day, very cold winter day

Performance controls such as hygrothermal stress and strain, cooling loads, overheating risk, etc demand hourly temperature, insolation and relative humidity values at a daily basis. Each day, however, is different in terms of outdoor conditions. A hold may be using a very hot summer day and very cold winter day with a return rate of $x\%$ per year which then are considered as being representative for extreme weather conditions and, handling them as if they belong to a spell of identical cold and hot days. That assumption allows treating both days as periodic boundary condition, allowing an analytic solution of the transient assembly response. Table 1.16 give such days for Uccle.

1.3 Indoor conditions

Building usage normally defines the indoor conditions. Tenants wish temperatures at comfort level. Relative humidity and water vapour pressure might be adjusted for several reasons. That and the air pressure difference between rooms and between rooms and the outside define what is called the environmental load inside.

1.3.1 Dry bulb (or air) temperature

1.3.1.1 In general

The dry bulb or air temperature is of decisive importance when thermal comfort is considered. The value further ranks high as parameter impacting the net and gross energy demand and the end energy consumption for heating and cooling, whereas she also influences the moisture response of a building.

Table 1.17. Operative temperatures needed according to DIN 4701 and EN 12831.

Building Type	Room	Temperature	
		DIN 4701	**EN 12831**
Dwelling	Daytime rooms	20	20
	Bathroom	24	24
	Bedroom	20	
Hospital	Nursery	22	20
	Surgery	25	
	Premature births	25	
Office building	Single office	20	20
	Landscape office		20
	Conference room		20
	Auditorium		20
	Cafeteria, restaurant		20
	Hall	20	
	Corridors, rest rooms	15	
Indoor swimming pool	Natatorium	28	
	Showers	24	
	Cabins	22	
Schools	Classroom		20
Department store			20
Church			15
Museum, gallery			16

Most standards yet focus on the resulting or operative temperature, see Table 1.17, though in many buildings the dry bulb temperature is the quantity controlled. Anyway, in well insulated buildings, both hardly differ.

1.3.1.2 Measured data

Residential buildings

In Figure 1.18 334 weekly mean indoor dry bulb temperature records as measured in Belgian dwellings and apartments between 1972 and 1992 are collected as a function of the weekly mean outdoor dry bulb temperatures logged simultaneously. Table 1.18 gives the number of records, the least square regression and the correlation coefficient for daytime, bath- and bedrooms. The weekly means in the daytime rooms near the comfort temperature and seem hardly influenced by the weekly mean outside temperature, proving they are well heated. Bedrooms and bathrooms instead show significantly lower weekly means which depend explicitly on the weekly mean outside temperature. Whereas bathrooms are intermittently kept warm, sleeping rooms are hardly heated. Their indoor temperature reflects the balance between transmission and ventilation gains from adjacent rooms, transmission and ventilation losses to the outside and solar and internal gains.

1.3 Indoor conditions

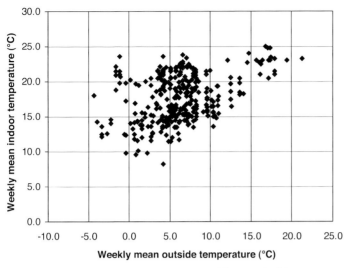

Figure 1.18. Dwellings and apartments: weekly mean inside temperature in living room, sleeping rooms and bathrooms in relation to the weekly mean outside temperature.

Table 1.18. Dwellings and apartments, measured weekly mean air temperatures (1972–1992).

	Number of weekly records	$\overline{\theta}_i = a + b\,\overline{\theta}_e$		Correlation coefficient
		a (°C)	b (–)	
Daytime rooms	126	19.8	0.12	0.08
Bedrooms	179	12.7	0.45	0.47
Bathrooms	29	16.5	0.34	0.47

Table 1.19. Measured daily mean air temperature in dwellings (2002–2005).

Room	$\overline{\theta}_i = a + b\,\overline{\theta}_e$	
	a (°C)	b (–)
Daytime rooms	20.0	0.22
Bedrooms	16.0	0.45
Bathrooms	19.0	0.32

Between 2002 and 2005, the data of Table 1.18 were reassessed during a new measuring campaign in 39 Belgian dwellings. The conclusions were different. Daytime rooms remain well heated but bedrooms looked somewhat warmer, probably because of a better thermal insulation, although they still seemed either intermittently or hardly heated at all. Bathrooms were warmer with weekly means less depending on the outside temperature (see Table 1.19).

Indoor swimming pools

Figure 1.19 collects 162 records of weekly mean indoor air temperatures measured in 16 Belgian indoor swimming pools, together with the weekly mean outdoor air temperature logged simultaneously. The least square line and the correlation coefficient are given in Table 1.20.

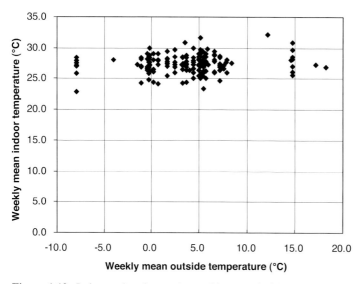

Figure 1.19. Indoor swimming pools: weekly mean inside temperature as function of the weekly mean outside temperature.

Table 1.20. Indoor swimming pools, measured weekly mean air temperatures.

Number of weekly records	$\theta_i = a + b\,\theta_e$		Correlation coefficient
	a (°C)	b (–)	
162	27.3	0.045	0.02

1.3.2 Relative humidity and (partial water) vapour pressure

1.3.2.1 In general

Indoor relative humidity and indoor vapour pressure are prime players when it comes to the moisture tolerance of building components and whole buildings. That is mirrored in their usage as pivot in the indoor climate class listing of buildings. Both have also a significant impact on indoor environmental quality. In most buildings they remain uncontrolled as a quantity. Actual value therefore depends on the whole building heat-air-moisture balance. Only in specific cases such as museums, archives, clean room spaces, computer rooms, surgical units, intensive care units etc., a stricter control is necessary, demanding complete HVAC-systems.

1.3.2.2 Measured data

Residential buildings

In Figure 1.20 334 weekly mean indoor/outdoor vapour pressure excesses measured in Belgian dwellings and apartments between 1972 and 1992 are collected as a function of the weekly mean outdoor air temperature logged simultaneously. Table 1.21 gives the number of records, the regression line, the 95% line and the correlation coefficient for the daytime, bath- and

1.3 Indoor conditions

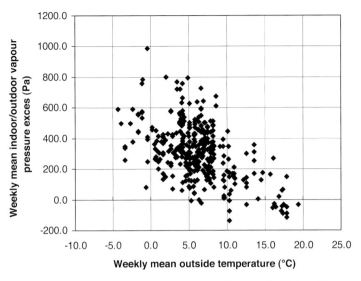

Figure 1.20. Dwellings and apartments: weekly mean inside/outside vapour pressure excess in relation to the weekly mean outside temperature.

Table 1.21. Dwellings and apartments, measured weekly mean indoor/outdoor vapour pressure excess and relative humidity (1976–1982).

	Number of records	$\Delta p_{ie} = a + b\,\theta_e$		Correlation coefficient r^2	$\varphi_i = a + b\,\theta_i + c\,\theta_e$			Correlation coefficient r^2
		a Pa	b Pa/°C		a %	b %/°C	c %/°C	
Daytime rooms	124	640	−35.4	0.50	99.4	−2.6	0.87	0.40
95% line		891	−35.4					
Bedrooms	177	323	−14.2	0.20	94.7	−3.1	1.9	0.47
95% line		503	−13.6					
Bathrooms	30	468	−24.5	0.64	94.5	−2.5	1.2	0.58
95% line		678	−24.5					

bedrooms. Also a multiple regression between relative humidity at one side and air temperature inside and air temperature outside is listed. Important as a first observation is that the weekly mean indoor/outdoor vapour pressure excess decreases with increasing weekly mean outdoor temperature. There are two reasons for that: first, residential buildings are more intensely ventilated in summer than in winter and secondly, the long term hygric inertia of the building enclosure and all furnishings which introduces a limited damping of the annual amplitude and some time shift compared to the outside. A second observation is that not the bathrooms but the daytime rooms are the most humid ones in residential buildings. As third observation one sees the relative humidity decreasing in all rooms with increasing air temperature indoors, whereas the outdoor temperature has the opposite effect, an increase.

Table 1.22. Dwellings and apartments, measured daily mean indoor/outdoor vapour pressure excess (2002–2005).

	$\Delta p_{ie} = a + b\, \theta_e$		Correlation coefficient r^2
	a Pa	B Pa/°C	
Daytime rooms	299	−12.4	0.53
95% line	594	−18.0	
Sleeping rooms	261	−10.7	0.18
95% line	497	−13.1	
Bathrooms	338	−11.3	0.64
95% line	658	−16.1	

Between 2002 and 2005, the data of Table 1.21 were reassessed on lasting validity by a new measuring campaign in 39 Belgian dwellings, see Table 1.22. The results were close-by in sleeping rooms and bath rooms but very different in daytime rooms. These were much dryer. The reason for that is unclear. Perhaps cooking hoods became standard equipment in kitchens.

Indoor swimming pools

Figure 1.21 collects 162 records of weekly mean indoor/outdoor water vapour pressure excesses measured in 16 different indoor swimming pools, together with the weekly mean outdoor air temperature. A least square analysis and the correlation coefficients are given in Table 1.23. Again, the weekly mean decreases with increasing outside air temperature. Noticeable is the large difference between the mean and the 95% line, indicating that indoor swimming pools range from humid to quite dry. The relative humidity for a temperature indoors of 27.5 °C lies also far below the comfort value of 60 to 70%, see Figure 1.22. Apparently the air is too dry in natatoriums. A possible explanation is that most indoor swimming pools have an enclosure which is not designed to tolerate a relative humidity at comfort level without severe moisture problems. Anyhow, keeping relative humidity as low as measured, demands intensive ventilation and leads to a high end energy consumption.

Table 1.23. Indoor swimming pools, measured weekly mean indoor/outdoor water vapour pressure excess and relative humidity.

Number of records	Line	$\Delta p_{ie} = a + b\, \theta_e$		Correlation coefficient r^2	$\varphi_i = a + b\, \theta_i + c\, \theta_e$			Correlation coefficient r^2
		a Pa	b Pa/°C		a %	b %/°C	c %/°C	
162	Mean	1229	−27.7	0.13	125	−2.8	0.36	0.17
	95%	1793	−27.7					

1.3 Indoor conditions

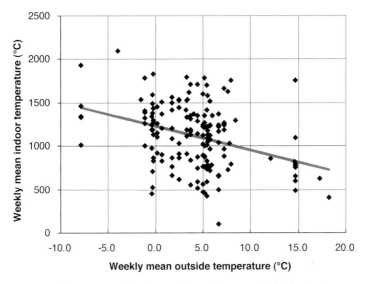

Figure 1.21. Indoor swimming pools: weekly mean inside/outside vapour pressure excess in relation to the weekly mean outside temperature.

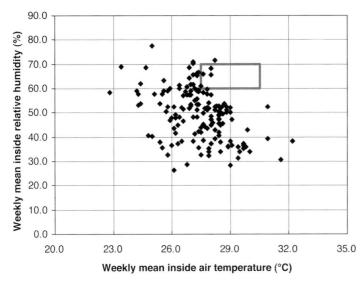

Figure 1.22. Indoor swimming pools: weekly mean inside relative humidity in relation to the weekly mean inside temperature, the rectangle delimitates the comfort zone.

1.3.3 Water vapour release indoors

The tables 1.24 to 1.26 list several literature data that apply for humans, plants and human activities.

Table 1.24. Water vapour release, in general.

	Activity	Vapour release g/h
Adults, metabolism	Sleeping	30
	At rest (depends on temperature)	33–70
	Light physical activity	50–120
	Moderate physical activity	120–200
	Heavy physical activity	200–470
Bathroom	Bath (15′)	60–700
	Shower (15′)	≤ 660
Kitchen	Breakfast preparation (4 people)	160–270
	Lunch preparation (4 people)	250–320
	Dinner preparation (4 people)	550–720
	Dish cleaning by hand	480
	Daily average release	100
Laundry drying	After dry spinning	50–500
	Starting from wet laundry	100–1500
Plants (per pot)		5–20
Jung trees		2000–4000
Full-grown trees		$2 \cdot 10^6 – 4 \cdot 10^6$

Table 1.25. Water vapour release in residences, depending on the number of family members.

Number of family members	Average water vapour release in kg/day		
	Low water usage[1]	Average water usage[2]	High water usage[3]
1	3 to 4	6	9
2	4	6	11
3	4	9	12
4	5	10	14
5	6	11	15
6	7	12	16

[1] Dwelling frequently unoccupied
[2] Average families with children
[3] Teenager kids, frequent showers, laundry dried indoors, etc.

1.3 Indoor conditions

Table 1.26. Water vapour release for an average family, no kids, the two partners working.

Hour	Number of people	Vapour release in g/h				Total
		Persons	**Cooking**	**Hygiene**	**Laundry**	
1–5	2	120				120
6	2	120	240	720		1080
7	2	120	240			360
8–17	0					0
18	2	120				120
19	2	120	480			600
20	2	120	480			600
21–24	2	120				120
Total		1680	1440	1200		4320

1.3.4 Indoor climate classes

The indoor/outdoor water vapour pressure excess is such important parameter when it comes to moisture tolerance that buildings have been grouped in indoor climate classes with the excess as pivot. Table 1.27 summarizes the classification as used in Belgium since 1982. The boundaries between the 4 classes reflect the following situations:

Class 1/2 Vapour diffusion does not cause interstitial condensation on a monthly mean basis in a continuously shaded air-tight low-sloped roof without vapour retarder, composed of non-hygroscopic materials and covered by a perfectly vapour tight roofing felt

Class 2/3 Vapour diffusion does not give accumulating condensate in a north-oriented, non-hygroscopic airtight wall without vapour retarder at the inside, covered by a vapour tight outside cladding

Class 3/4 Vapour diffusion results in accumulating condensate in the roof, used for the class 1/2 boundary, now radiated by the sun.

Especially in indoor climate class 4 a correct heat-air-moisture design of the building enclosure is crucial to avoid damage. In climate class 1, 2 and 3, it suffices to respect some simple design rules for avoiding interstitial condensation. The boundaries have been recalculated since, using transient modelling. That resulted in some limited differences in annual means except for the boundary between class 3/4. There an important shift was seen, confer the values between brackets in column 2 of Table 1.27.

EN-ISO introduced the indoor climate class concept in 2001, grouping buildings in five classes with as boundaries inbetween the values of Table 1.28 and Figure 1.23.

Important to know is that, when water vapour egress through envelope parts is governed by air egress and inside air washing, the indoor climate classes loose their applicability. They in fact were constructed to evaluate the slow phenomenon of interstitial condensation by diffusion, not to judge fast events such as interstitial condensation by air carried water vapour. Even climate class 1 buildings may experience problems then.

Table 1.27. Belgium, indoor climate classes ($\bar{\theta}_{e,m}$: monthly mean outside temperature).

Indoor climate classes	Annual mean $\Delta\bar{p}_{ie}$ Pa	Applies to
1–2	$150 - 8.9\,\bar{\theta}_{e,m}$ $(87 - 8.9\,\bar{\theta}_{e,m})$	Building with hardly any vapour release (dry storage rooms, sport arenas, garages, etc.)
2–3	$540 - 29\,\bar{\theta}_{e,m}$ $(550 - 29\,\bar{\theta}_{e,m})$	Buildings with limited vapour release per m³ of air volume and appropriate ventilation (offices, schools, shops, large dwellings, apartments)
3–4	$670 - 29\,\bar{\theta}_{e,m}$ $(1030 - 29\,\bar{\theta}_{e,m})$	Buildings with higher vapour release per m³ of air volume but still appropriately ventilated (small dwellings, hospitals, pubs, restaurants)
4	$> 670 - 29\,\bar{\theta}_{e,m}$ $(> 1030 - 29\,\bar{\theta}_{e,m})$	Buildings with high vapour release per m³ of air volume (indoor swimming pools, several industrial complexes, hydro-therapy spaces)

Table 1.28. EN-ISO 13788, indoor climate classes, boundaries.

Indoor climate classes	Annual mean $\Delta\bar{p}_{ie}$ Pa	
	$\bar{\theta}_{e,m} < 0$	$\bar{\theta}_{e,m} \geq 0$
1–2	270	$270 - 13.5\,\bar{\theta}_{e,m}$
2–3	540	$540 - 27.0\,\bar{\theta}_{e,m}$
3–4	810	$810 - 40.5\,\bar{\theta}_{e,m}$
4–5	1080	$1080 - 54.0\,\bar{\theta}_{e,m}$

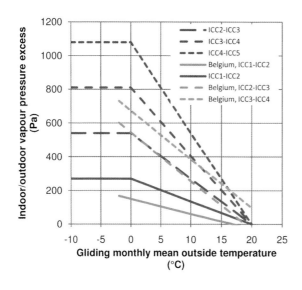

Figure 1.23. The indoor climate classes of EN ISO 13788 compared with the indoor climate classes as defined for Belgium in 1982.

1.3.5 Inside/outside air pressure differentials

Air pressure differentials by stack, wind and fans act as the driving force behind air in- and egress, air looping and air carried vapour and enthalpy displacement. That way, they cause thermal transmittance and thermal transient response to loose significance, induce interstitial condensation worries, highen energy consumption, activate drafts and degrade sound insulation.

Stack gives overpressure indoors in winter and possibly under-pressure in summer. Pressure differences by wind depend on its speed, related pressure coefficients along the building envelope and the air permeance distribution over the building fabric. When for example the envelope is most air permeable at the windside, than the air pressure in the building will track the mean overpressure there resulting in air egress at the leward side. Also forced-air heating and mechanical ventilation cause pressures differentials between spaces and across the envelope. Through it, industrial buildings may experience extreme situations. A clean room laboratory for example showed overpressures up to 150 Pa in the plenum above, while on the top-floor of the oast of a brewery even 300 Pa was registered.

Measurements in homes show stack as being the dominant force. Wind has mainly an impact on peak pressure differentials while differentials caused by mechanical ventilation are often negligible compared to stack and wind except when the building is very airtight.

1.4 References

[1.1] Poncelet, L., Martin, H. (1947). *Main characteristics of the Belgian climate*. Verhandelingen van het K.M.I. (in Dutch).

[1.2] Best, A. C. (1950). *The size distribution of rain drops*. Quart. J. Royal Meteor. Soc., Vol. 76, p. 16–36.

[1.3] WTCB (1969). *Study of the heat gains by natural illumination of buildings*. Part 1, Solar irradiation, Research report 10 (in Dutch).

[1.4] BRE (1971). *An index of exposure to driving rain*. Digest 127.

[1.5] Hens, H. (1975). *Theoretical and experimental study of the hygrothermal behaviour of building- and insulating materials by interstitial condensation and drying, with application on low-sloped roofs*. Doctoral thesis, KU-Leuven (in Dutch).

[1.6] Dogniaux, R. (1978). *Recueil des données climatologiques exigentielles pour le calcul des gains solaires dans l'habitat et l'estimation de la consommation de l'énergie pour le chauffage des bâtiments*. K.M.I. (Contract EG) (in French).

[1.7] ISO (1980). *Climatologie et industrie du bâtiment*. Rapport, Genève (in French).

[1.8] Hens, H. (1981). *Building Physics 2, practical problems and applications 2a*. Acco, Leuven (in Dutch).

[1.9] Davies, J., Mc Kay, D. (1982). *Estimating solar irradiance and components*. Solar Energy, Vol. 29, No. 1, pp. 55–64.

[1.10] Berdahl, P., Fromberg, R. (1982). *The thermal radiance of clear skies*. Solar Energy, Vol. 29, No. 4, pp. 299–314.

[1.11] WTCB tijdschrift (1982). *Moisture response of building components*, 1 (52 pp.) (in Dutch and in French).

[1.12] Anon. (1983). *Draft European Passive Solar Handbook*. Section 2 on Climate, CEC.

[1.13] Vaes, F. (1984). *Hygrothermal behaviour of light-weight ventilated roofs*. Onderzoek IWONL-KULeuven-TCHN, final report (in Dutch).

[1.14] Uyttenbroeck, J. (1985). *Exterior climate data for energy consumption and load calculations*. Report WTCB-RD-Energie (in Dutch).

[1.15] Uyttenbroeck, J., en Carpentier, G. (1985). *Exterior climate data for building physics applications*. Report WTCB-RD-Energie (in Dutch).

[1.16] Costrop, D. (1985). *The use of climate data for simulating of thermal systems*. Report KU-Leuven-RD-Energie (in Dutch).

[1.17] Liddament, M. W. (1986). *Air Infiltration Calculation Techniques*. An Application Guide, AIVC.

[1.18] Wouters, P., L'Heureux, D., Voordecker, P. (1987). *The wind as parameter in ventilation studies, a short description*. Report WTCB (in Dutch).

[1.19] IEA-Annex 14 (1991). *Condensation and Energy*. Sourcebook, Acco Leuven.

[1.20] Sanders, C. (1996). *Environmental Conditions*. Final report IEA-ECBCS Annex 24 'Heat, Air and Moisture Transfer in Insulated Envelope Parts', Vol. 2, ACCO, Leuven, 96 pp.

[1.21] Straube, J. (1998). *Moisture control and enclosure wall systems*. PhD-thesis, University of Waterloo (Canada).

[1.22] Blocken, B., Carmeliet, J., Hens, H. (1999). *Preliminary results on estimating wind driven rain loads on building envelopes: A numerical and experimental approach*. K. U. Leuven, Laboratory of Building Physics, 78 p.

[1.23] Straube, J., Burnett, E. (2000). *Simplified prediction of driving rain on buildings*. Proceedings of the International Building Physics Conference, TU/e, pp. 375–382.

[1.24] Blocken, B., Carmeliet, J., Hens, H. (2000). *On the use of microclimatic data for estimating driving rain on buildings*. Proceedings of the International Building Physics Conference, TU/e, pp. 581–588.

[1.25] ASHRAE (2001). *Handbook on Fundamentals*. Chapter 27, Climatic Design Information.

[1.26] EN ISO 13788 (2001). *Hygrothermal performance of building components and building elements-Internal surface temperature to avoid critical surface humidity and interstitial condensation*. Dalculation method.

[1.27] Van Mook, F. (2002). *Driving rain on building envelopes*. PhD-thesis, TU/e, 198 p.

[1.28] ASHRAE (2005). *Handbook on Fundamentals*. Chapter 27, Climatic Design Information.

[1.29] EN ISO 15927 (2002). *Hygrothermal performance of buildings*. Calculation and presentation of climatic data.

[1.30] Cornick, S., Rousseau, M. (2003). *Understanding the Severity of Climate Loads for Moisture-related Design of Walls*. IRC Building Science Insight, 13 p.

[1.31] Cornick, S., Djebbar, R., Dalgliesh, W. (2003). *Selecting Moisture Reference Years Using a Moisture Index Approach*. IRC Research Paper, 19 p.

[1.32] Blocken, B. (2004). *Wind-driven rain on buildings, measurement. Numerical modelling and applications*. PhD-thesis, K. U. Leuven, 323 p.

[1.33] Anon. (2006). *Energy Performance Decree*. Text and Addenda (in Dutch).

[1.34] Kumaran, K., Sanders, C. (2008). *Boundary conditions and whole building HAM analysis*. Final report IEA-ECBCS Annex 41 'Whole Building Heat, Air, Moisture Response, Vol. 3, ACCO, Leuven, 235 p.

[1.35] ASHRAE (2009). *Handbook of Fundamentals*. Chapter 27, Climatic Design Information.

[1.36] CD with weather data from K. U. Leuven, Laboratory of Building Physics, 1996–2008.

2 Performance metrics and arrays

2.1 Definitions

A building's physical properties and qualities plus possible functional qualities become performance metrics when expressible in an engineering way, predictable during design and controllable during and after construction. The difference between properties and qualities is quite subtle. The first do not reflect a graded judgment whereas the second do. Area and thickness of a wall for example are properties. The thermal transmittance figures instead as a quality. A low value is seen as beneficial, a high value as undesirable. The statement 'in an engineering way' paves the way for performance requirements and reference values. The latter may be imposed legally or demanded by the principal. Some are so obvious that they are imperative.

Typical for performance values is their hierarchical structure, top down, with as highest level the built environment (level 0), then the building (level 1), building components (level 2) and finally layers and materials (level 3). A correct performance rationale starts with requirements at the levels 0 and 1 to descend to the levels 2 and 3, a track that gives designers the greatest leeway.

2.2 Functional demands

Performance metrics emanate from sets of functional demands. That is called the fitness for purpose approach. At the built environment level the demands include accessibility by public transport, quality of the environment, proximity of shops, parking infrastructure, etc. At the building level, the functional demands include safety, accessibility for disabled, communication infrastructure, flexibility, sanitary facilities, intensity of maintenance, primary energy consumption, durability, sustainability, functionality, ergonomics, thermal comfort, acoustical comfort, visual comfort, indoor air quality, etc. Of course, form, space and architectural quality are also important assets.

2.3 Performance requirements

Performance requirements turn the functional demands into the engineering metrics mentioned. These metrics are predicted during design, when the building exists on paper, which is why calculation, computer simulation and prototype testing are used. Performance requirements in fact only make sense when such methodologies exist.

Control during and after construction demands testing methods and protocols. For some requirements such as bending of a floor or beam, air-tightness, reverberation time, sound and contact noise reduction, thermal comfort, acoustical comfort and visual comfort, testing is quite simple. For others, such as thermal transmittance, end energy consumption or moisture response, testing is complex and time-consuming. Therefore, prototype testing and quality guaranty often replace control during and after construction, which means that a well established agreement policy should exist. For that purpose, certification institutes have been founded at the national level, Europe-wide (EOTA, European Organisation for Technical Approvals) and worldwide (WFTAO, World Federation of Technical Assessment Organisations). EOTA delivers the Euromark for free traffic of building materials and building components between the 27 member states, while WFTAO fosters worldwide cooperation between continents.

2.4 Some history

The interest in a performance-based design approach dates from the seventies of last century. The driving forces were a better knowledge in fields such as structural mechanics, material science, building physics and building services but also the energy crisis of the seventies, increased complaints about sick buildings, a growing interest in sustainable construction and the search for upgraded safety. The process got accelerated by a growing quality consciousness and increased quality insurance demands. However some barriers continue to retard that evolution. Many specifications remain descriptive, imposing what materials are to be used and how they should be used. Too many designers also know only one performance requirement 'according to the art of construction'. That was acceptable in times building was based on tradition but today, with waves of new materials, new components, new systems and new demands engulfing the markets, nobody exactly knows what the art of construction means anymore. The sentence also does not permit enforcing correct and objective quality metrics by contract.

In the nineteen seventies, a Belgian inter-industrial study group wrote a first 'Performance Guide for Buildings', which today can be seen as an early trial to produce coherent sets of performance-based specifications. The guide formulates requirements at the whole building level, followed by requirements at the facade, roof, inside partition walls, floors, staircases, water distribution system, waste pipe installations, heating system, ventilation system, electrical system, transport system and communication level. The metrics applied were based on the functional demands advanced by ISO DP 6241:

1. Structural integrity
2. Fire safety
3. Safety at use
4. Tightness (water and air)
5. Thermal comfort
6. Indoor air quality
7. Acoustical comfort
8. Visual comfort
9. Contact comfort
10. Vibratory comfort
11. Hygiene
12. Functionality
13. Durability
14. Economy

Sustainability and energy were not on the list. Sustainability was not yet an issue in the seventies whereas energy was considered as economics, rather than being a requirement on its own.

The international interest in a performance-based approach grew from the eighties on. Within CIB (International Council for Research and Innovation in Buildings) a working group 'Performance Concept in Buildings' was established. A real break-through came with the European CPD's (Construction Products Directive) which differentiated between six groups of functional demands:

1. Structural safety
2. Fire safety
3. Health, hygiene, environment
4. Safety at use
5. Acoustical comfort
6. Energy efficiency

Building systems, building components and building materials have to comply with these demands before they can be traded freely among member states. A full application of a performance approach was introduced in the Netherlands, with their Building Decree, first published in 1992 and upgraded in 2000. Later-on, also Canada, Australia and New Zealand made their building regulations mainly performance based.

2.5 Performance arrays

2.5.1 Overview

In the framework of a research project by the International Energy Agency's Executive Committee on Energy Conservation in Buildings and Community Systems, quite complete performance metrics were proposed at level 1, the building, and level 2, building components, see Table 2.1 and Table 2.2. Both arrays reflect the different knowledge fields upon which construction is based.

Table 2.1. Performance array at the building level (level 1).

Field		Performances
Functionality		Safety when used
		adapted to usage
Structural adequacy		Global stability
		Strength and stiffness against vertical loads
		Strength and stiffness against horizontal loads
		Dynamic response
Bilding physics	Heat, air, moisture	Thermal comfort in winter
		Thermal comfort in summer
		Moisture tolerance (mould, dust-mites, surface condensation)
		Indoor air quality
		Energy efficiency
	Sound	Acoustical comfort
		Room acoustics
		Overall sound insulation (more specific: flanking transmission)
	Light	Visual comfort
		Day-lighting
		Energy efficient artificial lighting
	Fire safety[1]	Fire containment
		Means for active fire fighting
		Escape routes
Durability		Functional service life
		Economic service life
		Technical service life
Maintenance		Accessibility
Costs		Total and net present value, life cycle costs
Sustainability		Whole building life cycle assessment and evaluation

[1] In countries like The Netherlands, Germany and Austria fire safety belongs to building physics. In other countries, it doesn't.

Table 2.2. Performance array at the building component level (level 2).

Field		Performances
Structural adequacy		Strength and stiffness against vertical loads
		Strength and stiffness against horizontal loads
		Dynamic response

Building physics	Heat, air, moisture	Air-tightness
		    Inflow, outflow
		    Venting
		    Wind washing
		    Indoor air venting
		    Indoor air washing
		    Air looping
		Thermal insulation
		    Thermal transmittance (U)
		    Thermal bridging (linear and local thermal transmittance)
		    Thermal transmittance of doors and windows
		    Mean thermal transmittance of the envelope
		Transient response
		    Dynamic thermal resistance, temperature damping and admittance
		    Solar transmittance
		    Glass percentage in the envelope
		Moisture tolerance
		    Building moisture and dry-ability
		    Rain-tightness
		    Rising damp
		    Hygroscopic loading
		    Surface condensation
		    Interstitial condensation
		Thermal bridging
		    Temperature factor
		Others (i.e. the contact coefficient)

	Acoustics	Sound attenuation factor and sound insulation
		Sound insulation of the envelope against noise from outside
		Flanking sound transmission
		Sound absorption
	Lighting	Light transmittance of the transparent parts
		Glass percentage in the envelope
	Fire safety[1]	Fire reaction of the materials used
		Fire resistance
Durability		Resistance against physical attack (mechanical loads, moisture, temperature, frost, UV-radiation, etc.)
		Resistance against chemical attack
		Resistance against biological attack
Maintenance		Resistance against soiling
		Easiness of cleaning
Costs		Total and net present value
Sustainability		Life cycle analysis profiles

[1] In countries like The Netherlands, Germany and Austria fire safety belongs to building physics. In other countries, it doesn't.

2.5.2 In detail

2.5.2.1 Functionality

Functionality belongs to the CPD's safety at use. As a performance value, it has an important impact on floor lay-out, room dimensions and auxiliary space inclusion, on the kind of finishes used, etc. A bedroom for example must be high enough to accommodate a wardrobe with upper shelf. Balustrades are 0.9 m high or more. A doorway measures 0.9×2.1 meters. Enough space should be left under the windows for radiators. A kitchen floor may not suck fat. Ventilation grids have to be burglary-save and cleanable.

A detailed discussion on functional requirements can be found in the appropriate literature.

2.5.2.2 Structural adequacy

For the adequacy of a structure three conditions have to be met:

- **Stable.** Foundations and retaining walls must respect sliding equilibrium. A building may neither slide nor turn over under extreme wind. An envelope panel must remain vertical under extreme wind, a roof truss may not sag, etc
- **Allowable stresses, no buckling.** Here we look to own weight, death load and meaningful combinations of live load, static and dynamic wind, snow and accidental loads by earthquakes and skewedness. After a first design, a control starts with equilibrating forces and moments (normal forces, bending moments, shear forces and torsion moments), transposing these into stresses, evaluating if allowed, judging buckling equilibrium, etc. If problems arise, the design should be corrected and controlled again
- **Allowable settings and deformations.** Setting follows from the compressibility of the soil, while deformations relate to the stiffness of the load bearing structure. Depending on function and finishes, allowable deformations are usually small in buildings, although floor tiles for example demand a floor slab with higher stiffness than a plastered ceiling does. A false ceiling is even more forgiving, etc.

Besides, there are specific requirements. To give an example, acceleration in high rises due to wind gusts must remain below allowable limits.

2.5.2.3 Building physics related quality

This coincides with what is called health, hygiene and environment, acoustical comfort and energy efficiency in the CPD's. Building physics related performances covers a whole set of heat-air-moisture, acoustical and visual requirements, as listed in Table 2.1 and 2.2. The heat-air-moisture requirements are discussed in detail in Chapters 3 and 4.

2.5.2.4 Fire safety

Although everyone hopes fire will never happen during the lifespan of a building, the consequences are so devastating that many countries impose measures to keep fire risk as low as achievable by law 'as achievable' indicating that zero risk is not affordable.

The performance metrics consider three levels: (1) building, (2) building fabric and (3) materials and furnishings. Materials and furnishings are what burns. As far as the fabric is concerned, fire resistant building elements should curtail the fire, while load bearing parts such as columns, beams, floor slabs and stiffening walls should withstand collapse long enough to allow evacuation. A correct lay-out and the possibility for active fire fighting must help in containing the flames and facilitate evacuation.

Safe evacuation is at the core of any fire policy. All requirements in fact must safeguard a long enough evacuation time. A second concern of course is keeping fire damage as low as possible. Specific requirements are:

- Level 3, materials and furnishing

  A distinction is made between inflammable and flammable materials. Burning is an exothermal oxidation reaction. Inflammable materials hardly oxidize or oxidize so slowly that the heat produced is too insignificant to cause trouble. When it comes to fire reaction, the following aspects are looked for: flammability, flame spread, smoke release and napalm behaviour. The European standard differentiates between six classes, A1 and A2 for inflammable materials and B, C, D and E for flammable materials, with the steps from B to E designating increasing flammability. Untested materials get a classification F, fire reaction unknown.

- Level 2, building fabric components

  Fire resistance is the governing property. It gives the time in minutes between the start of a standardized fire test and the moment that one of the following events occur: (1) temperature at the other side becomes high enough to propagate fire by radiation, (2) the element is no longer smoke-tight and (3) the element loses its structural integrity. The three are controlled according to standardized procedures. Fire tests on walls and floors consider all three events with a score in hours per event, while columns are tested for structural integrity only.

- Level 1, building

  The goals here are:

  1. Keep local fire risk as low as achievable
  2. Keep fire from spreading to neighbourhood buildings as best as can be achieved.
  3. Guarantee structural integrity during a sufficiently long time span
  4. Guarantee enough time for building users to be evacuated
  5. Create optimal working conditions for the fire brigade

  That requires a large enough fire spread length between the windows on successive floors and defines the extent of fire zones in the building. It also fixes the number, location and width of escape routes, stair cases included, and the presence of active means for fire-fighting.

Regulations start with the fire reaction of the finishes and furnishings and end with zoning, escape routing and active fire fighting. Severity of the requirements depends on building height, fire load per m^2, assumed familiarity of the user with the building lay-out, ease of evacuation, the value of the furnishings, etc.

In the last decade CFD opened up possibilities to simulate fire and its consequences numerically. The calculations start from the fire load per m^2, its distribution over the zone, the area, form and dimensions of the zone, thermal properties of the separation walls, structural characteristics of the load-bearing part and the nature and magnitude of the ventilation flows (which supply oxygen). Simulation opened the door for a more advanced fire safety design.

2.5.2.5 Durability

One of the boundary conditions for a long lasting durability is a correct heat-air-moisture design. Of course, there are other important requirements. Some of them are:

Material	Rquirements other than moisture tolerance
Concrete	Frost resisting, correct water/cement factor, well compacted, reinforcement correctly covered with concrete, limited diffusion of Cl-ions
Plaster	Frost resisting, sufficient ductility, good resistance against fatigue
Natural stone	Frost resisting, good resistance against salt attack
Bituminous felts	Sufficiently UV-resistant, flexible, good resistance against fatigue
Plastics	Sufficiently UV-resistant

Some building components demand specific requirements. The hinges of operable windows and doors must continue to function properly even after opening and closing them tens of thousands times. Also, windows and doors may not bulge.

2.5.2.6 Maintenance

Nobody doubts that easy maintenance is an important asset when talking about durability and costs. 'Easy' points to accessibility, correct dimensions, form and finish. A nice example is given by those spaces where strict hygienic and immunological rules have to be accounted for (surgery wards, production spaces in the micro-electronic and pharmaceutical industry, centres for haemic treatment, etc). All surfaces should be hard with high resistance against impact load. Surface finishes must be easily disinfected (water-tight and no chemical attack by the disinfectant). Corners are rounded, electrical switches and sockets are water-tight and pipes and wires are mounted in shafts.

Of course maintenance should be correctly managed. Soiling loads differ between building parts and there are areas that demand more regular cleaning whereas others need less.

2.5.2.7 Costs

Reference is made to Chapter 3, where some of the performance metrics at building level are discussed in detail.

2.6 References

[2.1] IC-IB (1979). Performance Guide for Buildings, 9 tomes (in Dutch and in French).

[2.2] Building decree (1991). Staatsblad, 680, Staatsdrukkerij, Den Haag (adapted version: 2000) (in Dutch: Het bouwbesluit).

[2.3] Lovegrove, K., Borthwick, L., Bowen, N. (1995). *New Draft Performance Based Building Code Based on a Pyramid of Principles.* Australian Building Codes Board News, September, pp. 6–9.

[2.4] Hens, H. (1996). *The performance concept: a way to innovation in construction.* Proceedings of the CIB-ASTM-ISO-RILEM 3rd International Symposium on Application of the Performance Concept in Buildings, Tel-Aviv, December 9–12, p. 5–1 to 5–12.

[2.5]   CERF (1996). *Assessing Global Research Needs.* Report #96-5061, Washington DC.

[2.6]   CERF (1996). *Construction Industry: Research Prospectuses for the 21th Century.* Report #96-5016, Washington DC.

[2.7]   Lstiburek, J., Bomberg, M. (1996). *The Performance Linkage Approach to the Environmental Control of Buildings.* Part I, Construction Today. Journal of Thermal Insulation and Building Envelopes, Vol. 19, January, pp. 244–278.

[2.8]   Lstiburek, J., Bomberg, M. (1996). *The Performance Linkage Approach to the Environmental Control of Buildings.* Part I, Construction Tomorrow. Journal of Thermal Insulation and Building Envelopes, Vol. 19, April, pp. 386–402.

[2.9]   Hendriks, L., Hens, H. (2000). *Building Envelopes in a Holistic Perspective.* Final Report IEA, EXCO ECBCS Annex 32, ACCO, Leuven.

[2.10] Vitse, P., Vandevelde, P., Thylde, J. (2003). *European test methods and classification of the fire reaction of building products.* WTCB-tijdschrift, 2^e trimester, p. 27–36 (in Dutch and in French).

[2.11] Hens, H, Rose, W. (2008). *The Erlanger House at the University of Illinois_A Performance-based evaluation.* Proceedings of the Building Physics Symposium, Leuven, October 29–31, pp. 227–236.

[2.12] Becker, R. (2008). *Fundamentals of Performance-Based Building Design.* Building Simulation, Vol. 1, No. 4, December.

3 Functional requirements and performances at the building level

Chapter 1 looked at the environmental loads on a building and its components. In Chapter 2 performance metrics were focussed on, resulting in two arrays, one at the building level, called level 1, and one at the building parts level, called level 2.

Chapter 3 concentrates on part of the metrics at the building level. Six of those listed in the array are discussed in detail: (1) thermal comfort, (2) health and indoor air quality, (3) energy efficiency, (4) durability and service life, (5) economics with total and net present value as key elements and (6) sustainability with green building assessment as main element. Existing knowledge is broad enough to formulate requirements for these six,.

3.1 Thermal comfort

3.1.1 In general

Feeling comfortable is typically defined as a condition of mind that expresses satisfaction with the environment. A prerequisite is healthiness and satisfaction with the living, housing and working conditions, whereas a series of environmental parameters must have values that satisfy the individual: air temperature, radiant temperature, air speed and relative humidity when thermal comfort is at stake, sound pressure levels when acoustical comfort is at stake, luminance, glare and light colour when visual comfort is at stake and odours when olfactory comfort is at stake

Evaluation of an indoor space on comfort satisfaction demands knowledge of the impact of each of these environmental parameters. Relationships of that kind are well established for thermal, acoustical and visual comfort. while research indicated that dissatisfaction with one field of comfort is hardly compensated by a better adjustment of the other two. Of the four fields mentioned, thermal comfort is the most critical one, as satisfaction is directly linked to human physiology. The other three involve one of the senses only. Although the lack of a sense because of long-term overload or injury is highly undesirable, it is not as threatening as adverse thermal conditions may be.

3.1.2 Physiological basis

Human sensitivity to the thermal environment originates from two physiological facts found with all warm-blooded creatures: being exothermic and homoeothermic.

Exothermic means that a human needs energy for his metabolism (M), i.e. the whole of food-consuming chemical activity enabling cell growth, dead cell replacement, respiration, heart and liver activity, digestion, brain activity and others. Metabolism forces humans to eat regularly. If not, the body first consumes its own fat reserves and after that muscle tissue to sustain metabolism. The blood takes up the metabolic heat (Φ_M in W) resulting from that chemical activity and helps in transmitting it to the environment.

Table 3.1. Metabolic rates.

Activity	Air speed m/s	Metabolic rate W/m²	Heat produced W	Power produced W
Rest				
Sleeping	0	41	41	
Lying	0	46	46	
Sitting	0	58 (= 1 Met)	58	
Standing	0	70	70	
Light activity				
Laboratory	0	93	93	
Teaching	0	93	93	
Car driving	0	58–116	58–116	
Cooking	0	96–116	96–116	
Typing	0	62–68	62–68	
Studying	0	78	78	
Normal activity				
Walking on flat terrain				
3.2 km/h	0.9	116	116	
4.8 km/h	1.3	150	150	
Climbing				
3.2 km/h	0.9	174	156	18
4.8 km/h	1.3	232	206	26
Assembling	0.05	128	128	0
Dancing	0.2–2.0	139–255	139–255	0
Heavy activity and sport				
Spadework	0.5	348	279	69
Tennis	0.5–2.0	267	240–267	0–40
Basketball	1.0–3.0	441	397–441	0–44

The term basic metabolism (M_o), ≈ 73 W, relates to the energy needed by a 35 years old male, 1.7 m tall, weighting 70 kg, who is sleeping in a thermally neutral environment, 10 h after his last meal. When waking up metabolism increases. A metabolic rate (M_A) of 58 W per m² body surface is called **1 Met**. For an overview of rates at different activity levels, see Table 3.1.

Physical labour turns part of the metabolism into mechanical power (P_M in W). The metabolic equilibrium so becomes: $M = P_M + \Phi_M$. With η_{mech} the mechanical efficiency of the body, power P_M may be written as: $P_M = \eta_{mech} M$, resulting in a metabolic heat rate (q_M), given by:

$$q_M = (1-\eta_{mech})\frac{M}{A_{body}} = (1-\eta_{mech}) M_A \quad (W/m^2) \qquad (3.1)$$

3.1 Thermal comfort

where A_{body} is the body surface in m², a function of weight (M_{body} in kg) and stature (L_{body} in m):

$$A_{body} = 0.202 \, M_{body}^{0.425} \, L_{body}^{0.725} \tag{3.2}$$

For an adult male, 1.7 m tall and weighting 70 kg, the formula gives a body area of ≈ 1.8 m². Between metabolic heat rate (q_M) and mechanical power (P_M) one has as relations:

$$q_M = \left(\frac{1-\eta_{mech}}{\eta_{mech}}\right) P_M \qquad P_M = \left(\frac{\eta_{mech}}{1-\eta_{mech}}\right) q_M \tag{3.3}$$

Homoeothermic in turn indicates that the body temperature has to remain close to a set value of 36.8 °C. That set point rises somewhat after eating, excessive alcohol consumption or heavy activity, while also being subjected to a daily cycle: lowest in the morning and highest in the afternoon. Yet, the body temperature can only remain constant when equilibrium is guaranteed between the average metabolic heat produced and the average heat lost, or:

$$(1-\eta_{mech}) M = \sum \Phi_j \tag{3.4}$$

with $\sum \Phi_j$ the sum of all heat flows between the body and the environment.

Without the feed back between metabolism and heat loss, equilibrium would only exist accidentally, which is why human beings possess an autonomic control system that balances average metabolism and average heat loss using the hypothalamus as relay. That relay notes blood temperature continuously and receives temperature signals from the skin, the spinal cord and the mucous membrane in nose and bronchi. Out of these it distils a core temperature. The difference between set value and core temperature stimulates the physiological actuators. If positive, meaning that the environment is too cold for a given metabolism, then the subcutaneous blood stream slows down (may vary from 25 ml/(m² · s) down to 1.7 ml/(m² · s)), diminishing the heat loss through the skin, and people start shivering and chattering with a higher metabolism as a result. Subcutaneous blood stream of course cannot slow down indefinitely. When the average 'core' temperature drops below 28 °C, progressive hypothermia develops, first at the feet, hands, ears and nose but finally reaching the vital organs and the brain.

In case of a negative difference, meaning that the environment is too hot for a given metabolism, the subcutaneous blood stream boosts and sweating starts, increasing both the sensible and latent heat loss. The larger the absolute difference between set and sensed value, the more sweat is produced. People simultaneously feel languid. Again, there is a limit. When the core temperature passes 46 °C, hyperthermia starts with irreversible brain damage as a possible consequence. The core temperature the hypothalamus registers can be checked by measuring the temperature of the tympanum in the ear.

Thanks to their autonomic control system, humans can survive in thermal extremes. Too intense control interventions, however, are not perceived as comfortable. Absolute thermal comfort therefore confines with situations where metabolism and heat loss are equilibrated without control action, meaning it is only achievable at rest. Any activity automatically activates autonomic control, resulting in what may be called relative thermal comfort with limited subcutaneous blood stream and sweating changes.

The physiological approach suggests individuals react equally from a thermal point of view. This is not true. Subtle differences in core temperature, a more or less sensitive autonomic control system, thermal environments experienced induce variation. Illness also impairs the hypothalamus, while extreme activity restrains its functionality, etc.

3.1.3 Global steady state thermal comfort

Basic assumption behind the global steady state thermal comfort theory as developed by P. O. Fanger is that the clothed body reacts as a volume, whose surface temperature in a thermally homogeneous environment is uniform. Such an environment is one in which air temperature, radiant temperatures and relative humidity have a constant value all over the body while air speed is identical all-around the body. In such case the physiological reality is transposable in a system of algebraic equations with activity and clothing as adaptable personal parameters.

3.1.3.1 Clothing

Clothing helps in staying comfortable in environments that are too cold for the activities going-on. It in fact increases the thermal resistance of the skin, limit the heat loss that way and keep up the average skin temperature. The thermal resistance they add has clo as a unit, i.e. the ratio between the thermal resistance of an individual garment and the typical business dress (underwear, shoes, socks, straight trousers, shirt, tie and vest) with a thermal resistance of 0.155 m$^2 \cdot$ K/W, and I_{cl} as symbol. Table 3.2 gives clo-values for some garments and a series of dresses. The data listed are subjected to uncertainty. Values in fact vary between sources. The reason is that clo-values are quantified by several hours of measuring in a controlled climate chamber using a thermal manikin. These manikins slightly differ between test institutes.

The clo-value of a piece of clothing is assumed equal to the sum of the clo-values of the separate garments:

$$I_{clo,T} = \sum I_{clo,j} \tag{3.5}$$

Table 3.2. Clo-values.

Garments and clothing	I_{Clo}	f_{cl}
Nude	0	1
Brief	0.14	1
Long underwear top, long underwear bottom	0.35	1
Bra	0.01	1
Undershirt	0.08	1
Socks	0.02	1
Pantyhose	0.03	1
Short or swimming trunk	0.1	1
Office chair (is an extra, add to the garment)	0.15	1
Bed clothes (mattress, sheets and blankets)	> 1.6	1
Short, summer shirt, socks, sandals	0.3–0.4	1.05
Light summer clothing (straight trousers, shirt)	0.5	1.1
Female dress (bra, slip, pantyhose, skirt, shirt,)	0.7–0.84	1.1
Typical business dress	1.0	1.15
Typical business dress with raincoat	1.5	1.16
Polar outfit	> 4	–

3.1 Thermal comfort

A chair or seat counts for an additional clo-value, given by:

$$\Delta I_{clo} = 0.748 \, A_{ch} - 0.1 \tag{3.6}$$

where A_{ch} is the contact area between the chair or seat and the human body.

3.1.3.2 Heat flow between the body and the environment

Let us return to Equation (3.4). The right part contains the heat flows between the body and the environment. They consist of:

Sensible heat

Convection and radiation

Heat transmission across the clothing is given by:

$$q_{cl} = \frac{\theta_{skin} - \theta_{cl}}{0.155 \, I_{clo,T}} \tag{3.7}$$

with θ_{skin} uniform skin temperature and θ_{cl} uniform surface temperature of the clothed body. In practice, uniformity is replaced by averages. At the clothed body surface, transmission becomes convective exchange with the air (q_c) and radiant exchange with all surrounding surfaces (q_R), the convective part being given by:

$$q_c = h_c \, f_{cl} \, (\theta_{cl} - \theta_a) \tag{3.8}$$

where θ_a is the air temperature assumed uniform along the clothed body, h_c the mean convective surface film coefficient along the clothed body and f_{cl} a multiplier which account for the larger surface a clothed body has compared to the nude body (see Table 3.2, third column). For natural convection the mean convective surface film coefficient looks like:

$$h_c = 2.4 \, (\theta_{cl} - \theta_a)^{0.25} \tag{3.9}$$

In case of forced convection the expression becomes:

$$h_c = 12 \, \sqrt{v_a} \tag{3.10}$$

where v_a is the mean relative air speed along the clothed body (i.e. included any body movement).

The radiant part becomes:

$$q_R = 5.67 \, e_L \, F_T \, f_R \, f_{cl} \, (\theta_{cl} - \theta_r) \tag{3.11}$$

with f_R being a reduction factor as parts of the clothed body radiate to one another. e_L represents the average long-wave emissivity of the clothed body whereas θ_r is the radiant temperature of the environment. For a standing individual f_R equals 0.73. For a sitting individual that value reduces to 0.7.

Convection, radiation and linked heat transmission across the clothes may result in heat gain or heat loss at the skin. There is a gain when the difference between a weighted average of the air and radiant temperature ($a \, \theta_a + (1-a) \, \theta_R$) and the mean skin temperature is positive and a loss when that difference is negative. In cases of thermal comfort outdoors, solar gains and long wave losses to the sky should be accounted for at the clothed body surface.

Respiration

The sensible part in the heat exchanged by respiration is given by:

$$q_{bs} = 0.143 \cdot 10^{-2} \, M_A \, (34 - \theta_a) \qquad (3.12)$$

That formula combines the respirated sensible heat, equal to $c_a \, G_{a,b} \, (34 - \theta_a)$, where $G_{a,b}$ is the respirated airflow in kg/s and c_a the specific heat capacity of air in J/(kg · K), with the relation between respirated airflow and metabolism ($G_{a,b} = 1.43 \cdot 10^{-6} \, M_A \, A_{skin}$). At air temperatures above 34 °C, respiration results in sensible heat gain.

Latent heat

Perspiration and sweating

Perspiration and sweating induce a latent heat exchange calculated as

$$q_{PS} = l_b \, (0.06 + 0.94 \, f_w) \underbrace{\left(\frac{p_{sat,skin} - p}{\dfrac{1}{f_{cl} \, \beta} + Z_{clo,T}} \right)}_{B} \qquad (3.13)$$

where f_w is a factor representing the skin fraction moistened by sweat, l_b is the heat of evaporation for water in J/kg, p is the vapour pressure in the surrounding air in Pa, $p_{sat,skin}$ the mean vapour saturation pressure at the skin in Pa, β the surface film coefficient for diffusion at the clothed body surface in s/m and $Z_{clo,T}$ is the diffusion resistance of the clothing in m/s. The factor f_w becomes 0 for a metabolic rate below 58 W/m² and gets a value proportional to the ratio between the amount of sweat and the term called B in Equation (3.13) for a rate beyond 58 W/m². The water vapour flow rate lost by perspiration and sweating is given by the ratio between the latent heat released and the latent heat of evaporation (kg/(m² · h)).

Latent heat flow rate by respiration

The latent part in the heat exchanged by respiration is given by

$$q_{bl} = 3.575 \, M_A \, (0.033 - 6.21 \cdot 10^{-6} \, p) \qquad (3.14)$$

with 0.033 the water vapour ratio in the exhaled air in kg/kg.

3.1.3.3 Comfort equations

Dividing the left part of Equation (3.4) by the skin surface, transposing all heat flow rates into the formula and moving all latent and the sensible heat flow rate by respiration to the left as respiration, perspiration and sweating are not directly linked to clothing, results in:

$$(1 - \eta_{mech}) M_A - 0.143 \cdot 10^{-2} M_A (34 - \theta_a) - 3.575 M_A (0.033 - 6.21 \cdot 01^{-6} \, p)$$

$$- l_b \, (0.06 + 0.94 \, f_w) \left(\frac{p_{sat,skin} - p}{\dfrac{1}{f_{cl} \, \beta} + Z_{clo,T}} \right) = \frac{\theta_{skin} - \theta_{cl}}{0.155 \, I_{clo,T}} \qquad (3.15)$$

3.1 Thermal comfort

As mentioned, the right hand side of that equation equals the convective and radiant heat flow rate at the clothed body surface:

$$\frac{\theta_{skin} - \theta_{cl}}{0.155\, I_{clo,T}} = f_{cl}\left[h_c \left(\theta_{cl} - \theta_a\right) + e_L\, 5.67\, F_T\, f_R \left(\theta_{cl} - \theta_r\right)\right] \quad (3.16)$$

The two Equations (3.15) and (3.16) are a prerequisite for a globally comfortable environment.

However, for comfort as well the acceptable sweat rate (g_S in kg/(m^2 · s)) as the highest allowable mean skin temperature should be observed:

$$g_S = \max\left\{1.68 \cdot 10^{-7}\left[(1-\eta)\,M_A - 58\right], 0\right\} \quad (3.17)$$

$$\theta_{skin} = 35.7 - 0.0276\,(1-\eta)\,M_A \quad (3.18)$$

The combination shows sweating to lower the skin temperature. The reason is the latent heat exchanged. The Equations (3.15) to (3.18) make evaluation of overall steady state thermal comfort possible.

3.1.3.4 Comfort parameters

The named equations contain two types of variables: human being dependent and environment dependent. The first, which are handled as parameters, include activity and clothing with its specific long-wave emissivity – typically 0.9 although other values are possible – and the vapour resistance it adds to the skin. The second, which are the true variables, include air temperature, radiant temperature, relative humidity and air speed relative to the moving human body. In reality, (partial water) vapour pressure in the air and not relative humidity is the governing parameter. At any rate, as air temperature already intervenes, relative humidity is usable as a substitute for vapour pressure. Relative air speed in turn fixes the convective surface film coefficient in case of forced convection.

In theory, all combinations of the four environmental variables that satisfy the system of equations (3.15) to (3.18) define environments which are comfortable for a given activity and clothing. Since the system counts three equations for four variables, an infinite number of comfortable thermal environments can be found for any given metabolism and clothing.

3.1.3.5 Equivalent environments and comfort temperatures

Manipulating four environmental variables is complex. Therefore, since thermal comfort research started, those involved tried to replace the four by one using the concept of 'thermally equivalent environments'. As was mentioned, for given activity and clothing, an infinite number of comfortable thermal environments exist. These are called equivalent. Their infinite number must include environments with three preset variables. Assume for example equal radiant and air temperature, a relative humidity 100% and a relative air speed zero. Since at 100% relative humidity wet and dry bulb temperature coincide, only one unknown temperature value is left. That value is called the resulting or operative temperature, symbol θ_o. Its value characterizes all environments that are perceived as comfortable for a given activity and clothing. The value is found by solving the system of comfort Equations (3.15) to (3.18) for the conditions imposed. Other rules for equivalency are of course possible. If for example the relative humidity is assumed to be 50%, then the operative temperature turns into an effective temperature, symbol θ_{eff}.

In practice, thermal comfort is most critical in environments where people perform light manual labour or are intellectually active. These environments are rather steady state. Temperature fluctuations, if any, are moderate and slower than the time needed by the human body to adapt, whereas in moderate climates the following conditions are typically met:

Clothing	$I_{clo,T} \leq 1.2, \quad e_L = 0.9$
Inside air temperature	$15\ °C \leq \theta_a \leq 27\ °C$
Radiant temperature indoors	$\theta_a - \theta_r \leq 8\ °C$
Relative humidity indoors	$30 \leq \phi \leq 70\%$
Relative air speed indoors	$v_a \leq 0.2$ m/s

In such environments, respiration, perspiration and sweating dissipate some 30% of the metabolic heat. That makes convection and radiation the only environmental variables to be adjusted, simplifying the Equations (3.15) and (3.16) to ($F_T = 0.98, f_R = 0.7$):

$$0.7\ M_A = f_{cl} \left[h_c \left(\theta_{cl} - \theta_a \right) + 3.5 \left(\theta_{cl} - \theta_r \right) \right] \quad (3.19)$$

Assuming an equivalent environment with equal air and radiant temperature (θ_o) that expression reduces to:

$$0.7\ M_A = f_{cl} \left[h_c + 3.5 \right] \left(\theta_{cl} - \theta_o \right) \quad (3.20)$$

Equalizing both gives:

$$\theta_o = \frac{h_c\ \theta_a + 3.5\ \theta_r}{h_c + 3.5} \quad (3.21)$$

meaning that in those environments the operative temperature equals the weighted average between the air and the radiant temperature. For an average convective surface film coefficient equal to $2.4(6)^{0.25} \leq h_c \leq 2.4(14)^{0.25}$ (i.e. a temperature difference between the clothed body and the air between 6 and 14 °C) that weighted average hardly differs from the arithmetic mean:

$$\theta_o = \frac{(3.8\ à\ 4.6)\ \theta_a + 3.5\ \theta_r}{(3.8\ à\ 4.6) + 3.5} \approx 0.55\ \theta_a + 0.45\ \theta_r \approx 0.5 \left(\theta_a + \theta_r \right) \quad (3.22)$$

The result shows that in environments where people live or perform light manual labour or are intellectually active the air and radiant temperature are of equal importance. A too low value of the one may be compensated by a higher value of the other.

What comfort value the operative temperature should have, follows from the Equations (3.20)–(3.22). In winter, for an individual who is moderately active and whose dress has a clo-value of 1.2, about 21 °C seems optimal whereas in summer, with a clo-value down to 0.6–0.8, the operative temperature may reach 26 °C.

Of course, one should not forget that the operative temperature is only a true comfort parameter in environments where the conditions listed above are met. If not, one should return to the set of comfort Equations (3.15) to (3.18). Such exercises show that the relative humidity is the least important of the environmental variables. Only at air temperatures beyond 25 °C, does a too high relative humidity feels sultry. Of course, relative humidity has a broader impact than thermal comfort only. Humid air for example is typically perceived as less fresh. Another lesson is that high operative temperatures can be compensated by an increased air speed, which is why in hot climates ceiling fans ameliorate thermal comfort.

3.1.3.6 Comfort appreciation

Predicted Mean Vote versus Predicted Percentage of Dissatisfied

The fact that individuals quote thermal comfort differently was already mentioned. These differences emerge when thermal comfort appreciation is at stake. One in fact could think that, knowing the activity and the dress, all environments that receive a 'comfortable' evaluation according to the mathematical model will be perceived as such by everyone. Yet, even then, a percentage of individuals judge the environment as too cold, much too cold, too warm or much too warm. That turns comfort appreciation into a statistical reality and not a deterministic fact. The statistical approach that was developed by P. O. Fanger combines three concepts:

1. The <u>L</u>oad (L). Equals the difference between the metabolic heat produced and the heat lost (($1 - \eta_{mech}$) $M - \Sigma \Phi_j$)). The larger the load, the more the conditions drift away from a comfortable situation. A negative load marks environments that are colder than desired for a given activity and dress. A positive load instead marks environments that are warmer than desired for a given activity and dress. A load zero marks environments that fit with the comfort equations for a given activity and dress.

2. The <u>P</u>redicted <u>M</u>ean <u>V</u>ote (PMV). Through extended laboratory experiments simulating typical steady state office environments, involving thousands of individuals and testing thermal comfort as perceived using the thermal sensation scale of Table 3.3, the following statistical relation between load and mean vote, now called predicted, has been advanced:

$$\text{PMV} = \left[0.303 \exp(-0.036\, M_A) + 0.28 \right] L \tag{3.23}$$

If PMV > 3, then 3, if PMV < −3, then − 3

Table 3.3. The ASHRAE thermal sensation scale.

Vote	Thermal sensation
3	Much too warm
2	Too warm
1	Somewhat too warm
0	Neutral
−1	Somewhat too cold
−2	Too cold
−3	Much too cold

3. The <u>P</u>redicted <u>P</u>ercentage of <u>D</u>issatisfied (PPD). Using the results of the laboratory tests, a relationship was proposed between the predicted mean vote and the percentage of dissatisfied, again called predicted. Dissatisfied were those judging the environmental conditions as being too warm, much too warm, too cold or much too cold while wearing a standard dress with known clo-value and doing typical office jobs. As a formula:

$$\text{PPD} = 100 - 95 \exp\left[-\left(0.03353\, \text{PMV}^4 + 0.2179\, \text{PMV}^2 \right) \right] \tag{3.24}$$

See Figure 3.1. Even at PMV zero 5% of all individuals will statistically speaking still express dissatisfaction with perceived thermal comfort, half of them citing the conditions as too warm or much too warm and half of them as too cold or much too cold.

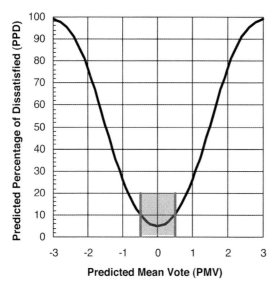

Figure 3.1. PPD versus PMV as proposed by P. O. Fanger and included in ISO 7730.

That 5% dissatisfied at PMV zero is an ongoing subject of discussion. Some researchers claim the real percentage lays statistically higher with a minimum of 15% dissatisfied at a PMV 0.5, whereas the curve should be asymmetric with more dissatisfied giving a negative vote than dissatisfied giving a positive vote (Figure 3.2).

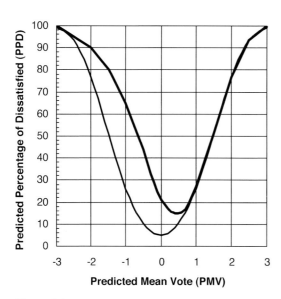

Figure 3.2. PPD versus PMV as proposed by E. Mayer of the Fraunhofer Institut für Bauphysik (Germany).

3.1 Thermal comfort

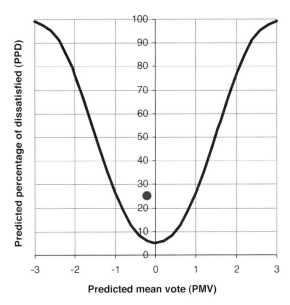

Figure 3.3. Comfort enquiry in an air conditioned office building: mean vote close to zero. The dot gives the number of dissatisfied counted.

Global thermal comfort is typically judged acceptable when PPD does not exceed 10%. According to the Fanger curve, that convenes with a PMV between –0.5 and +0.5. According to the Mayer curve less than 10% dissatisfied is not achievable.

Experiences with comfort surveys in air conditioned office buildings tended to confirm the Mayer results rather then agreeing with the Fanger curve. As Figure 3.3 underlines for one of the building surveyed, perceived comfort gave even more dissatisfied for a mean vote of zero than Mayer found in his experiments.

The adaptive model

P. O. Fanger published his book on Thermal Comfort with the PMV/PPD model in 1970. Already in the early eighties the model was questioned. It should not consider the psychological and adaptive aspects in perceiving conditions as thermally comfortable or not. It should overestimate the impact of the environment on everyone's comfort evaluation and underestimate expectation and adaptation by changing posture, slowing down activity or varying clothing. It should overlook the fact that other advantages such as energy economy may accommodate people with their thermal environment even when not comfortable according to the PMV/PPD-model. The importance of adaptation was clearly seen in the comfort response of employees who could adjust their thermal environment. On the average they complained less than employees who could not even when the conditions these experienced were in the comfort zone.

Field studies in naturally ventilated buildings showed a relationship between the operative temperature the employees preferred and the outside temperature, a fact which led to other values perceived as comfortable as the PMV/PPD-model predicted. Each author however advances different equations. Some for example couple the preferred operative temperature to the monthly mean outside temperature:

$$\theta_{o,comfort} = 11.9 + 0.534\, \theta_e$$
$$10\,°C \leq \theta_e \leq 34\,°C \quad (3.25)$$

Others refine the definition of outside temperature, for example by using a weighted value over a period of 4 days:

$$\theta_{e,ref} = \frac{\theta_{e,today} + 0.8\, \theta_{e,today-1} + 0.4\, \theta_{e,today-2} + 0.2\, \theta_{e,today-3}}{2.4} \quad (3.26)$$

That results in comfort equations, such as:

$$\theta_{o,comfort} = 20.4 + 0.06\, \theta_{e,ref}$$
$$\theta_{e,ref} \leq 12.5\,°C$$

As usual, the spread around the least square lines drawn through the measured values is large. Recent studies confirm that the PPD/PMV-model does very well in air conditioned buildings but is less accurate in naturally ventilated buildings. This, however, is contradicted in a German project on adaptive comfort, resulting in the following equations for the measured operative temperatures:

Office buildings	$\theta_e \leq 10\,°C$	$\theta_e > 10\,°C$
Heated only, natural ventilation	$\theta_o = 22.5 - 0.2\, \theta_e$	$\theta_o = 17.9 - 0.42\, \theta_e$
Heated, balanced ventilation	$\theta_o = 23.1 - 0.05\, \theta_e$	$\theta_o = 20.0 - 0.23\, \theta_e$
Air conditioning, window airing impossible	$\theta_o = 22.2 - 0.1\, \theta_e$	$\theta_o = 22.2 - 0.1\, \theta_e$
Air conditioning, window airing possible	$\theta_o = 23.4 - 0.05\, \theta_e$	$\theta_o = 23.4 - 0.05\, \theta_e$

The report did not mention what outside temperature was used (daily mean, weighted four days mean, monthly mean). A comfort enquiry, using the ASHRAE scale, revealed that the naturally ventilated offices were perceived as being too warm in summer but pleasing in winter. The other three cases scored neutral in both seasons.

Whether the adaptive model stays closer to reality then the PMV/PPD-model has still to be shown. As mentioned, some particular comfort studies seem to contradict the statement it is closer. In later publications, Fanger included adaptation in his PMV/PPD-model by multiplying PMV with an expectancy factor (e):

Expectancy	Boundary conditions	Value of e
High	Naturally ventilated office building in regions where air-conditioning is the norm Regions where heat waves are limited in duration	0.9–1.0
Moderate	Naturally ventilated office building in regions where air-conditioning is less common Regions where the summers are warm	0.7–0.9
Low	Naturally ventilated office building in regions where air-conditioning is the true exception Regions where it is hot the whole year	0.5–0.7

3.1.4 Local discomfort

Two of the assumptions behind global thermal comfort were: the human body is a system at a constant temperature and the environment is thermally homogeneous, characterized by one air temperature, one radiant temperature, one relative humidity and one relative air speed. Both assumptions misrepresent reality. The human body is not a system at a constant temperature and its skin does not sense the same operative temperature everywhere. On the contrary, the human body behaves as a complex thermal system: inert, partly clothed, blood pumped around continuously, other skin temperatures at the extremes than at the central part, etc. The air temperature differs between head and ankles, each part of the body senses other radiant temperatures, relative humidity changes somewhat along the clothed body and relative air speed varies considerably, depending on how the body and body parts move. Although computer modelling may help in mastering that complexity, still, for practical applications, the differences are bundled under the heading 'local discomfort'.

3.1.4.1 Draft

Draft complaints are quite common in buildings with air conditioning. The term reflects unwanted local cooling by air movement. Especially the neck, the lower back and the ankles are draft sensitive. A reason is the high convective surface film coefficient (h_c) noted there. Research showed their value not only depends on the mean relative air speed (in m/s), but also on the ratio between mean and standard deviation, called the turbulence intensity (T_u in %):

$$h_c = h_{co} + 0.27 \sqrt{v T_u} \quad (W/(m^2 \cdot K)) \tag{3.27}$$

where h_{co} is the convective surface film coefficient for natural convection and v is local mean air speed. Flow direction also intervenes, with flow touching the back being worse than one touching the chest, just as a flow coming from above is worse than a flow coming from below. The percentage of dissatisfied, called the draft rate (DR), equals (T_u in %):

$$DR = (34 - \theta_a)(\bar{v} - 0.05)^{0.62} (0.37 \bar{v} T_u + 3.14) \tag{3.28}$$

with $\bar{v} = 0.05$ m/s for $\bar{v} < 0.05$ m/s and DR = 100 for DR > 100. The equation does not distinguish between cold air flow directions. The ANSI/ASHRAE standard 55-2004 advances an acceptable percentage of dissatisfied due to local discomfort from draft of 20%.

3.1.4.2 Vertical air temperature difference

Figure 3.4 gives the percentage of dissatisfied (PD) as a function of the difference in air temperature between head and ankles for seated people. The ANSI/ASHRAE standard 55-2004 advances an acceptable percentage of dissatisfied due to local discomfort from vertical air temperature difference of 5%, which allows a difference of maximum 3 °C.

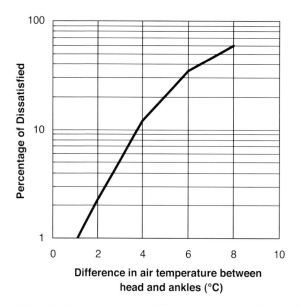

Figure 3.4. Air temperature difference between head and ankles, percentage of dissatisfied for seated people.

3.1.4.3 Radiant temperature asymmetry

The human body looks to two half-spaces, each with its own radiant temperature. Asymmetry causes problems when the difference between both passes certain values. Figure 3.5 gives the percentage of dissatisfied for radiant temperature asymmetry in the vertical direction.

When the upper half is colder, that percentage passes the 5% statistically called acceptable by the ANSI/ASHRAE standard 55-2004 for a difference with the lower half beyond 14 °C.

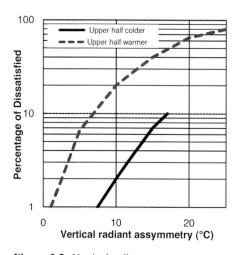

Figure 3.5. Vertical radiant temperature assymmetry, percentage of dissatisfied.

3.1 Thermal comfort

Table 3.4. Permitted mean temperature of a radiant ceiling.

$F_{h,c}$	0.06	0.08	0.10	0.12	0.14	0.16	0.18	0.20	0.22	0.24	0.26	0.28
θ_{pl} (°C)	50	43	38	35.2	33	31.3	30	28.6	27.5	26.8	26.0	25.8

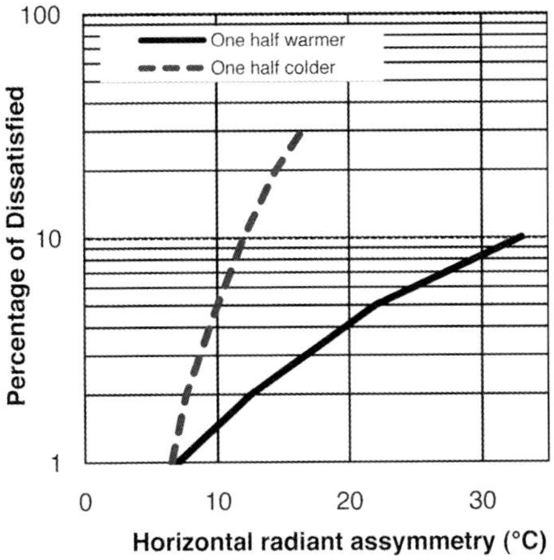

Figure 3.6. Horizontal radiant temperature assymmetry, percentage of dissatisfied.

For a warmer upper half, the limit is 4.5 °C. That number allows fixing the temperature of a heated ceiling that on the average should not be exceeded, depending on the view factor with the head (equal to the view factor between a dot and a surface), see Table 3.4.

Short peaks above are allowed. For example, at the design outside temperature (θ_{eb}) a heated ceiling may have a temperature $\theta_{pl,max} \leq 1.67\,\theta_{pl} - 0.67\,\theta_{o,comfort}$.

Figure 3.6 shows the percentage of dissatisfied for radiant temperature asymmetry in the horizontal direction. If a cold surface causes the asymmetry, the percentage passes 5% for a difference beyond 10 °C. A single glazed wall may be responsible for that, which is why Figure 3.6 is called the glass effect. A warmer surface hardly gives complaints.

3.1.4.4 Floor temperature

At the global comfort level transmission gain or loss to the floor via the feet were not included in the balance. From that, one may correctly conclude that its share in the heat exchanged with the environment is very limited. Still, feet comfort is a critical obstacle. The parameters of influence are: (1) floor temperature (θ_{fl}) and (2) contact coefficient with the floor cover (b_{fl}). The importance of both depends on (1) how long one's sole contacts the floor and (2) if the individual wears shoes or not. As an array:

Duration of the contact	Very short	Short	Long
Bare feet	$\theta_{fl} = F(b_{fl})$	$\theta_{fl} = F(b_{fl})$	$\theta_{fl} = F(b_{fl})$
Wearing shoes	θ_{fl}	$\theta_{fl}, (b_{fl})$	θ_{fl}

With bare feet, floor temperature perceived as comfortable will depend on the contact coefficient of the floor cover:

Floor cover	Floor temperature Duration of the contact		
	Very short	**Short**	**Long**
Carpet	21	24.5	21–28
Sisal carpet	23	25	22.5–28
Cork	24	26	23–28
Parquet (RNG)	25	25	22.5–28
Parquet (Oak)	26	26	24.5–28
Vinyl tiles	30	28.5	27.5–29
Vinyl on felt	28	27	27.5–28
Linoleum on planks	28	26	24–28
Linoleum on concrete	28	27	26–28.5
Concrete	30	28.5	27.5–29
Marmora	30	29	28–29.5
Tiles	30	29	28–29.5

For feet in shoes, Figure 3.7 gives the percentage of dissatisfied in locations where long lasting floor contact is the rule. 10% of dissatisfied allows a floor temperature between 19 and 28 °C. The highest value limits floor heating while floor cooling should respect the lowest value. In circulation zones, the limits are less severe, from 17 to 30 °C.

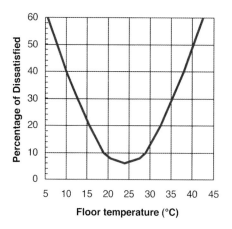

Figure 3.7. Feet with shoes, long lasting floor contact. Percentage of dissatisfied as function of floor temperature.

3.1 Thermal comfort

3.1.5 Transient conditions

A couple of rules should be respected. ASHRAE standard 55-2004 considers three factors: (1) the rate (°C/h) at which the temperature is increasing, (2) the rate (°C/h) at which the temperature is decreasing, and (3) the peak to peak amplitude of the fluctuations (cycling).

Temperature drifts refer to passive temperature changes in the indoor environment, while ramps refer to controlled temperature changes. For a steady state starting operative temperature between 21 and 23.3 °C, drifts and ramps should not pass the values given below over a time period between 0.25 h to 4 h.

Time period	0.25 h	0.5 h	1 h	2 h	4 h
Maximum operative temperature change allowed	1.1°	1.7°	2.2°	2.8°	3.3°

If peak cyclic variation in operative temperature during a period less than 15 minutes exceeds 1.1 °C, then temperature change shall not exceed 2.2 °C/h. If the peak to peak difference remains below 1.1 °C, then no restrictions exist on temperature change over 1 hour.

3.1.6 Comfort-related enclosure performance

Equation (3.22) with weighting factors 0.5 for the air and radiant temperature allows establishing a simple relation between the operative temperature and the mean thermal transmittance of a room enclosure needed for thermal comfort. All surfaces are assumed to be grey bodies. Then, the radiant temperature in the centre of the room is close to:

$$\theta_r = \frac{\sum (A_j \, \theta_{s,j})}{\sum A_j} \tag{3.29}$$

where A_j is the area and $\theta_{s,j}$ the temperature of surface j. The steady state relation between surface temperature and the thermal transmittance per wall looks like:

$$\theta_{s,j} = \theta_o - \frac{U_j}{h_i}(\theta_o - \theta_k) \tag{3.30}$$

In that equation θ_k is the reference temperature at the other side of the wall, equal to the operative temperature in the neighbour room for inside walls and the sol-air temperature outdoors for outside walls. Per outside wall l that equation may be reorganized as:

$$\theta_{s,l} = (1 - U_l/h_i)\theta_o + (U_l/h_i)\theta_e^*$$

For inside wall k separating the room under consideration from room k, reorganization gives:

$$\theta_{s,m} = (1 - a_k \, U_k/h_i)\theta_o + (a_k \, U_k/h_i)\theta_{o,k}$$

where a_k equals $(\theta_o - \theta_{o,k})/(\theta_o - \theta_e^*)$. That ratio turns zero when both rooms are at same operative temperature and approaches one when the separation wall is well insulated and the neighbour room is intensely ventilated with outside air.

Implementing both surface temperature relations in Equation (3.30) results in:

$$\theta_r = \theta_o - \frac{U_m}{h_i}(\theta_o - \theta_e^*) \tag{3.31}$$

where U_m is the mean thermal transmittance of the room enclosure:

$$U_m = \frac{\sum(A_l U_l) + \sum(a_k A_k U_k)}{\sum A_l + \sum A_k}$$

In that equation $\sum A_l$ refers to all exterior and $\sum A_k$ to all interior parts. Introducing (3.31) in Equation (3.22) then gives:

$$\theta_o = \frac{h_i \theta_a + U_m \theta_e^*}{h_i + U_m} \tag{3.32}$$

The lower the mean thermal transmittance, i.e. the better insulated the outside walls and the larger the inside surface share in the total enclosure, the smaller the difference between operative and air temperature. Writing (3.32) as $\theta_a = (1 + U_m/h_i)\theta_o - (U_m/h_i)\theta_e^*$ shows that a better thermal insulation allows lowering the air temperature without jeopardizing thermal comfort whereas end energy consumed by ventilation will be less. The envelope performance metric consists of fixing an upper limit for the mean thermal transmittance in case a maximum gap between air and radiant temperature is imposed.

3.2 Health and indoor environmental quality

3.2.1 In general

Habitation and health are so basic that their interaction has always inspired alternative theories. Quite popular during the last decades was the so-called bio-ecological construction practice, a faith based on a set of statements that are advanced as paradigms. To mention some:

- Everyone is subjected to terrestrial radiation. The lines of equal radiation form quadrants, which local water arteries disturb. These disturbances are found using the dowsing rod. Living inside a quadrant is healthy, dwelling on the quadrant lines may cause sleeplessness, headache and, in extreme cases, cancer
- Static electrical fields are curative while low frequency alternating field are unhealthy
- One's psychological and physiological equilibrium benefits from negative ions in the air. These are produced by an apparatus called the magic pyramid
- Receiving cosmic radiation assures wellbeing

Construction rules advanced are:

- Use healthy building, insulating and finishing materials. The adjective healthy refers to all materials where the basic substances are still recognizable. Materials that are manufactured chemically are unhealthy. They shield the curative cosmic rays, act as electrostatic bodies and emit all kind of noxious substances. Examples of healthy materials are loam, lime, brick (if hand-moulded), straw, timber, cork, cellulose, wax, etc. Unhealthy are concrete, polymers based on mineral oil, most metals, bitumen, etc.

- Walls must breath. Only healthy materials assure this will happen. Vapour retarding materials and vapour barriers should be excluded
- Radiant heat is healthier than convective heat. Heavy tile-stoves assure a perfect balance between both. This is the best way of heating.

Happily, their former aversion for thermal insulation and air-tightness that spoiled the early discussions on energy efficiency disappeared from the list. Today, bio-ecologists defend both options though only healthy insulation materials should be used and plastic air barriers may not be applied. They forward as alternatives for thermal insulation cellulose, sheep's wool, sea grass, straw, etc. That most of them behaves badly when humid and lack fire resistance if not treated chemically, does not concern them.

None of these bio-ecological paradigms and rules stand up to the laws of physics. But still they affect the public discussion about indoor environment and health. Many people simply prefer a fairy tale to a sober-minded scientific approach.

Yet, that there is a relation between habitation and health will not be denied by any scientist active in the field. Pollutants in the air people breath are found outdoors as well as indoors. To mention some: viruses, bacteria and fungal spores, house dust, volatile organic components, bio-odours, tobacco smoke, radon, nitrous oxides, carbon monoxide, vapours, fine dust and fibres. Moisture has a special status as it acts as a necessary boundary condition for fungi to germinate and grow, while humidity also increases out-gassing of several contaminants. Complaints about sick buildings and the sick building syndrome (SBS) even brought indoor air quality to the forefront of concern. SBS figures as catch-all for all ailments that could be the result of air pollution indoors. Moderate SBS-troubles include irritation of the eyes, the nose and the throat. Worse are all kinds of allergic reactions. Things turn dramatic when a polluted indoor environment causes infections and becomes carcinogenic. The sick building syndrome of course also bears a psychological component. Those who dislike their jobs or an individual who has to work with people she/he don't like, may seek SBS as escape route.

Since people in moderate climates spent up to 80% of their lifetime indoors, caring for health presumes that (1) the emission of harmful contaminants indoors should be minimized as much as achievable, (2) the concentrations left in the indoor air should be diluted by ventilation so they remain at acceptable levels and (3), if none of these help, supply air filtration must be considered or personal protective measures taken.

3.2.2 Health

The World Health Organization (WHO) defines health as a condition of physical, psychological and social well-being rather than an absence of disease or handicap. Related to the indoor environment, not only pathology but also uneasiness, perception and discomfort are therefore to be considered. The thermal aspects in this have been discussed above.

3.2.3 Definitions

Looking to pollutants and their effects, two notions surface: concentration and exposure time. Although not a paradigm, harmfulness generally aggravates at higher concentration and longer exposure time. All definitions therefore link both.

Emission	Pollutants' release. Is typically a function of time.
Immission	Pollutants' load. Immission outdoors and indoors depends on the distribution of pollutants in the air, where the receiving person is in space and his breathing rhythm and volume.
Ppm	Parts per million. The number of molecules of a component per million molecules in the mixture.
Radioactive radiation	Radiation emitted by decaying atom nuclei. Radioactivity consists of particle rays (α and β) and electromagnetic rays (γ). α-rays stand for emission of helium atoms, β-rays for emission of electrons.
Becquerel (Bq)	Unit of radioactivity. One Bq means the decay of one nucleus per second.
Sievert (Sv)	Unit of radioactive dose equivalent. The risk to develop biological effects is proportional to the Sievert received. One Sievert equals a concentration of 20 Bq/m^3.
AMP (USA)	Peak concentration allowed during 15'.
ACC (USA)	Highest concentration allowed during 15' along an exposure of 8 hours per day (with exclusion of the periods AMP applies).
MAC, TWA8 (USA)	Mean concentration allowed during 8 hours a day and 5 days a week.
DNEL (EU, REACH)	Maximum concentration allowed during a given exposure period (from peak to 24 hours a day).
AIC, NOAEL (EU, REACH)	Concentration a pollutant may reach without negative health effects.

3.2.4 Relation between pollution outdoors and indoors

Pollution indoors depends on what happens outdoors. No building is perfectly airtight and, if it was, entering and leaving people still allow unclean outside air to mix with inside air. Also ventilation uses outside air. As a result, each outdoor pollutant will contaminate the indoor air although not in the same concentration. Indeed, on the way inside the building, concentration decreases thanks to the filters in the ventilation system or to adsorption by the surfaces delimiting the leaks. Once inside, house dust and the surfaces present adsorb some of the pollutants. Some pollutants even react with one-another to form new compounds. And finally, dust from outdoors may deposit while migrating through complex envelope leaks.

The air from outside also has to fill a large space inside. That induces inertia in the build-up of pollutant concentrations indoors. That means that advice such as closing windows when a spike in outdoor pollution occurs makes sense. Yet, many pollution sources interact inside buildings! These are discussed below.

3.2.5 Physical, chemical and biological contaminants

3.2.5.1 Process related

Dust, vapours, smoke, mist and gaseous clouds

Process-related pollutants like dust, vapour, smoke, mist and gaseous clouds are typically emitted in industrial environments. Dust is produced during mechanical processing of materials. Vapours and gaseous clouds are the result of thermal treatment and chemical processes. Mist is caused by vaporisation of liquids. Smoke is linked to burning. In non-industrial environments traffic causes most of the dust that may enter buildings along busy roads.

Looking to dust, health effects mainly depend on the particle size, the nature and toxicity of the dust-forming agents, their solubility in biological liquids and the inhaled doses. Particles with sizes below 2.5 µm are the most feared. They penetrate into the lung alveoli, while dust from 2.5 to 10 µm in size is caught by the mucous membranes. In non-industrial environments most of the minus 2.5 µm particles consist of polycyclic aromatic hydrocarbons (PAH's) emitted by car traffic. This was confirmed by an investigation in four Chinese hospitals situated along busy roads. The concentration indoors and outdoors of PAH-loaded dust with particle size below 2.5 µm was consistently related, while by moving people whirled up deposited dust resulted in higher concentrations indoors than outdoors. PAHs are feared as carcinogens. They could be an important mortality cause although the proof is called flimsy by some.

For some dusts the following limits are set for a day-long exposure:

Dust	Limit mg/m^3
Cadmium	0.05
Manganese	1.0
Plaster	10.0
Corn	10.0
Silica fume	0.1
All dust together	15.0
Breathable dust	5.0

For vapours, smoke, mist and gaseous clouds, the diversity is impressive. Inhaling is hard to avoid, meaning that the allowable concentration should be strictly respected, be it by source control, by ventilation or by wearing masks.

Fibres

Fibres are mineral particles whose length to diameter ratio exceeds a value of 3, whose section remains constant along the length and whose diameter is less than 3 mm.

Asbestos, a natural silicate that splits along its length in ever thinner filaments, is a feared fibre. It passes the bronchia and enters the lung alveoli where the immune system gets activated, resulting in inflammation. Long time exposure to high concentrations may cause asbestosis or pleural asbestos calcification, which in turn initiates heart insufficiency. Occasional exposure to small concentrations of blue asbestos may cause mesotheliom, a deadly type of lung cancer. Very strict concentration limits are therefore imposed in indoor spaces:

	AIC Fibres per cm^3
Blue asbestos	0.0002 Directive CEN-IHE $L/D > 5$, $0.2 < D < 3$ mm, $L > 5$ mm
Other types of asbestos	0.001

Source control is the only strategy allowed. Using asbestos is forbidden. Buildings where asbestos has been used as thermal insulation or fire protection must be cleared. Asbestos cement as a product is not manufactured anymore. Very restrictive requirements are imposed when processing existing asbestos cement products.

Thanks to the large diameter of the fibres (3 to 5 µm), the great length (some centimetres), the fact they don't split in ever thinner fibres and the solubility in biologic liquids, glass and mineral wool, also silicate fibres but now called MMMF (man-made mineral fibres), are not carcinogenic. They however irritate the skin and the mucous membranes, may cause itchy dermatitis and eye irritation and give a temporary inflammation of the mucous membranes at high concentrations. OSHA imposes a TWA-value of 5 mg/m^3. When installing glass or mineral fibre insulation, workers must wear whole-body protecting garments and a dust mask

3.2.5.2 Material related

Volatile and semi-volatile organic compounds ((S)VOC's)

Volatile and semi-volatile organic compounds have boiling points between 50 and 250 °C. They are out-gassed by building and insulation materials, finishes, furniture, ventilation systems, consumption articles, polluted soils, micro-organisms and present in tobacco smoke. More than 1000 compounds are described in literature. These include benzene, toluene, *n*-alkenes, aromatic compounds, alkenes, cycloalkenes, pentane, heptanes, halogens, alcohols, ethers, aldehydes, ketenes, amides, etc. The amounts emitted depend on the source and the micro-

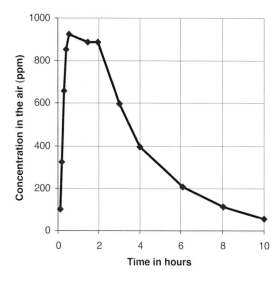

Figure 3.8. VOC concentrations after application of a wet finish.

3.2 Health and indoor environmental quality

climate in its vicinity (temperature, relative humidity, air speed). Some dry materials such as the binder in mineral fibre, polystyrene foam, polyurethane foam, timber, polymer floor covers, carpets and textiles show a limited but decreasing emission with time. The decrease is largest for VOC's with lowest boiling point. Wet finishing materials, such as paints, varnish and lacquers, have a boost in emission followed by a fast decrease, see Figure 3.8.

Part of the primary emission is at any rate adsorbed by house dust and by all inside surfaces and is re-emitted afterwards depending on the VOC-concentration in the air. This so-called secondary emission increases at higher surface temperatures. For certain materials, emission is also augmented when relative humidity goes up.

(S)VOC cause olfactory nuisances. They may irritate the eyes, the mucous membrane and the bronchi. Some are believed to be carcinogenic for people exposed to high concentrations during very long periods of time (the so-called painter's disease). The effects seem to be additive, which is why for safety reasons allowed concentrations should obey $\Sigma\ C_i/C_{accept,I} \leq 1$, with C_i the concentration and $C_{accept,i}$ the acceptable concentration for VOC i. The following array is used to judge possible harmful effects:

Total VOC concentration $\mu g/m^3$	Effects
$C_{VOC} < 200$	No irritation expected
$200 \leq C_{VOC} < 3000$	Possible irritation in combination with smoking
$3000 \leq C_{VOC} < 25000$	Irritation or nuisance expected
$C_{VOC} \geq 25000$	Neurotoxin effects possible

Investigations show that (S)VOCs are a problem worldwide though the mixture emitted differs between geographical regions. One of the reasons is the difference in materials applied. Yet, in most buildings the concentration per (S)VOC remains far below the olfactory limit and the concentration which could cause irritation. Remarkably, this does not exclude complaints. An assumption is that a summing-up effect may exist, though no objective data have been published that support that hypothesis. The summing-up effect should not hold for (S)VOCs only but also for the combination with other pollutants.

Formaldehyde

The VOC formaldehyde (HCHO) is a colourless gas, soluble in water, which is released by materials bonded with formalin or a phenol resin (such as chipboard, plywood and OSB) and by textiles. Emission increases with temperature and relative humidity. Formaldehyde irritates the mucous membranes of the upper and lower bronchi and evokes allergic reactions in sensitive people. The MAC-value equals 0.5 ppm.

Phthalates

Phthalates are organic compounds used in synthetic materials to keep them pliable and deformable. Their low boiling point is a problem, because it allows for release and absorption by house dust. They are suspected of irritating the bronchi and causing asthmatic complaints.

PCP's

Pentachlorinephenols were the active substance in wood preservation products. Investigation showed that they are highly carcinogenic and therefore no longer allowed.

3.2.5.3 Ground related

Radon

Radon is a radioactive inert gas belonging to the decay sequence of Uranium 238.

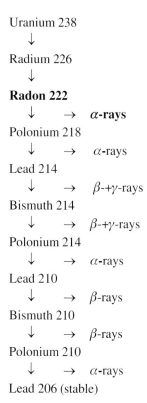

Uranium 238
↓
Radium 226
↓
Radon 222
↓ → α-rays
Polonium 218
↓ → α-rays
Lead 214
↓ → β-+γ-rays
Bismuth 214
↓ → β-+γ-rays
Polonium 214
↓ → α-rays
Lead 210
↓ → β-rays
Bismuth 210
↓ → β-rays
Polonium 210
↓ → α-rays
Lead 206 (stable)

Together with the cosmic rays and bottom-related γ-rays radon effused into the outside air causes background radioactivity outdoors,

Uranium 238 is found in small concentrations in primary rock formations. This makes the ground the most important radon source, with phosphoric gypsum, groundwater and the outside air a distant second. Thanks to its inert character and moderately long half life of 3.7 days, radon can effuse out of the bottom and spread in the atmosphere. The emission rate depends on the concentration in the ground, the diffusion resistance to be crossed and the presence of crevices purged by convection. Wind lowers the concentration outdoors to some 5 a 10 Bq/m³. Indoors instead, due to the limited ventilation rates, concentrations may reach much higher values. Measurements in Belgian dwellings gave:

Geographic region	Radon concentrations Bq/m³		
	Median	Mean	90%-value
Flanders, northern Wallonia, Southern Luxemburg	32	39	66
Remaining Wallonia	52	77	135

3.2 Health and indoor environmental quality

In problem buildings, concentrations above 1000 Bq/m^3 were measured. Radon is inhaled by the building users. While decaying in their lungs it emits α-rays, increasing lung cancer risk after long exposure times with the percentages given below:

Mean concentration Bq/m^3	Lung cancer risk %
100	1.5
200	3.0
400	6.0
800	12.0

In combination with smoking, lung cancer risk increases more than the sum of both risks. Also here some doubt anyhow exists about that direct cause/consequence model.

Allowable concentrations are:

Organisation	Buildings	
	Existing	New
International Commission for Radiological Protection, European Union and WHO	400	200

The difference between new and existing, reflects the fact that curing existing buildings with too high radon concentrations indoors is more difficult than preventing problems in new construction. It is possible to minimize immission by applying a radon-tight foundation floor in new buildings or radon-tightening the foundation floor in existing buildings. In both cases, diluting the concentration by correct ventilation and under-pressurization of the crawlspace is a possibility. In new buildings a solution is to under-pressurize a gravel bed below the foundation floor.

3.2.5.4 Combustion related

Combustion of fossil fuels releases water vapour, carbon dioxide (CO_2), carbon monoxide (CO), nitrous oxides (NO_x) and sulphur dioxide (SO_2). The dust formed is loaded with VOCs en PAHs. Water vapour and carbon dioxide are not harmful for health. Carbon monoxide and nitrous oxides are.

Carbon monoxide

Carbon monoxide is an extremely toxic, colourless, non-smelling gas released during incomplete combustion, though combustion under air excess will also produce it. Incomplete combustion points to lack of ventilation or a badly drawing chimney. When inhaled, carbon monoxide attaches to the haemoglobin of the red blood cells preventing them from adsorbing oxygen and disturbing the transport of oxygen to the cells that way. A person affected first loses consciousness and then dies. Even at very small concentrations, carbon monoxide is lethal. The MAC-value is 8.6 ppm, the AMP-value 86 ppm.

Nitrous dioxide

Nitrous dioxide (NO_2) is produced by combustion at such high temperatures that the oxygen in the air reacts with the nitrogen in the air. Long lasting inhalation at high concentration may inflame the lung tissue. At low concentrations nitrous dioxide aggravates breathing. The MAC-value is 0.08 ppm and the allowed concentration during a one hour exposure 0.2 ppm.

3.2.6 Bio-germs

3.2.6.1 Bacteria

Legionella pneumophila is a feared microorganism. The bacterium develops in 20 to 40° centigrade warm water, a temperature found in air humidifiers, cooling towers, domestic hot water pipes and spray installations. Legionella pneumophila propagates via water droplets in the air. When inhaled by persons with weakened immune response, it may cause a deadly lung inflammation known as the 'legionnaire's disease'. Combating the bacterium is quite simple. It suffices to boost domestic hot water to temperatures above 60 °C, to clean air handling units, to disinfect cooling towers regularly and to avoid stagnant water in long domestic hot water pipes and to mix hot and cold water close to the taps.

3.2.6.2 Mould

Moulds belong to the family of the micro-fungi. As all fungi, they have no leaf green activity, digest their food externally and reproduce by releasing spores. Once deposited on a suitable substrate, the life cycle includes spore germination, mycelium growth, spore formation and spore release. The following mould families are typically colonizing indoor surfaces (all deuteromycetes):

Family	In %.of the cases (14 studies)
Aspergillus's	93
Penicillium	85
Cladosporium	71
Aureobasidium	64
Alternaria	57

Mould spores are present in the air at all times. The risk of surface infestation depends on the conditions present close to and on that surface: some oxygen, the right temperature, carbon, nitrogen and salts in the substrate and the right humidity. Oxygen below 0.14 %m^3/m^3 suffices for germination and growth. The optimum temperature covers a range from 5 to 40 °C, which can be found on the inside surface of badly insulated exterior walls. More than minimal amounts of carbon, nitrogen and salts are not needed. The right humidity instead is critical. How high, differs between mould families and depends on temperature. That dependency and its impact on growth is reflected in mould specific isopleths. Figure 3.9 shows a simplified isopleth representing the lowest growth rate for aspergillus versicolor. As a formula:

$$\phi = 0.033\,\theta^2 - 1.5\,\theta + 96 \tag{3.33}$$

3.2 Health and indoor environmental quality

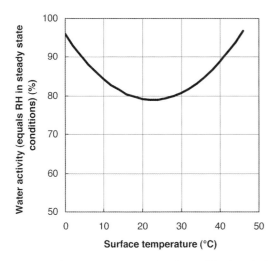

Figure 3.9. Lowest isopleths for Aspergillus Versicolor.

For isopleths closer to the optimal temperature/water activity combination, growth rate increases and the period before germination shortens. Beyond the optimum temperature/water activity combination, growth rate decreases again. Globally, at surface temperatures between 17 and 27 °C a monthly mean surface relative humidity of 80% suffices to see Aspergillus, Cladosporium and Penicillium start germinating on substrates with limited nutritive value. Other moulds, among them Stachybotrys, demand much higher surface relative humidity. Or, a direct relationship exists between long lasting surface relative humidity and mould risk.

Risk becomes 1 when the vapour pressure at a surface (p_s) passes the product of surface temperature related saturation pressure ($p_{sat,s}$) with the critical surface relative humidity that causes germination of the mould considered, in transient conditions called its water activity (a_w):

$$p_s \geq a_w \, p_{sat,s} \tag{3.34}$$

As mentioned, the lowest surface relative humidity on a monthly mean basis that enables germination to start is 80%, or, on a scale from 0 to 1 : 0.8. A higher value shortens the period (T) for germination:

$$\phi \geq \min\left[100, 100 - 5.8 \ln(T)\right] \tag{3.35}$$

Moulds may grow on any surface: inside, outside, even on filters in HVAC-systems and air handling units. In that case spores get distributed over the indoor space more intensely than when growing on inside surfaces.

In the last two decades intensive investigations have refined the tools for evaluating mould on different substrates. Research in Finland allowed predicting the critical surface relative humidity and the evolution in time of the mould infestation on pine and spruce through calculation of a growth index by a mixed population of moulds once surface relative humidity passes the critical value. That mould growth index was defined as:

Index	Growth rate	Description
0	None	Spores not active
1	Small spots of mould on the surface	Initial growth stages
2	Less than 10% of the surface covered	
3	10–30% of the surface covered	New spores produced
4	30–70% of the surface covered	Moderate growth
5	> 70% of the surface covered	Plenty of growth
6	Very heavy and tight growth	Coverage 100%

German researchers in turn proposed a classification of substrates in four classes according to mould sensitivity:

Class	Type
0	Optimal substrate (agar)
1	Bio-degradable materials
2	Typical building materials
K	Critical mould on an optimal substrate

They then considered the mould as an additional layer inside, with thickness defined by the growth rate, with a mould-specific vapour diffusion resistance and a mould-specific sorption isotherm.

Mould on inside surfaces has several annoying consequences. It damages the aesthetics and causes economic loss. In fact, people tend to clean the infected surfaces again and again, knowing that, if not, the market value of the dwelling decreases. Mould also upsets people and, although not convincingly proven, may harm health. Some assume that the mycotoxines moulds emit aggravate house dust allergy in sensitive people. Exposure to very high concentrations may impair the immune system. According to the John Hopkins School of Medicine the relation between mould on inside surfaces and allergies of the upper bronchi is not substantiated, while the literature on health effects is more agitational than scientific.

What is proven is that at high concentrations the spores of Aspergillus Fumigatus may colonize the lungs, causing an illness called aspergillosis. Spores of stachybotrys are so toxic that inhaling high concentrations could cause a deadly lung illness, known as IDH. That was found in tests using laboratory animals. At low concentrations, however, the white blood cells are perfectly able to eliminate the strachybotris spores. And, the probability of encountering the concentrations, used in the laboratory tests, in buildings is close to zero.

3.2.6.3 Dust mites

Dust mites are colourless, 0.5 mm large octopods, who colonize carpets, textiles, cushions and beds in large numbers and feed on human and animal hairs and scales. Forty-six different dust mite species are known. Thirteen of them colonize buildings. Of these thirteen, three are cosmopolitan, i.e. live worldwide.

Dust mites only feed, they do not drink. Respiration and perspiration happens through the skin at the back, the abdomen and the legs. That way, their moisture balance depends on

3.2 Health and indoor environmental quality

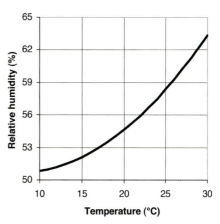

Figure 3.10. Critical relative humidity for the dustmite Dermatophagoides farinae (picture left).

the temperature and relative humidity in their direct environment. 10 to 30 °C is a preferred temperature interval while the critical relative humidity for dehydration differs between species. For dermatophagoides farinae its value is:

$$\phi_{critical} = 52.1 - 0.375\,\theta + 0.025\,\theta^2 \tag{3.36}$$

Also see Figure 3.10.

For other species the critical relative humidity is around 65%. Beyond that value, depending on temperature, the growth from egg to mature mite accelerates substantially (from 122.8 ± 14.5 days at 16 °C to 15 ± 2 days at 35 °C).

A direct relation exists between dust mite density and allergic reactions affecting some 20% of the sensitive population. A risk analysis in Denmark showed that the probability a random inhabitant will suffer from asthma could be explained in 60% of the cases by exposure to the allergens released by dust mites, consisting of digestive enzymes in the excrements suspended in the air. The risk to develop hypersensitivity starts at a concentration of 2 µg excrements per gram of house dust. In order for a sensitive person to get an allergic attack, the concentration should exceed 10 µg per gram of house dust.

3.2.6.4 Insects

The presence of insects is a sign that certain spots in a building show elevated humidity. Mosquitoes for example breed and hibernate in moist, warm crawl spaces. Moving into the building from there is not a problem given the the many leaks around pipes passing through the ground floor. At night the bloodthirsty female mosquitoes part via these paths for their sleeping victims.

Humid cavities behind baseboards and rotting timber sills are preferred shelters for cockroaches, fleas and other pests. Extermination demands quite some pesticides. Many times the unsuspecting inhabitant uses spray cans for that. That way the pesticide pollutes the indoor air.

In buildings with air heating systems cockroaches may become a nuisance as they nestle in the air ducts. Sensitive people react allergically to some enzymes in their excrements. These are injected indoors together with the warm air.

3.2.6.5 Rodents

Rodents are not welcome in buildings because they cause damage. They are combated using blood-diluting remedies. A better methods is to prevent them from entering. All openings wider than their skull – see Table 3.5 – should be sealed.

Table 3.5. Mean skull width of rodents and insectivores.

Species	Skull width mm
Common pipistrelle (protected)	7–8
Shrew mouse	9–10
House mouse	10–12
Black rat	16–22
Brown rat	17–27
Hedgehog (protected)	28–38

3.2.6.6 Pets

Cats release substances which provoke allergic and asthmatic reactions in sensitive people. An investigation in 93 office buildings in the USA showed that nearly all dust on the floor contained cat allergens. The average per sample was 0.3 µg/g; the maximum 19 µg/g. The risk for developing hypersensitivity starts at a concentration of 8 µg/g. That value was exceeded in 7 of the 93 office buildings. However, hardly any cats entered those office buildings. The allergen source was the dust from the clothes the employees brought with them from home.

3.2.7 Human related contaminants

Also humans contaminate their environment. They pick up oxygen during inhaling and emit carbon dioxide and water vapour when exhaling. At the same time perspiration and transpiration results in water vapour release, while bio-odours are out-gasses through the skin and by respiration. After the discovery of America and the cultivation of the tobacco plant people started smoking, first in Europe and later worldwide.

3.2.7.1 Carbon dioxide (CO_2)

Healthy humans inhale a volume of 4 M litres of air per hour with M metabolism in W. The lungs pick up some 27% of the oxygen present in the inhaled air and add 4 $\%m^3/m^3$ of CO_2. That way, the inhaled and exhaled air consists of:

	Inhaled air $\% \ m^3/m^3$	Exhaled air $\%m^3/m^3$
Oxygen	20.9	15.3
Nitrogen	79.6	74.5
Carbon dioxide	0.04	4.0
Water vapour	?	6.2

3.2 Health and indoor environmental quality

A human so emits 0.16 M norm-litres of carbon dioxide per hour (M in W). Until a concentration of 50,000 ppm, carbon dioxide is not poisonous. Above 10,000 ppm some physiological effects such as drowsiness are noted. Beyond 40,000 ppm people start complaining about headaches.

3.2.7.2 Water vapour

Water vapour cannot be called a contaminant. Its presence however activates out-gassing of volatile organic compounds and helps bio-germs to develop. At normal activity, the quantity released by a human does not exceed 40 to 60 g/h. As Figure 3.11 shows, emission increases proportionally with metabolism.

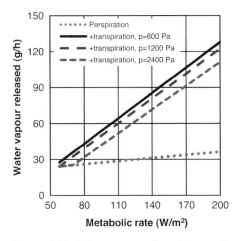

Figure 3.11. Water vapour released by people.

3.2.7.3 Bio-odours

Bio-odours like isoprene and 2-propanon are volatile organic components. Their release is best sensed when entering a badly ventilated room occupied by a bunch of people shortly before. It smells stuffy. That stuffiness has been the motor behind the olf/decipol model, see perceived indoor air quality.

3.2.7.4 Environmental tobacco smoke

A slow process of combustion like tobacco smoking is very efficient if only hydrocarbons are involved. When reacting, they form carbon dioxide and water vapour. Tobacco however contains much more complex compounds than hydrocarbons. Through that, even when burning slowly, an amalgam of combustion products is emitted, turning the smoke into a cocktail of thousands of airborne contaminants with the most important ones being:

Component	Concentration mg per cigarette	MAC ppm
NO_x	1.801	
NO	1.647	
NO_2	0.198	
CO	55.10	50
Ammoniac	4.148	25
Acetaldehyde	2.500	100
Formaldehyde	1.330	
Acetone	1.229	1000
Acetonitrile	1.145	
Benzene	0.280	
Toluene	0.498	
Xylene	0.297	
Styrene	0.094	
Isoprene	6.158	
1.3 Butadiene	0.372	5000
Limonene	1.585	
Nicotine	0.218	
Pyridine	0.569	
Dust	13.67	
Hydrogen cyanide	1600 ppm	10
Methyl chloride	1200 ppm	100

Smoking is also a source of olfactory irritation whereas nicotine acts as an habituating substance. A primary contaminant flow relates to the smoke inhaled by the smoker, a secondary flow to the smoke released by the burning cigarette. The active smoker is mainly exposed to a mixture of both. The passive smoker instead inhales a mixture of secondary and exhaled primary flow. A considerable amount of the contaminants present in the primary flow remain in the respiratory tract of the smoker. He is the one most exposed to the health risks smoking causes. Although the mixture of exhaled primary flow and secondary flow is different from the mixture the smoker inhales, the secondary flow will in the long run also subject the passive smoker to the same health risks.

For chain smokers, health risks include heart disease, vascular disease, pulmonary complaints, stomach complaints and lung cancer. Health damage is so great that pregnant women are firmly discouraged to smoke. Smoking more than one packet of cigarettes a day increases the probability of premature death with 15% meaning that 1 to 6 chain smokers die of their habit.

3.2.8 Perceived indoor air quality

3.2.8.1 Odour

People typically judge the indoor air quality by smelling. Yet, evaluating malodour objectively is a difficult task, which is why at the end of the eighties Fanger proposed a human perception based methodology called the olf/decipol model, a source/field model with a human odour panel as reference instrument.

The unit of odour emission is called the olf. One olf stands for the bio-odours emitted by an adult male with a body area of 1.8 m^2, who is lightly active, takes a shower five times a week, puts on fresh underwear every day and uses deodorant moderately. The unit of odour intensity is called the decipol (dP). One decipol characterizes the olfactory pollution caused by the one olf male adult in a room aired with 36 m^3 an hour of zero olf fresh air.

The human odour panel quantifies the actual decipol value in a room. Such a panel consists of ten individuals who are trained in estimating the decipol scale at the laboratory. During a short visit that panel is asked to scale the odour in a room in terms of decipol. The mean of the values noted represents the perceived malodour in decipol. When the ventilation flow is known, that perceived decipol value allows calculating the number of olfs present. In fact, if dP_e is the decipol value outside, dP_i the decipol value inside, n the ventilation rate in h^{-1} and V the air volume of the room in m^3, the number of olfs is:

$$P_{Olf} = \frac{(dP_i - dP_e) n V}{36} \tag{3.37}$$

In nearly all cases, more olfs are found than people present. They typically come from out gassing materials and poorly maintained ventilation systems. Such investigations allowed quantifying some of the sources:

Source	Olf
Reference person	1
Active person, 4 Met[1]	5
Active person, 6 Met	11
Smoker with cigarette	25
Smoker without cigarette	6
Materials and ventilation systems	0–0.4 Olf/m^2

[1] 1 met: a metabolism of 58 W per m^2 body area.

The many investigations also resulted in a statistical relation between the perceived malodour in decipol (dP_i) and the number of dissatisfied (PD, %):

$$G_{a,j} = \left(\frac{G_{P,j}}{x_{aj} - x_{oj}} \right) \frac{1}{\varepsilon} \tag{3.38}$$

ASHRAE treats 20% as an acceptable number of dissatisfied. That fixes 1.4 dP as being the limit for perceived malodour (Figure 3.12). In so called 'sick buildings' a decipol value beyond 10 is currently found.

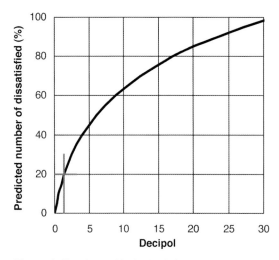

Figure 3.12. The PD/decipol relation.

The olf/decipol model has been criticised since. Using a human odour panel as measuring instrument induces random doubt on the decipol values perceived. Also questions such as 'How to calibrate the sources used to train the panels in terms of decipols?', 'How to evaluate nice odours?' create doubts about the true value of the model.

3.2.8.2 Inside air enthalpy

Investigations showed that complaints about bad indoor air quality multiply with increasing enthalpy (h) of the air, the enthalpy per kg of dry air being given by:

$$h = 1008\,\theta + x_v\,(2\,500\,000 + 1840\,\theta) \tag{3.39}$$

where θ is the air temperature in °C and x_v the water vapour ratio in kg/kg, which in turn is linked to relative humidity:

$$\phi = \frac{x_v\,P_{atm}}{p_{sat}\,(0.621 + x_v)} \quad (0 \leq \phi \leq 1) \tag{3.40}$$

with P_{atm} the atmospheric pressure in Pa ($\approx 100\,000$ Pa) and p_{sat} the water vapour partial saturation pressure at temperature θ in Pa. The statistical relation found between enthalpy and number of dissatisfied looks like:

$$PD = 100\,\frac{\exp\left[-0.18 - 5.28\,(-0.033\,h + 1.662)\right]}{1 + \exp\left[-0.18 - 5.28\,(-0.033\,h + 1.662)\right]} \tag{3.41}$$

Also here, 20% of dissatisfied is handled as the limit. Investigations in class rooms showed values during class hours largely surpassing that limit, see Figure 3.13. Outdoor air ventilation of classes in fact is typically largely insufficient to guarantee acceptable indoor air quality.

Again, not everyone agrees with the relation between perceived indoor air quality and enthalpy. Some presume that enthalpy only gains importance when people do not feel thermally comfortable. Then, individuals will perceive high air temperatures as an element of bad indoor

3.2 Health and indoor environmental quality

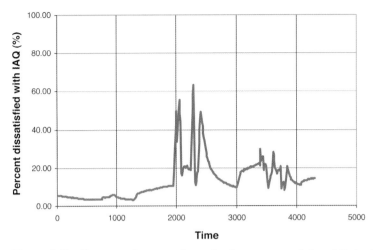

Figure 3.13. Class room in a secondary school, percentage of dissatisfied with perceived indoor air quality based on measured air enthalpy.

air quality, although low air temperatures will not be perceived better in comparison. When relative humidity is high it is perceived as unpleasant and when low as more pleasant.

3.2.9 Sick building syndrome

A much-discussed risk linked to bad environmental quality in office buildings is the sick building syndrome (SBS). In literature, an office building is called sick when:

1. Visual, acoustical and sensorial contact with the outdoors is lacking
2. Space, privacy and quietness are lacking
3. Visual, acoustical and thermal comfort are lacking
4. Unwanted pollutants are present in the air

SBS complaints typically disappear during weekends and hollidays but return when office work is resumed. The most common symptoms are: eye irritation, nose irritation, sinus irritation, throat irritation, difficulties with breathing, heavy chest, coughing, hacking, headache and dry skin.

The impact of the air quality on the magnitude of SBS complaints has been researched intensively, among others by looking to the ventilation system used and the size of the ventilation flows. Many of these studies however are one dimensional; they do not consider other possible causes. These that looked broader concluded that fully climatized office buildings showed 30 tot 200% more SBS-complaints than naturally or mechanically ventilated buildings. Especially adiabatic or temperature controlled air humidification came out as awkward factor, mainly because the ductwork was poorly cleaned and the filters got biologically contaminated.

Also a possible link between the CO_2-concentration inside and the number of SBS-complaints has been investigated. A positive correlation was found in a sense that each 100 ppm CO_2 extra resulted in a 10 to 30% increase in complaints about dry mucous membranes, irritated throat, irritated nose and coughing. As will be shown, the CO_2 excess inside forms a direct indication of the fresh air flow per employee.

Yet, another cause of SBS complaints, which deserves increasing interest in recent years, seems to be workstress. In fact, recent studies underline that employees who perform duties without power of decision complain more than executives. In all cases investigated the complainants experienced their job as stressing and missed encouragement. At the same time, in the buildings evaluated, no relationship could be found between the complaints and ventilation effectiviness, the complaints and CO_2 plus VOC-concentration in the air, the complaints and noise level inside. Moreover, employees that could adapt their work environment by adjusting heating or operate winows, complained less. All that allowed to conclude that SBS not only is the result of a failing building and HVAC performance, but rather the consequence of psychosocial stress. Own experiences confirm these observations, though the kind of office building and its HVAC-system also intervene.

3.2.10 Contaminant control by ventilation

The first and most effective strategy to combat indoor air contamination is by source control. Concentration dilution by ventilation forms a second line of defence. Source control must be the strategy followed in case of process related contaminants – fibres, volatile organic compounds, radon, combustion related pollutants and smoking – with ventilation as additional measure. However, for pollution by humans, ventilation is the only way-out to upgrade perceived indoor air quality and keep carbon dioxide concentration below acceptable. For bio-germs, ventilation is part of the policy.

3.2.10.1 Ventilation effectiveness

Ventilation effectiveness concerns the way the supply air mixes with the indoor air. Effectiveness always lies between two extremes: perfect mixing ventilation and perfect displacement ventilation. In case of mixing ventilation the objective is realizing a room-wide uniform mixture of supply and 'polluted' air. With displacement ventilation the objective is complete removal of the 'polluted' air by creating a laminar moving front with the fresh air. That way, the polluted air is replaced by fresh air. A worst mixing situation is given by a ventilation short circuit, when the supply and return grids sit too close to each other. Then, the ventilation air is immediately sucked by the return. Displacement ventilation is distorted by the presence of furniture, apparatus and people. Every individual in fact sees stack flow developing around his body. That creates turbulence and a local break in the displacement front.

3.2.10.2 Ventilation needs

In general

Assume n pollutants are emitted. The emission per pollutant is $G_{P,j}$ (kg/s). The concentration in the supply air is x_{oj} (kg/kg). The concentration allowed is x_{aj} (kg/kg). In case of mixing or displacement ventilation, the steady state ventilation flow ($G_{a,j}$ in kg/s) needed is:

$$G_{a,j} = \left(\frac{G_{P,j}}{x_{aj} - x_{oj}} \right) \frac{1}{\varepsilon} \qquad (3.42)$$

where ε is the ventilation effectiveness. For ideal displacement ventilation, effectiveness equals 2. Real displacement ventilation scores between 1 and 2, ideal mixing ventilation has a

3.2 Health and indoor environmental quality

ventilation effectiviness 1 while real mixing ventilation stays below 1. Formula (3.42) applies for each pollutant. The ventilation flow to be imposed is the largest of those calculated, or:

$$G_a = \max(\in G_{a,j}) \qquad (3.43)$$

Under transient conditions and assuming perfect mixing, the concentration x_{aj} of contaminant j with source strength $G_{P,j}$ is given by:

$$\rho_a V_a \frac{dx_{aj}}{dt} = G_{a,j}(x_{oj} - x_{aj}) + G_{P,j} \qquad (3.44)$$

a differential equation of first order which is solved analytically or numerically depending on the initial and boundary conditions imposed (concentration outside, ventilation flow and source strength (x_{oj}, $G_{a,j}$, $G_{P,j}$) as a function of time).

Carbon dioxide as a tracer

Carbon dioxide is an effective tracer when human presence is the main contaminant source. Although Table 3.6 shows differences in AIC-values between countries, the Pettenkofer number of 1500 ppm seems an acceptable average limit ($c_{CO_2,i}$). With an outside concentration of some 380 ppm ($c_{CO_2,e}$) and an emission per individual of 0.16 M liters per hour (G_{CO_2}), the ventilation flow needed per person and per hour to keep 1500 ppm is:

$$\dot{V}_{a,pers} = \frac{1000\, G_{CO_2}}{c_{CO_2,i} - c_{CO_2,e}} = \frac{160\, M}{1120} = 0.143\, M \quad (m^3/(h \cdot pers)) \qquad (3.45)$$

Table 3.6. MAC- and AIC-values for carbon dioxide ($c_{CO_2,i}$).

Country	MAC ppm	AIC ppm
Canada, USA	5000	1000
Germany, The Netherlands, Switzerland	5000	1000 a 1500
Finland	5000	2500
Italy	5000	1500

Results are shown in Table 3.7. In the future, if carbon dioxide concentration outdoors increases, more air shall be needed per person. Anyhow, the average ventilation flow needed at low activity seems to be some 20 m³ of fresh air per person and per hour. The European standard EN 13779 refines that approach by introducing carbon dioxide excess classes, see Table 3.8. IDA 1 represents a high, IDA 2 a mean, IDA 3 a moderate and IDA 4 a low indoor air quality.

Carbon dioxide is also used as tracer for judging ventilation effectiveness. Figure 3.14 shows results for 19 class rooms, all with a different numbers of scholars attending the class. The maxima were measured at the end of a teaching hour. Their mean largely surpasses the 1500 ppm with a worst case of 5000 ppm, i.e. the MAC value. The reason for these alarming results was a total lack of adequate class room ventilation during teaching hours.

Table 3.7. Ventilation flow as a function of activity.

Activity	Metabolic rate W/m²	Ventilation flow AIC = 1500 ppm m³/(h · pers)
Sleeping	41	10.0
Lying	46	11.5
Seating	58	14.6
Standing	70	17.6
Laboratory work	93	23.4
Teaching	93	23.4
Typing	62–68	15.6–17.1
Studying	78	19.7
Dancing	139–255	35.0–64.3
Playing tennis	240–267	60.5–67.2
Playing basketball	397–441	100.0–111.1

Table 3.8. Indoor air quality classes according the EN-standard 13779, based in carbon dioxide excess indoors.

Class	CO_2-concentration excess indoors	
	Interval	Mean
IDA 1	≤ 400	350
IDA 2	400–600	500
IDA 3	600–1000	800
IDA 4	≥ 1000	1200

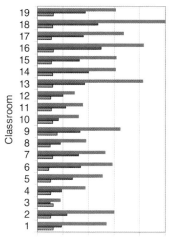

Figure 3.14. Carbon dioxide concentrations measured in 19 class rooms (lowest, mean and maximum values measured).

3.2 Health and indoor environmental quality

Avoiding mould

Whether a building will suffer from mould or not depends on many factors. The most important ones are: outside temperature and relative humidity, overall thermal insulation of the building and each envelope part, the presence of thermal bridges, the amounts of water vapour released indoors, the way the building is heated and how good the ventilation is.

The starting-point is no visible mould on inside surfaces. For that the monthly mean relative humidity at any inside surface should not exceed 80%. In case of no surface condensation on the coldest surfaces (normally single glass if any), that mean is given by:

$$\phi_s = 100 \left(\frac{p_i}{p_{sat,s}} \right) = \frac{100}{p_{sat,s}} \left(p_e + \frac{462 \, T_i \, G_{v,P}}{\dot{V}_a} \right) \quad (3.46)$$

where p_i is vapour pressure inside, $p_{sat,s}$ the saturation pressure at the surface, T_i inside temperature in K, $G_{v,P}$ the vapour release indoors and $\dot{V}_a$ ventilation flow (all monthly averages in SI). Between saturation pressure and surface temperature (θ_s) the following relation exists:

$$p_{sat,s} = 611 \exp\left(\frac{17.08 \, \theta_s}{234.18 + \theta_s} \right) \quad (3.47)$$

Monthly mean surface temperature at the coldest spot of the opaque envelope becomes:

$$\theta_s = \theta_e^* + f_{h_i} \left(\theta_o - \theta_e^* \right) \quad (3.48)$$

with f_{h_i} the temperature factor, θ_e^* the equivalent outside temperature in °C and θ_o the operative temperature inside in °C. The temperature factor f_{h_i} is given by:

$$f_{h_i} = \frac{\theta_s - \theta_e^*}{\theta_i - \theta_e^*} \quad (3.49)$$

If all other parameters are known, the set of Equations (3.46)–(3.49) allows calculating the ventilation needed. Figure 3.15 summarizes some results. What is needed strongly depends on the temperature indoors. The higher it is, the less ventilation is demanded. At the same time

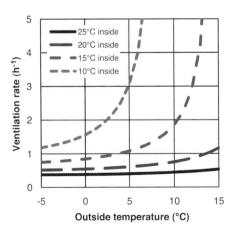

Figure 3.15. Ventilation rate needed to avoid mould in a dwelling with a volume of 200 m³, a vapour release of 14 kg/dag and a lowest temperature ratio of 0.7. The month considered is January.

the minimum ventilation rate is augmented at higher outside temperatures, when more water vapour is released and when the lowest temperature ratio encountered on opaque inside surfaces drops. In other words, avoiding mould in cool and cold climates is not a matter of ventilation only. The question is better posed as follows: what is the lowest acceptable temperature ratio to reduce mould risk to 5% in dwellings where ventilation and heating guarantee good indoor environmental quality and vapour release is not excessive. Calculations show that threshold is 0.7. In hot and humid climates only one measure helps in reducing mould risk: drying the ventilation air.

Controlling dust mite population

The best way to combat dust mite overpopulation is by keeping the hygroscopic moisture in carpets, mattresses, sheets, blankets and wall textiles low. During winter, that can be achieved by excellent thermal insulation, hardly any thermal bridging, good ventilation and correct heating habits. Also mattress finishes, sheets and blankets with a low specific surface help as does a sudden drop in temperature (for example by airing mattresses, sheets and blankets during cold weather) and washing bedding at high temperature ($\approx 60°$). In summer and in warm and humid climates only one measure has an effect: drying the ventilation air.

Guaranteeing good perceived indoor air quality

As was mentioned, 20% dissatisfied demands a decipol value 1.4. If the olf load inside (P_{Olf}) and decipol value outside (dP_e) are known, then less than 1.4 inside requires a ventilation flow in m^3/h, equal to:

$$\dot{V}_a = \frac{36\, P_{Olf}}{1.4 - dP_e} \qquad (3.50)$$

For the decipol value outside, the following numbers are used:

	dP_e
Mist	> 1
Urban environment	0.05 tot 0.35
At the coast, in the mountains	0.01

In an urban environment, the ventilation flow needed per person thus becomes:

$$36\, P_{Olf}/(1.35\, m) \leq \dot{V}_{a,pers} \leq 36\, P_{Olf}/(1.05\, m)$$

with m the number of people inside. In a typical office, building materials and HVAC give a load of ≈ 0.4 Olf per m^2 of floor area (A_{fl}). Each building user represents ≈ 1 Olf. That gives as lower and upper ventilation threshold per person:

$$27 + 10.7\, A_{fl}/m \leq \dot{V}_{a,pers} \leq 34 + 13.7\, A_{fl}/m \qquad (3.51)$$

For a floor area of 10 m^2 per person, the result is $134 < \dot{V}_{a,pers} < 171$ m^3/(h · pers), i.e. a multiple of the value based on allowable carbon dioxide concentration. Use of low out-gassing materials and good maintenance of the HVAC system may lower the olf-load to less than 0.1 Olf/m^2, reducing the ventilation needed to a value between 54 and 65.8 m^3/(h · pers). The olf/decipol-

3.2 Health and indoor environmental quality

Table 3.9. The four indoor air quality classes according to EN-standard 13779.

Class	Perceived indoor air quality dP		
	Interval	Mean	% dissatisfied
IDA 1	≤ 1.0	0.8	13
IDA 2	1–1.4	1.2	18
IDA 3	1.4–2.5	2.0	26
IDA 4	≥ 2.5	3.0	33

approach was used in the European standard 13779 to further differentiate between the four IDA-classes. See Table 3.9.

Smoking

Ventilation should be coupled to the number of cigarettes smoked per hour. Typically each cigarette demands 120 m³/h for the number of dissatisfied with perceived indoor air quality in a smokers room not to exceed 20%. Smoking a cigarette takes some 7.5 minutes. An average chain smoker consumes 3 to 6 cigarettes an hour, demanding 360 to 720 m³ of fresh air per hour. Through that, a smoking room with only chain smokers requires an amazing ventilation flow. An alternative could be to apply an olf/decipol approach, adding 0.05 times the percentage of smokers present as extra olfs per person in the smoking room. That way the additional ventilation flow equals:

$$\frac{36 \cdot 0.05 \cdot \%\text{smokers}}{1.4 - 0.05} \leq \Delta \dot{V}_{a,\text{pers}} \leq \frac{36 \cdot 0.05 \cdot \%\text{smokers}}{1.4 - 0.35} \quad (\text{m}^3/(\text{h} \cdot \text{pers}))$$

or:

$$1.33 \cdot \%\text{smokers} \leq \Delta \dot{V}_{a,\text{pers}} \leq 1.71 \cdot \%\text{smokers} \quad (\text{m}^3/(\text{h} \cdot \text{pers}))$$

If only chain smokers are present and their number equals 1 per 10 m², than 268 to 328 m³ of fresh air per person per hour is needed, which is much less than calculated above but still a good amount. Therefore current practice imposes a ventilation flow double the value needed in a non smoking environment. That can result in quite a few complaints about perceived air quality. Investigations however show that smokers are more tolerant from that point of view than non-smokers. At any rate the best policy is to ban smoking in buildings.

3.2.11 Ventilation requirements

The basics discussed above are not practical for every day use, which is why standardization bodies impose ventilation requirements. The EN-standard 13779 for example, which applies for non-residential buildings, defines 4 IDA-classes, requiring per class the ventilation flows shown in Table 3.10.

The ASHRAE standard 62.2-2007 proposes as equation for whole-building ventilation in residences:

$$\dot{V}_a = 0.18 \, A_{\text{fl}} + 12.6 \left(n_{\text{br}} + 1 \right) \quad (\text{m}^3/\text{h}) \tag{3.52}$$

Table 3.10. Ventilation in non-residential buildings according to EN 13779.

Class	Ventilation flow per person present in m³/h			
	Smoking not allowed		Smoking allowed	
	Interval	Mean	Interval	Mean
IDA 1	> 54	72	> 108	144
IDA 2	36–54	45	72–108	90
IDA 3	22–36	29	44–72	58
IDA 4	< 22	18	< 44	36

Table 3.11. Design ventilation flows in residential buildings according to NBN D50-001.

Room	Design ventilation flow
Living room	3.6 m³/h per m² of floor surface, with a minimum of 75 m³/h and a maximum of 150 m³/h
Sleeping room, playroom	3.6 m³/h per m² of floor surface with a minimum of 25 m³/h and not exceeding 36 m³/h per person
Kitchen, bathroom, washing room, etc.	3.6 m³/h per m² of floor surface, with a minimum of 50 m³/h and a maximum of 75 m³/h
Toilet	25 m³/h
Corridors, staircases, day and night hall and analogous spaces	3.6 m³/h per m² of floor surface
Garage	Should be ventilated separately from the dwelling

with A_{fl} the whole dwelling floor area in m² and n_{br} the number of bedrooms. The assumptions are two persons at minimum in a studio or a one-bedroom dwelling and an additional person for each additional bedroom. If higher occupancies for a given number of bedrooms are noted, 12.6 m³/h per person should be added.

Table 3.11 lists the design ventilation flows in residential buildings as imposed by the Belgian standard NBN D 50-001. 'Design' should be read as: the ventilation components must be dimensioned for these values, though less ventilation will be measured under circumstances that differ from those used by the norm to define the 'design flows'.

3.3 Energy efficiency

3.3.1 In general

Three main users define nationwide annual end energy consumption: industry, traffic and residential plus equivalents. While the industry includes several big consumers who need energy for production, traffic and residential plus equivalents represent a huge number of small consumers who appreciate the service energy provides but don't care about energy.

Only when energy prices skyrocket and the services become more expensive, do they make themselves heard.

Yet, from a societal point of view, energy is a permanent challenge with availability, affordability and sustainability as main concerns. Many industrialized countries lack energy sources. They have to import coal, oil and gas, which not only impacts their balance of payments but is also a problem because many imports come from countries that are politically unstable.

The most alarming sustainability-related concern today is world-wide global warming, caused by the CO_2 emitted by fossil fuel combustion. In 1997 that concern lead to the Kyoto protocol, demanding the industrialized world to cut global warming gas emission (GWG) by 5% in 2008–2012, compared to 1990. The protocol was ratified in 2004. Whether the objective will be reached is questionable because the GWG emissions still rise in several countries. The post Kyoto era even looks more difficult. To stabilize global warming, world-wide, cuts from 50 and 80% will be needed in 2050 compared to business as usual. This will change the economy, demand another kind of end energy production and impact town planning, housing and transportation. As an intermediary step the EU is aiming for a 20/20/20 target in 2020: 20% of its energy demand covered by renewables, 20% less GWG-emissions compared to 1990 and 20% less energy consumed compared to business as usual.

Progress in 'residential plus equivalents' will require a consequent application of the 'trias energetica': first minimizing the end energy demand, secondly using renewables as far as economically feasible, and thirdly, covering the demand left with fossil fuels as efficiently as possible. A problem with renewables like wind and sun is the absence of a power guarantee. In society today, where energy keeps things running, such a guarantee is a prerequisite. Of course, smart grids with demand-side management and energy storage may help, but even that will not eliminate the need for back-up power plants to align production and demand.

In Europe, governments actively promote rational use of energy. In the building sector, with its millions of non-knowledgeable decision makers, legal obligations are a most effective way to proceed. There are many arguments in favour of that. If correctly designed and constructed, energy efficient buildings offer better thermal comfort, a healthier indoor environment and longer service life than more energy consuming ones. If designed in a way the volume to enclosure ratio is high and if the HVAC-system is well chosen and correctly dimensioned, energy efficient buildings may even be cheaper then more energy consuming ones. And finally, each fossil MJ avoided helps in combating global warming.

3.3.2 Some statistics

According to the OESO, 'residential plus equivalents' represent some 40% of the annual end use in developed countries. That fits in with the data for Flanders, Belgium (Figure 3.16), those for Germany shortly after reunification (Table 3.12) and the ones for the USA (Table 3.13).

The share 'buildings' have is even larger if the energy needed for manufacturing materials and components, the energy used during construction and demolition, and, the energy consumed for acclimatizing industrial premises is added. Buildings are thus a greater factor than the industry, a fact not always recognized. As most energy used in buildings comes from burning fossil fuels or from electricity produced in fossil fuel powered plants, the share in CO_2-emission is almost as important.

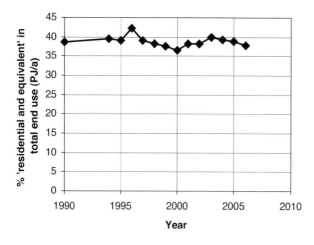

Figure 3.16. Flanders, Belgium, %-share of 'residential and equivalent' in the annual end energy consumption.

Table 3.12. End use in Germany.

End use	%
Industry	28
Traffic	28
Households plus equivalents (= buildings)	23
Army	1

Table 3.13. End use in the USA

End use	%
Industry	32.6
Traffic	32.3
Buildings	33.7

3.3.3 End energy use in buildings

End energy use in buildings includes (1) lighting and appliances (F), (2) domestic hot water (DHW) and (3) HVAC (mainly heating in moderate climates). Also part of the end use for traffic could be added. Distances travelled in fact also depend on settlement patterns with ribbon development and scattered habitation as less energy efficient choices. Table 3.14 summarizes consumption data for a moderate climate, collected in 201 dwellings.

On the average function represents 16% of the total. Domestic hot water has an equal or smaller share, while heating consumes the remaining 68 to 75%. The three together represent an annual average of 180 kWh per m^2 of floor surface. That number is high knowing that 'low energy (LEB)' focuses on less than 50 kWh/($m^2 \cdot$ a) for heating and less than 85 kWh/($m^2 \cdot$ a) for heating, domestic hot water and function.

3.3 Energy efficiency

Table 3.14. Annual end energy consumption in 201 dwellings built between 1910 and 1995.

		Mean	Standard deviation	Minimum	Maximum
Gross floor surface	m²	242	89	98	485
Total end use (H + DHW + F)	kWh/(m² · a)	180	68	66	488
End use for H and DHW	kWh/(m² · a)	149	71	66	428
End use for F	kWh/(m² · a)	28	14	8	74

3.3.3.1 Lighting and appliances (function)

Household activity, office work, teaching, nursing, all demand appliances. These in turn use energy, in most cases electricity. Everyone also wants task-appropriate, comfortable lighting, again consuming a substantial amount of electricity. Statistically, annual electricity use for appliances and lighting in dwellings may be calculated as:

$$Q_{function,a} = \sum Q_{basis} + \sum_{j=1}^{m}\left(\frac{n_j}{168}\right)\sum Q_{extra,P_{eff}} + (V-540)\sum Q_{extra,V} \quad \text{(kWh/a)} \quad (3.53)$$

where Q_{basis} is the annual electricity use listed in the second column of Table 3.15, m the number of inhabitants, n the average number of hours per week each inhabitant is present, $Q_{extra,P_{eff}}$ a correction per effective person present and $Q_{extra,V}$ a correction that accounts for the out to out protected volume of the dwelling.

Table 3.15. Annual end energy for appliances and lighting in detail.

Appliance	End use Q_{basic} kWh/a	$Q_{extra,P_{eff}}$ kWh/a	$Q_{extra,V}$ kWh/a
Stove	650	152	0.08
Dish washer	360	86	0.03
Washing machine	300	86	0.03
Dryer	440	57	0.03
Lighting	700	190	0.69
Fridge	270	85	0.06
Freezer	400	133	0.06
Computer	150	0	0
Pump, central heating	300	0	0.97
Others	400	161	2.08

Formula (3.35) was tested using the field data summarized in Table 3.16. The deviation between the calculated and measured mean was –4.5%, the difference in standard deviation between both collections –21.7%, the deviation for the maximum measured –16.6% and the deviation for the minimum measured –44.5%

Table 3.16. Electricity use for appliances and lighting measured in 26 dwellings.

		Mean	Standard deviation	Minimum	Maximum
End energy	kWh/a	5804	1995	2025	9925
Primary energy[1]	kWh$_{th}$/a	14510	4990	5060	24800

[1] the end to primary conversion factor for electricity used is 2.5

A least square fit between the measured electricity use and the out to out protected volume V in m³ gave:

$$Q_{appl+light} = (201 + 0.242\ V)\Delta t/1\,000\,000 \quad (MJ) \tag{3.53}$$

with Δt the time step in s.

In poorly insulated domestic buildings in cool climates, primary energy for lighting and appliances is not a first concern. Heating dominates. However, when moving to 'low energy' and beyond, lighting, appliances and domestic hot water may become more important than heating in terms of primary energy used. The annual cost is surely higher as heating uses fossil fuel whereas lighting and appliances consume expensive electricity. Energy efficient appliances thus become a prerequisite in 'low energy' and beyond. To give some examples: well insulated refrigerators and freezers consume half the energy than less insulated ones; coupling washing machines and dish washers to gas-fired domestic hot water lowers primary energy consumed; compact fluorescent lamps halve electricity use for lighting and LED's may even do better. Briefly, lowering the post function is not utopia, although a higher efficiency in power consumed lowers the internal heat gains and may be an incentive for more intense use.

In non-domestic buildings, appliances and lighting must be evaluated in full detail. What appliances are used? What power do they require? What is the average use in terms of hours per week? How many lamps are present? What is the luminous efficiency and how much power do they require? What switching systems are used? Is day lighting integrated? And so on.

3.3.3.2 Domestic hot water

End energy use for domestic hot water depends on the volume consumed, the water temperature and the total efficiency of the installation. Building use, the number of family members, their age and activities, their standard of living, and hygienic habits all define the volume. Water temperature differs according to usage: hygiene, recreation, industrial processes, etc. In condominiums and public buildings, domestic hot water must be regularly heated to 60 °C to avoid legionella. Total efficiency is determined by the way domestic hot water is prepared, stored and distributed. In domestic buildings, the formula below allows guessing related net energy demand:

$$Q_{dhv,net} = \max\left[80,\ 80 + 0.275(V-196)\right]\Delta t/1\,000\,000 \quad (MJ) \tag{3.54}$$

where V is the out to out protected volume of the dwelling in m³ and Δt the time step in seconds considered. Assumptions are: a statistical relation exists between the (protected) volume of a domestic building and the number of inhabitants; the water consumed is heated from 10 to 60 °C; on average people consume 33 l of domestic hot water per day.

3.3 Energy efficiency

3.3.3.3 Space heating, space cooling and air conditioning

In cool and cold climates space heating is the main energy user whereas in domestic buildings cooling is easily avoided thanks to passive measures such as limited glass surface, solar shading, massive construction and night ventilation. In buildings with high internal gains such as landscape offices, trading rooms and high care units in hospitals cooling has become a necessity. It is also gaining market share in shops, pubs and restaurants. Whether air humidity control is needed or not depends on the building's function – museum, clean room laboratory – and the sensitivity and value of furnishings, products manufactured, pieces of art exposed, etc.

3.3.4 Space heating

3.3.4.1 Terminology

Net energy demand	The energy needed to keep a building on set-point temperature, assuming a heating system with infinite capacity and an efficiency of one
Gross energy demand	The net energy demand over a given time span, divided by the mean building-coupled heating system efficiency over that time span
End energy use	The energy consumed by the boiler or heat pump to keep the building on set point temperature over a given time span. Encompasses also the energy used by the heating system to function (pumps, controls, fans)
Primary energy use	The end energy for heating plus the extra energy needed to produce and transport the energy source used to the building
Total net energy demand	The energy needed by a building for heating, cooling, air conditioning and domestic hot water assuming all systems have infinite capacity and an efficiency of one
Total gross energy demand	The energy needed by a building for heating, cooling, air conditioning and domestic hot water, divided by the mean building-coupled system efficiency per system over that time span
Total end energy use	The energy consumed by the HVAC-system to keep the building on set point temperature plus the energy used for producing domestic hot water and the energy consumed for lighting and appliances, all that over a given time span. Encompasses also the energy needed for all systems to function
Total primary energy use	Total end energy over a given time span, plus the extra energy needed to produce and transport the energy sources used to the building
Exergy	Describes what part of a given quantity of energy can be transformed into labour. The larger that part, the higher the quality of the energy
Steady state	Exists when all calculations are done for average boundary conditions. The period covered goes from 1 day to a heating season. Looking to the net energy demand, one month is typically taken as a unit period
Transient	Applies to calculations done using the instantaneous value of all boundary conditions. Instantaneous stretches from second to hour, the latter when the building is simulated, the first when systems and controls are simulated
Protected volume	The sum of all spaces in a building that demand thermal comfort conditions and need heating, sometimes cooling and accidental humidity control.

	The protected volume in residential premises includes all living spaces in the daytime and night time zone. Measuring is done out to out.
Zone	Each part of the protected volume on a different temperature. In domestic buildings, zones may correspond with rooms and corridors. Often it suffices to consider a day- and night time zone. It is also common to take the whole protected volume as one zone.
Envelope	The surface enclosing the protected volume. Includes roofs, façade walls, lowest floor and separation walls between the protected volume and adjacent spaces at different temperature. Separations with adjacent spaces at the same temperature are not part of the envelope. Area measuring is done out to out
Loss surface	Envelope surfaces with different orientation and/or slope
Separation	Each inside wall or floor between adjacent zones
Mid plane	Middle of a vertical separation between two zones, upper area of a horizontal separation between two zones
Envelope part	Each fraction of the envelope with different physical properties (thermal transmittance, transient response, long wave emissivity, short wave absorptivity)
Separation part	Each fraction of a separation surface with different physical properties (thermal transmittance, transient response, long wave emissivity, short wave absorptivity)
Inside temperature	The operative temperature in each building zone averaged over the time span considered. In case the protected volume figures as one zone, the inside temperature equals the uniform operative temperature the protected volume should have to give the same net energy demand for heating as it has with different zone temperatures.
Ventilation rate	Represents the average ratio between the outside air flow in m^3 per hour and the air volume in a zone over the time span considered. In case the protected volume acts as one zone, the ventilation rate equals the mean outside air flow entering the protected volume over the time span considered, divided by the air volume, typically set at 0.8 times the out to out protected volume

3.3.4.2 Steady state energy balance for heating at zone level

In general

Take a building containing j zones, each with volume V_j. Its protected volume is V ($V = \Sigma\ V_j$). Two quantities characterize a zone: a monthly mean inside (operative) temperature (θ_j) and a monthly net energy demand for heating ($Q_{\text{heat,net},j,\text{mo}}$). If the inside temperature is known, net energy demand is the unknown. Conversely, if the net energy demand is known, inside temperature becomes the unknown. Zones are represented by nodes that exchange heat and enthalpy with adjacent nodes (Figure 3.17). Transmission goes from adjacent nodes and the outside (together called the surroundings) to the node considered. Enthalpy instead flows in the same direction that the air moves. Solar and internal gains are directly injected in a node. The same happens with the enthalpy flow from a ventilation system. The monthly energy balance for node j so becomes (algebraic sum, Figure 3.17):

$$Q_{\text{heat,net},j,\text{mo}} + \left[\left(\Phi_{T,j,\text{mo}} + \Phi_{V,j,\text{mo}}\right) + \eta_{\text{use},j,\text{mo}}\left(\Phi_{\text{SLW},j,\text{mo}} + \Phi_{I,j,\text{mo}}\right)\right]\Delta t_{\text{mo}} = 0 \quad (\text{J}) \quad (3.55)$$

3.3 Energy efficiency

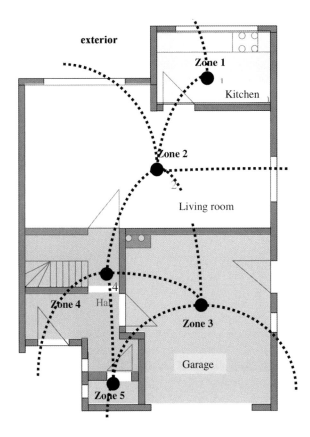

Figure 3.17. From zones to nodes.

where $\Phi_{T,j,mo}$ is the transmission flow from the surroundings, $\Phi_{V,j,mo}$ the total enthalpy flow by air exchange with the surroundings, included the enthalpy flow injected by an air-based heating system or a mechanical ventilation system, $\Phi_{SLW,j,mo}$ solar gain corrected for long wave losses and $\Phi_{I,j,mo}$ the internal gain, all for node j. The gains are multiplied with a utilization efficiency $\eta_{use,j,mo}$, whose value is 1 when no heating is provided. $Q_{heat,net,j,mo}$, the remaining net energy demand for heating in node j, closes the equation. That demand is positive when the losses are largest and zero the other way around. Δt_{mo} is a month long period in s. All terms are monthly means with Watt as a unit, except for $\eta_{use,j,mo}$, Δt_{mo} and $Q_{heat,net,j,mo}$.

Apparently, two tracks allow lowering the net energy demand for heating: limiting the losses or increasing the gains. In heating climates with little sun in winter, the first is the prime track to follow. The best of course is optimizing the algebraic sum.

Transmission

Transmission from the surroundings includes flow from the outside (air, bottom) and flows from adjacent nodes or neighbouring protected volumes at different temperature. As an equation:

$$\Phi_{T,j,mo} = H_{T,je,mo}\left(\theta_{e,mo} - \theta_{j,mo}\right) + \sum_{k=1}^{m} H_{T,jk,mo}\left(\theta_{k,mo} - \theta_{j,mo}\right) \tag{3.56}$$

where $H_{T,je,mo}$ is the transmission heat transfer coefficient between node j and the outside:

$$H_{T,je,mo} = \underbrace{\sum U_{e,op} A_{op}}_{\text{opaque}} + \underbrace{\sum U_{e,w} A_w}_{\text{windows}} + \underbrace{\sum a_{fl} U_{e,fl} A_{fl}}_{\text{ground}}$$
$$+ \underbrace{\sum \psi_e L}_{\text{linear TB}} + \underbrace{\sum \chi_e}_{\text{local TB}} \qquad (3.57)$$

and $H_{T,jk,mo}$ is the transmission heat transfer coefficient between adjacent node k and node j:

$$H_{T,jk,mo} = \underbrace{\sum U_{i,op} A_{op}}_{\text{opaque}} + \underbrace{\sum U_{i,w} A_w}_{\text{windows}} + \underbrace{\sum \psi_i L}_{\text{linear TB}} + \underbrace{\sum \chi_i}_{\text{local TB}} \qquad (3.58)$$

The U's are the thermal transmittances of all envelope and separation parts, A their surface, ψ the linear thermal transmittances of all linear thermal bridges, L their length and χ the local thermal transmittances of all local thermal bridges. The suffix e refers to the envelope, the suffix i to the separations. The sums are a reminder that all parts have to be considered. All transmission heat transfer coefficients are monthly means.

Ventilation and infiltration

Air is exchanged with the outside and adjacent nodes by infiltration and the purpose designed ventilation system. Moreover, a mechanical ventilation system may directly supply air to and/or extract air from a node. The result is a complex pattern of air flows G_{ajk} with the suffix order indicating flow direction, $_{jk}$ for a flow from node j to node k, $_{je}$ or $_{ej}$ for a flow to or from the outside, $_{in,j}$ for a flow supplied and $_{out,j}$ for a flow extracted by a mechanical ventilation system (Figure 3.18).

A sensible enthalpy balance for node j is represented by:

$$\Phi_{V,j,mo} = \left[-H_{V,je,mo}\,\theta_{j,mo} + H_{V,ej,mo}\,\theta_{e,mo} - \sum_{k=1}^{m} H_{V,jk,mo}\,\theta_j \right.$$
$$\left. + \sum_{k=1}^{m} H_{V,kj,mo}\,\theta_{k,mo} + H_{V,in,j,mo}\,\theta_{in,j,mo} - H_{V,out,j,mo}\,\theta_{j,mo} \right] \qquad (3.59)$$

where $H_{V,xy,mo}$ are the ventilation heat transfer coefficients, to the outside for xy = je, from the outside for xy = ej, to node k for xy = jk and from node k for xy = kj. Heat transfer coefficients in turn equal:

$$H_{V,xx,mo} = (1 - \eta_{rec,mo})\,c_a\,G_{a,xy,mo} \qquad (3.60)$$

with $\eta_{rec,mo}$ heat recovery efficiency ($\eta_{rec,mo} = 0$ without recovery), c_a specific heat capacity of air (1000 a 1030 J/(kg · K)) and $G_{a,xy,mo}$ the air flow in kg/s. Take heat recovery.

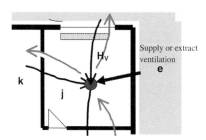

Figure 3.18. Air flows between node j and adjacent nodes, air flows between node j and the outside, supplied or extracted air flows.

3.3 Energy efficiency

When infiltrating flows pass leaks in the envelope some recovery occurs. Exfiltrating flows in turn temper transmission losses. However, combined, the effect is too minimal to be considered. Of course, purpose induced ventilation flows passing unheated spaces will pick up lost heat whereas a ducted balanced ventilation system with supply and extract fans may include a heat recovery unit. Remember that all quantities listed relate to monthly means.

In principle, infiltration and ventilation flows should follow from solving the system of nodal air balances. However, most of the time basic information on air permeances between zones and the surroundings is lacking, which is why default values for the flows are commonly used.

Solar gains and long wave losses

In addition opaque and transparent buildings provide solar gains and suffer from long wave losses ($\Phi_{S/LW,j,op,mo}$, $\Phi_{S/LW,j,w,mo}$):

$$\Phi_{S/LW,j,mo} = \sum_{opaque} \Phi_{S/LW,j,op,mo} + \sum_{window} \Phi_{S/LW,j,w,mo} \tag{3.61}$$

Resulting inflow into node j across the opaque parts is given by:

$$\Phi_{S/LW,j,op,mo} = \frac{U_{e,op} A_{e,op} J}{h_e} \left[\alpha_K r_{sT} \Phi_{sT,mo} - 100 r_{sd} e_L F_{op/sky} \right] \tag{3.62}$$

where $\Phi_{sT,mo}$ is the monthly mean total insolation on the opaque part under clear sky conditions, with 'total' referring to the sum of beam, diffuse and reflected radiation. J stands for the ratio between the insolation under real sky conditions and clear sky conditions while r_{sT} is the shadow factor for total insolation and r_{sd} the shadow factor for diffuse insolation. Both get as default value 0.9. $F_{op/sky}$ finally is the view factor between the surface considered and the sky. The equation shows that solar gains are larger when opaque parts are less insulated (higher U-value), the short wave absorptivity of the exterior surface lays higher, its long wave emissivity is lower and wind protection is better (surface film coefficient h_e lower).

Resulting inflow through the transparent parts mainly depends on the overall solar transmittance g of the glazing included the solar shading system:

$$\Phi_{S/LW,j,w,mo} = A_{e,w} J \left[0.95 \cdot 0.9 \, g \, f_{wgl} \, r_{sT} \, \Phi_{sT,mo} - 100 \frac{U_{e,w}}{h_e} r_{sd} e_L F_{w/sky} \right] \tag{3.63}$$

where f_{wgl} is the ratio between glass and window area, g the solar transmittance for normal beam insolation, 0.9 a conversion factor to account for the beam angle effect on the solar transmittance and 0.95 a multiplier accounting for dust deposit on the glass.

Transparent insulation, TIM, figures as separate case. TIM transmits shortwave and stops long wave radiation, whereas its structure is such that conduction and convection are impeded. Solar gains and long wave losses by a building part finished with TIM at the exterior become:

$$\Phi_{S/LW,j,TIM,mo} = A_{e,TIM} J \\ \cdot \left[\tau_{gl} \tau_{TIM} \alpha_K R'_{e,TIM} f_r r_{sT} \Phi_{sT,mo} - 100 \frac{U_{e,TIM}}{h_e} r_{sd} e_L F_{TIM/sky} \right] \tag{3.64}$$

with $R'_{e,TIM}$ the thermal resistance between the TIM's backside and the outside (equal to $R'_{e,TIM} = 1/h_e + R_{TIM}$, where R_{TIM} is the thermal resistance of the TIM-layer), $U_{e,TIM}$ the thermal transmittance of the whole assembly, τ_{gl} short wave transmissivity of the glass cover, τ_{TIM} short wave transmissivity of the TIM-layer, α_K short wave absorptivity of the surface behind the TIM-layer and f_r de ratio between the TIM covered area and the total area of the part.

When calculating insolation and long wave losses, the following simplifications are applied:

- The ratio J between real insolation and insolation under clear sky conditions is a constant, independent of the slope and azimuth of a building part
- The ratio between the insolation on a arbitrary surface and a horizontal surface for clear sky conditions during each month's central day is valid for that whole month
- The shadow factor r_{sT} combines horizon shielding with shielding by the building and its overhangs. Horizon shielding depends on the vertical angle α_e between the horizontal plane through the transparent part's centre and a horizontal line at the mean height of the real horizon above that centre. Shielding by horizontal building overhangs is quantified using a vertical overhang angle α_v, equal to the angle between the perpendicular line through the transparent part's centre, projected on a horizontal surface, and the connecting line from that centre to the mean overhang depth, a horizontal line bordering the average of the varying overhang depths. Building protrusions left and right are replaced by vertical overhang planes. Shielding is quantified by two horizontal overhang angles $\alpha_{h,l}$ en $\alpha_{h,r}$ equal to the angles between the horizontal projection of both the perpendicular line through the transparent part's centre and the connecting lines from that centre to the vertical lines bordering both average overhang depths (Figure 3.19).
- Beam solar radiation becomes zero for a solar height below the horizon angle α_e, a solar height above the vertical overhang angle α_v and for time angles beyond $a_s + \alpha_{h,l}$ and below $a_s - \alpha_{h,r}$, with a_s the azimuth of the loss surface containing the window. In all other cases, beam radiation reaches the window unhindered

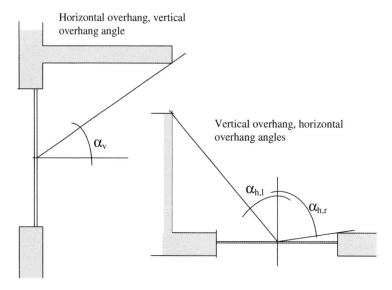

Figure 3.19. The two horizontal (α_{hl}, α_{hr}) and one vertical overhang angle (α_v).

3.3 Energy efficiency

- The shadow factor r_{sd} for diffuse solar plus long wave sky radiation is given by:

$$r_{sd} = \left[2\left(1-\frac{s_s}{\pi}\right)(1-\sin\alpha_e)-(1-\cos\alpha_v)\right]\left(\frac{\pi-\alpha_{h,l}-\alpha_{h,r}}{2(\pi-s_s)}\right)$$

with s_s the slope of the loss surface that contains the transparent part. Reflected solar gets a shadow factor 1.

Internal gains

Internal gains are stochastic in nature. They include the sensible heat released by the building occupants, the heat artificial lighting emits and the heat produced by appliances and apparatus. If no better data are available, internal gains in domestic buildings are quantified as:

$$\Phi_{I,j} = (220+0.67\,V)\frac{V_j}{V} \quad (W) \tag{3.65}$$

where V_j is the volume of the zone considered and V the protected volume of the building. In other buildings, gains by lighting should be calculated starting from the lighting design and projected lighting use and control. Human heat released follows from the usage pattern and mean occupancy (70 to 90 W per adult for office work). Guessing the heat produced by apparatus and appliances is more difficult. Data from buildings with similar function may be helpful. Otherwise, data from literature should be used.

Utilization efficiency

As said, not all gains are usable in heated zones. Especially on days with limited losses and quite important gains, the zone temperature may temporarily exceed the set point and turn part of the gains into unneeded heat. This is accounted for by the utilization efficiency.

First, the ratio between monthly gains and losses for the heated node j is calculated:

$$\gamma_{j,mo} = \frac{\Phi_{SLW,j,mo}+\Phi_{I,j,mo}}{\Phi_{T,j,mo}+\Phi_{V,j,mo}} \tag{3.66}$$

Then its time constant is quantified:

$$\tau_{j,mo} = \frac{C_j}{H_{T,j,mo}+H_{V,j,mo}}$$

where C_j is heat storage in the node, per loss and separation surface given by the heat capacity of all layers at the zone side of the thermal insulation within 10 cm from inside (Figure 3.20):

$$C_j = \sum_{\text{all surfaces}}\left[A\sum_{\text{inside},\,d\leq 0.1\,m}(\rho\,c\,d)\right] \tag{3.67}$$

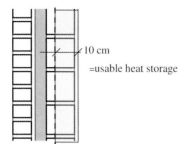

Figure 3.20. Usable hat stored in a loss or dividing surface (left is outside or the other side).

The monthly mean utilization efficiency than becomes:

$$\gamma_{j,mo} \neq 1: \quad \eta_{use,j,mo} = \frac{1 - \gamma_{j,mo}^{a_{ma,j}}}{1 - \gamma_{j,mo}^{a_{ma,j}+1}} \qquad \gamma_{j,mo} = 1: \quad \eta_{use,ma,j} = \frac{a_{j,mo}}{1 + a_{j,mo}} \qquad (3.68)$$

with $a_{j,mo} = 1 + \tau_{j,mo} / 54\,000$. In case a zone stays unheated or in case heating demand is zero, utilization efficiency equals 1.

3.3.4.3 Steady state energy balance for heating at building level

The balance per zone rewrites as:

$$-\left(H_{T,je,mo} + H_{V,je,mo} + \sum_{k=1}^{m}\left(H_{T,jk,mo} + H_{V,jk,mo}\right)\right)\theta_{i,j,mo}$$

$$+\sum_{k=1}^{m}\left(H_{T,jk,mo} + H_{V,kj,mo}\right)\theta_{i,k,mo} + \frac{Q_{heat,net,j,mo}}{\Delta t_{mo}}$$

$$= -\left(H_{T,je,mo} + H_{V,ej,mo}\right)\theta_{e,mo} - \eta_{use,j,mo}\left(\Phi_{S/LW,j,mo} + \Phi_{I,j,mo}\right)$$

Combining the transmission and ventilation heat transfer coefficients into the heat transfer coefficients $H_{j,mo}$, $H_{jk,mo}$ and $H_{ej,mo}$ allows simpler writing:

$$-H_{j,mo}\,\theta_{i,j,mo} + \sum_{k=1}^{m} H_{kj,mo}\,\theta_{i,k,mo} + \frac{Q_{heat,net,j,mo}}{\Delta t_{mo}}$$

$$= -H_{ej,mo}\,\theta_{e,mo} - \eta_{use,j,mo}\left(\Phi_{S/LW,j,mo} + \Phi_{I,j,mo}\right)$$

A building containing n zones produces a system of n equations with $2n$ unknown (n inside temperatures $\theta_{i,j,mo}$ and n net energy demands for heating $Q_{heat,net,j,mo}$):

$$\left|H_{j,mo}\right|\left|\theta_{i,j,mo}\right| + \left|1\right|\left|\frac{Q_{heat,net,j,mo}}{\Delta t_{mo}}\right| \qquad (3.69)$$

$$= -\left|H_{je,mo}\,\theta_{e,mo} - \eta_{use,j,mo}\left(\Phi_{S/LW,j,mo} + \Phi_{I,j,mo}\right)\right|$$

3.3 Energy efficiency

where $|H_{j,mo}|$ is an $[n \times n]$ array of heat transfer coefficients, $|\theta_{i,j,mo}|$ a column matrix of the n zone temperatures, $|1|$ a $[n \times n]$ unit array, $|Q_{heat,net,j,mo} / \Delta t_{mo}|$ the column matrix of the n net energy demands and $|H_{je,mo}\,\theta_{e,mo} - \eta_{use,j,mo}\,(\Phi_{SLW,j,mo} + \Phi_{I,j,mo})|$ the column matrix of the n known terms. If temperatures are known in $n - m$ zones and net energy demand in the m left, the system (3.69) can be solved. If one or more of the m zones show a negative net demand, these are set at zero, their utilization efficiency at 1 and the system is solved again. Summing up the final net demand for the n zones gives the monthly net demand for the whole building:

$$Q_{heat,net,mo} = \sum_{j=1}^{n} Q_{heat,net,j,mo}$$

Annual net demand follows from adding up all 12 months:

$$Q_{heat,net} = \sum_{mo=1}^{12} Q_{heat,net,mo}$$

In case one heating system warms the building, dividing each monthly net demand by a building-linked monthly mean system efficiency $\eta_{heat,sys,mo}$ gives the monthly gross demand:

$$Q_{heat,gross,mo} = \frac{Q_{heat,net,mo}}{\eta_{heat,sys,mo}} \quad (3.70)$$

End energy use for heating becomes:

$$Q_{heat,use} = \sum_{mo=1}^{12} \left(\frac{Q_{heat,gross,mo}}{\eta_{heat,prod,mo}} \right) \quad (3.71)$$

where $\eta_{heat,prod,mo}$ is the monthly mean production efficiency of the heat producing unit. In case the building has different heating systems, the Equations (3.70) and (3.71) hold per group of zones served by the same system. Adding the monthly totals give the overall end energy use per month. The twelve months together then equal the annual total. With an active solar heating system the annual end use writes as:

$$Q_{heat,use} = \sum_{mo=1}^{12} \left(\frac{Q_{heat,gross,mo} - \eta_{S,mo}\,Q_{S,mo}}{\eta_{heat,prod,mo}} \right) \quad (3.72)$$

where $Q_{S,mo}$ is the monthly insolation on the collector surface and $\eta_{S,mo}$ the monthly mean efficiency of the active solar system.

3.3.5 Parameters defining the net energy demand for heating in residential buildings

3.3.5.1 Overview

Four sets of parameters that shape the annual net energy demand, gross energy demand and end use for heating are contained in the equations of previous paragraph. Table 3.17 lists the three sets impacting the net demand: outside climate, building use and building design. Gross demand and end use add a fourth set: the heating and ventilation system used.

Table 3.17. Parameters influencing the annual net energy demand for heating.

Set	Parameter	Impact on net heating demand, all other parameters equal
Outside climate (heating season mean)	Temperature	Increases when lower
	Insolation	Decreases when more
	Wind speed	Increases when higher
	Precipitation	Increase when more
Building use (heating season mean)	Inside temperature	Increases when higher
	Temperature adaptation and control	Decreases when the principle 'heating a zone only when used' is better applied
	Ventilation	Increases when higher
	Internal gains	Decreases when higher
Building design	Protected volume	Increases when larger
	Compactness	Decreases when higher
	Floor organisation	Decreases when the floor organisation considers usability of solar gains and shielding of living zones by utility rooms.
	Surface area, type, orientation and slope of the transparent surfaces (typically windows)	Increases when the loss surface contains larger transparent parts. It decreases when those parts have a higher solar transmittance, a lower thermal transmittance, when more of them face the sunny sides and if close to vertical
	Thermal insulation	Decreases for a lower mean thermal transmittance of the envelope
	Thermal inertia	Thermal inertia hardly impacts net demand in cool and cold climates
	Envelope air-tightness	Decreases when better

3.3.5.2 In detail

Outside climate

Temperature

Net energy demand for heating increases with lower heating season mean outside temperature, all other parameters being equal.

The heating season mean outside temperature drops nearer to the poles, further inland and higher above sea level. Although situated at a higher latitude than the US, Northwestern Europe has a much more moderate climate thanks to the warm Gulf Stream that washes its coasts.

3.3 Energy efficiency

Insolation and clear sky long wave radiation

Net energy demand for heating drops when the insolation during the heating season increases, all other parameters equal.

Insolation during the heating season is quite low in countries with a maritime climate. Less cloudy skies and more sun occur in regions with inland climate and countries at latitudes closer to the tropics. Less cloudy skies however mean clearer nights and more clear sky long wave losses.

Wind

Higher wind speeds during the heating season increase the net energy demand for heating, all other parameters equal.

More wind has two effects: increased infiltration and ventilation losses and higher outside surface film coefficients. Higher speeds are typical for locations close to the sea, locations at the top of a hill and locations in valleys that are oriented parallel to the main wind direction. Also the lack of protection by nearby buildings and the absence of screening by trees may result in more wind.

Precipitation

More precipitation during the heating season may increase the net energy demand for heating, all other parameters equal.

The impact has to do with the way buildings are constructed. Contrary to North America where timber framing is preferred, brick buildings with cavity walls are a popular choice in Northwestern Europe. A cavity wall assures rain-tightness by drainage along the outside surface of the brick veneer, water buffering in the brick veneer and drainage at its cavity-side. Drying of that moisture demands energy, for a capillary saturated brick veneer up to 45 MJ/m^2. Part comes from outside, part from inside. Only the part from inside increases the net energy demand for heating. Also the lower thermal resistance of the wet veneer adds a little. Both effects decrease with the increasing insulation quality of the cavity wall.

Summary

Though climate is perceived as a given, mankind's activity nevertheless induces changes as global warming shows. Settlements also alter the local micro-climate. Cities are warmer, less sunny and less windy than rural regions. Open landscapes transform into less windy environments by planting trees. Broadleaf trees let the sun through in winter and protect from the sun in summer, etc.

Building use

Inside temperature

Net energy demand for heating drops when the inside temperature stays lower during the heating season, all other parameters equal.

Inside temperature is explicitly present in the model. In single zone tools, the average value for the protected volume is the parameter intervening:

$$\theta_{i,mo} = \sum_{j=1}^{n} \left(H_{je,mo}\, \theta_{i,j,mo} \right) \bigg/ \sum_{j=1}^{n} H_{je,mo} \qquad (3.73)$$

At first sight the drive for thermal comfort hardly leaves freedom in inside temperature setting. However, there is some elasticity. Habituation to high temperatures is not uncommon. Warmer dress allows for lower temperatures. Excellent thermal insulation increases the radiant temperature experienced, giving the same operative temperature at lower air temperature. Inhabitants accept lower temperatures for financial reasons as proven by the calculation of the average inside temperature to explain measured heating consumption data:

$$\theta_i = 15.7 - 0.00425\, H_T \quad (°C) \tag{3.74}$$

In moderate climates, the inside temperature has a large impact on net energy demand for heating: 1 °C more or less results in plus or minus 10 to 15% heating demand! The impact is less in cold climates.

Temperature control

Net energy demand for heating drops with better inside temperature control during the heating season, all other parameters being equal.

A good temperature control delivers substantial savings without harming thermal comfort. The basic rules are: (1) only heat the rooms in use, (2) during the hours they are used. Number (2) is called intermittent heating with night setback as a typical example. Comparison of the calculated annual end energy for heating, assuming 18 °C as protected volume inside temperature, with data measured in 964 dwellings showed that warming locally and intermittently is widespread (Figure 3.21). In fact, the least square straight line through the calculated data largely surpassed the measured exponential best fit:

$$\frac{E_h}{V} = 229.6 \left(\frac{H_T}{V}\right)^{0.84}$$

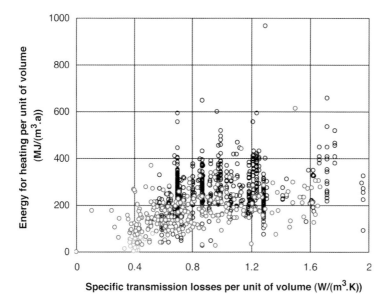

Figure 3.21. Measured annual end energy for heating in 964 dwellings as function of the specific transmission losses, both per unit of protected volume.

3.3 Energy efficiency

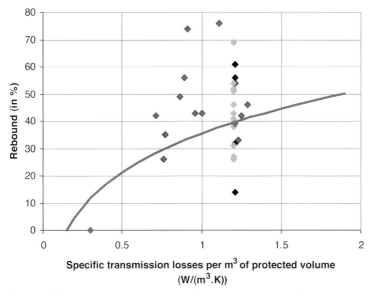

Figure 3.22. Random rebound factor. The dots show rebound factors measured for individual dwellings.

One reason for that is direct rebound behaviour: dwellers adapt the services desired to the price. The service heating delivers is thermal comfort, the price the energy cost. Figure 3.21 also reflects a physical reality: partial heating gives higher mean inside temperatures in dwellings that are better insulated, leaving less room for direct rebound in well insulated dwellings. From the figure a random direct rebound factor was deduced, the reference being the calculated end energy for heating at 18 °C inside (see Figure 3.22):

$$a_{\text{rebound}} = 1.355 \left(\frac{U_m}{C}\right)^{0.16} - 1 \qquad E_{\text{end,heat,a}} = (1 - a_{\text{rebound}}) E_{\text{end,heat,a,calc}}^{\theta_i = 18\,°C} \qquad (3.75)$$

Rebound has important effects. Better insulated designs save less energy compared to less insulated ones than assumed based on calculation with 18 °C inside. Insulation as a retrofit measure loses part of its benefit, make it less economic. And, energy projections for the residential sector overestimate annual heating consumption when direct rebound is not considered.

Ventilation

Net energy demand for heating drops with less ventilation during the heating season, all other parameters equal.

Ventilation habits are co-responsible for the air flows entering a building from outside. These have to be warmed to comfort temperature. A strict energy-based approach thus makes less ventilation attractive. However, ventilation is not a free lunch. Enough is a necessity for keeping the indoor environment healthy. The question is: what is enough? A formula used for residences is:

$$\dot{V}_a = m V \left[0.2 + 0.5 \exp\left(-\frac{V}{500}\right)\right] \quad (m^3/h) \qquad (3.76)$$

where V is the protected volume and m a factor reflecting the quality of the purpose designed ventilation system with value between 1 and 1.5. As mentioned before, ASHRAE standard 62.2-2007 advances as necessary ventilation flow for residences:

$$\dot{V}_a = 0.18\, A_{\text{floor}} + 12.6\,(n_{\text{br}} + 1) \quad (\text{m}^3/\text{h}) \tag{3.77}$$

where A_{floor} is the net floor area in the protected volume and n_{br} the number of bedrooms. Following EN 13779, the ventilation system in non-residential buildings for human activity must deliver an hourly airflow per person enough to maintain an indoor environment of quality IDA 3, IDA 2 or IDA 1, see Table 3.10.

Besides purpose designed ventilation, also uncontrolled infiltration intervenes:

$$\dot{V}_a = 0.04\, n_{50}\, V \quad (\text{m}^3/\text{h}) \tag{3.78}$$

Where n_{50} is the ventilation rate in ACH, measured with a blowerdoor for an inside/outside air pressure differential of 50 Pa. More recent research suggest as a more correct formula:

$$\dot{V}_a = (0.05 \text{ à } 0.06)\, n_{50}\, V \quad (\text{m}^3/\text{h}) \tag{3.79}$$

In the past, ventilation in domestic buildings relied on uncontrolled infiltration and window opening, which can be called airing. Although since the energy crises of the seventies and the trend towards more air-tight envelopes, infiltration could no longer guarantee basic ventilation with windows closed, that seemed no to be a reason for concern even when research showed that in cold and cool climates window opening was typical for the warm season, not for the winter season.

On average, window opening adds 0.21 to 0.34 ACH to the average infiltration rate. But excessive window airing in winter lifts net heating demand to unacceptable heights. Happily most heating systems cannot deliver the power needed to keep thermal comfort intact with open windows. That compels people to moderate winter window airing. Therefore, in practice, one may use the adventitious ventilation rates of Table 3.18.

Table 3.18. Adventitious ventilation rates in ACH's by window airing.

	Apartment buildings, per apartment					
Shielding	Two façade apartments Air-tightness envelope			One façade apartment Air-tightness envelope		
	Low	Average	High	Low	Average	High
None	1.2	0.7	0.5	1.0	0.5	0.5
Moderate	0.9	0.6	0.5	0.7	0.5	0.5
Strong	0.6	0.5	0.5	0.5	0.5	0.5
	Dwellings					
Shielding	Air-tightness envelope					
	Low	Average	High			
None	1.2	0.7	0.5			
Moderate	0.9	0.6	0.5			
Strong	0.6	0.5	0.5			

A purpose designed ventilation system assures more stable ventilation. The following systems are optional in residential buildings: (1) natural ventilation with wind and stack as driving forces, (2) mechanical supply ventilation, (3) mechanical extract ventilation and (4) balanced ventilation. The last allows heat recovery to be included. On condition the building is very airtight ($n_{50} < 1$ ACH) and the system is correctly designed and built (airtight ducts, ducts insulated where needed, recovery unit insulated) heat recovery may add quite some energy efficiency. Extract ventilation in turn creates the possibility to use the exhaust air as heat source for a heat pump. Peak ventilation by opening windows however must remain possible, which is why each room should have 0.064 times its floor surface in m^2 as m^2 operable window.

During the warmer months, purpose designed natural ventilation with operable windows remains the most energy efficient option in cool climates, which is why a balanced system should allow switching to natural ventilation. A recent development is mixed purpose designed natural/extract ventilation, with small fans on the exhausts in toilet, bathroom and storage room that run during and some time after that space is used.

Internal gains

Net energy demand for heating drops with higher internal gains during the heating season, all other parameters being equal.

Internal gains combine the occupant's heat with the heat released by artificial lighting and appliances. The drop mentioned in net heating demand yet should not be read as: more lighting and appliances save energy. In fact, the electricity needed to power the two requires much more primary energy in countries with thermal power plants than the fossil fuel saved by less heating.

Summary

Building use clearly is adaptable. Lower mean temperatures, better temperature control, correct ventilation, all help lowering the heating bill. In the seventies of last century, after the energy crises, authorities tried to change people's habits via the media. These campaigns were quite successful as long as energy was expensive. Once the prices dropped, people resumed their old habits especially when they didn't have to pay the heating bill. Habits are a complex boundary condition. Buildings and systems should therefore be designed so energy efficiently that wasteful behaviour does not require much extra energy. One advantage however must be recognized: wasteful behaviour in energy inefficient buildings makes the saving potential large.

Building design

Building design figures as key factor in energy efficiency, among others because most parameters of influence are under the control of the design team.

Protected volume

Net energy demand for heating increases with a higher protected volume, all other parameters being equal.

The impact of the protected volume (V in m^3) is directly visible when writing the heat loss by transmission (Φ_T), infiltration and ventilation ((Φ_V) for the building as single zone:

$$\Phi_T + \Phi_V = V\left[\frac{U_m}{C} + 0.34\left(n_V + n_{inf}\right)\right]\left(\theta_i - \theta_e\right) \quad (W) \tag{3.80}$$

where U_m is the mean thermal transmittance of the envelope in W/(m² · K), C the compactness in m (see below), n_V the ventilation rate and n_{inf} the infiltration rate, both in ACH. Heat loss and thus the net energy demand clearly increase in proportion to the protected volume, all other parameters being equal. Energy efficient buildings should thus have a volume dictated by the functions housed and not by the ambition of the principal. Yet, that conflicts with human nature. With higher prosperity also the net living area people wish expands and thus the protected volume, included the negative consequences for the net heating demand.

Compactness

Net energy demand for heating drops with higher compactness of the protected volume, all other parameters being equal.

The compactness (C) equals the ratio between the protected volume in m³ and the enclosing surface in m². As Equation (3.80) underlines, energy demand for heating changes in inverse proportion to compactness. A globe with compactness equal to one third of the radius ($R/3$ in m) is the densest volume known. Globes however lack functionality: lots of space is lost, and placing furniture is a problem. A cube instead is very usable, with compactness equal to one sixth of the side ($a/6$ in m). Yet, compared to a globe with same volume, a cube is 24% less compact.

Compactness increases with volume. In fact, a large volume always shows higher compactness than a smaller one (Figure 3.23). The absolute compactness instead, the ratio between the compactness of the volume (V) considered and a globe with same volume, remains constant:

$$C_{abs} = 4.836 \frac{V^{2/3}}{A_T} \quad (-) \tag{3.81}$$

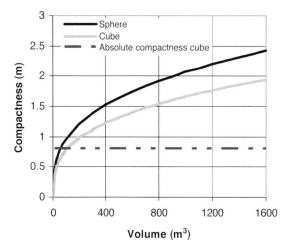

Figure 3.23. Compactness and absolute compactness as function of volume.

3.3 Energy efficiency

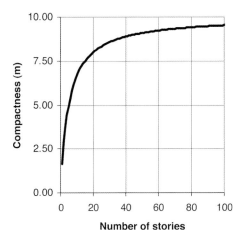

Figure 3.24. High rises, compactness.

Absolute compactness of detached houses lies below one whereas terraced houses with simple volumes are by far the most compact individual dwellings. A detached cube of 500 m^3 for example has compactness 1.32 m (absolute 0.806). Terracing five of them, increases compactness of the middle three to 1.98 m (absolute 1.21). Compact buildings are also more cost-conscious. In fact not the volume but the envelope and the partitions inside determine costs. Consider a building with a volume of 600 m^3. Compactness 1.405 means an envelope of 427 m^2. Fragmenting that volume to compactness 1 m demands 600 m^2 of envelope, included more partitions inside, extra foundations, extra roof edges, etc., i.e. clearly a more expensive choice. Moreover, compact buildings offer more thermal comfort than fragmented ones.

High compactness nevertheless is not a paradigm. High rises are very compact. In fact, one with rectangular floors has as compactness (see Figure 3.24):

$$C = 0.5 \bigg/ \left(\frac{L+B}{L\,B} + \frac{1}{H} \right) \tag{3.82}$$

where H is the height, L the depth and B the width of the building. The more stories the highrise has, the more compact it is, although above thirty stories the curve flattens. But more stories demand more space for elevators and more room for HVAC and sanitary systems. That diminishes the ratio between gross and usable floor area.

Floor organisation

Net energy demand for heating may drop when the floors are well organized, all other parameters being equal.

Floor organization first of all means lockable rooms. Otherwise, forget differentiation in temperature and ventilation flows, while solar gains usage may remain sub-optimal. A hall in dwellings with an inside door to the living room limits infiltration losses. If possible, windows in daytime rooms should look south-east over south to south-west for optimal usage of solar gains (north-east over north to north-west in the southern hemisphere). Anyhow, care must be taken in summer. Without solar shading, overheating and too strong differences in luminance may become a problem.

Area, orientation, slope and type of glazing

Net energy demand for heating could increase when more glass is used, all other parameters being equal. However, if the glazing has a low thermal transmittance and a high solar transmittance, if more glass is oriented south-east over south to south-west (north-east over north to north-west in the southern hemisphere) and if all glass is mounted vertically, than net energy demand for heating may drop, all other parameters being equal.

Glass area and glass type fix the transmission losses, while glass area, glass type, glass orientation and glass slope fix solar gains. Table 3.19 lists glass types available today. Where the thermal transmittance shows a ratio of 8.5 between worst and best, ratios for the solar transmittance and the visible transmittance are much smaller.

Table 3.19. Glass types.

Types	U-value W/(m² · K)	g-value –	$\tau_{visible}$ –
Single glass	5.9	0.81	0.90
Double glass	3.0	0.72	0.80
Low-e double glass	1.8	0.63	0.70
Argon filled low-e double glass	1.3	0.58	0.75
Krypton filled low-e double glass	1.0–1.1	0.58	0.75
Xenon filled low-e double glass	0.9–1.0	0.58	0.75
Krypton filled low-e triple glass	0.7	0.50	0.65
Xenon filled low-e triple glass	0.6	0.50	0.65

Despite that impressive decrease in thermal transmittance, large glass areas do not necessarily upgrade energy efficiency in the cloudy climate of Northwestern Europe. Even with the best types, more glass means replacing better insulating opaque assemblies by less insulating transparent ones. That difference in insulation quality is not brushed away by increased solar gains. With argon filled low-e double glass more area south is a neutral operation but more east over north to west increases the losses. Only krypton or xenon filled low-e double glass realizes neutrality there. But even then, the following three rules for dwelling remain valid in moderate climates:

1. Try to limit the glass area per room to 1/5 of the floor area.
2. Orient, if possible and advisable, a large part of that glass south-east over south to south-west (north-east over north to north-west in the southern hemisphere). Of these, south is preferred because at higher latitudes solar gains in winter are largest while in summer solar height makes shading easier. 'Advisable' underlines the rule has its limits. If for any reason sleeping room windows face south, do not make them larger than needed for daylighting. A living room facing north in turn loses attraction when equipped with such small windows that contact with outside is lost.
3. Mount glass vertically. In addition to being the easiest to maintain, winter solar gains at higher latitudes are largest.

Large glass areas oriented east over south (north in the southern hemisphere) to west augment overheating risk in springtime, summer and autumn. This shapes a fourth rule:

3.3 Energy efficiency

4. Provide large glass areas oriented east over south (north in the southern hemisphere) to west with solar shading. Otherwise, it may turn so warm indoors that inhabitants install active cooling! Exterior shading systems perform the best.

Thermal insulation

Net energy demand for heating drops with better thermal insulation, all other parameters being equal.

A well insulated envelope remains the most effective measure for lowering net energy demand for heating in cool and cold climates.

The whole thermal transmittance (U) quantifies insulation quality of an airtight envelope part:

$$U = U_o + \frac{\sum_{k=1}^{m} \psi_k L_k + \sum_{l=1}^{p} \chi_l}{A} \quad (W/(m^2 \cdot K)) \tag{3.83}$$

where U_o is the clear thermal transmittance of the part, A its surface, ψ the linear thermal transmittance of all linear thermal bridges in the part, L their length and χ the local thermal transmittance of all local thermal bridges in the part. The lower the whole thermal transmittance, the better the insulation quality. For the envelope the average thermal transmittance (U_m) becomes the quality mark:

$$U_m = \sum_{all} a\, U\, A \Big/ A_T, \quad A_T = \sum A \quad (W/(m^2 \cdot K)) \tag{3.84}$$

In that formula A_T is the envelope area in m² and a is a reduction factor for separations between heated and unheated spaces or parts in contact with grade or below grade.

Physically, proportionality exists between the net energy demand for heating and the specific transmission losses $H_T = U_m A_T$. A statistical analysis on measured heating data of 20 dwellings in a climate with on average 1923 degree days 15/15 ($D_{15}^{15} = 1923$) gave as a result:

$$Q_{heat,use} = (216 \text{ à } 360)\, U_m\, A_T \quad (MJ/a) \tag{3.85}$$

Multiplying with $D_{15}^{15}/1923$ extends that formula to other climates. Linearity is also seen in Figure 3.25, showing the impact of a step-wise retrofit of an end of a row house on the end energy used for heating. Linearity allows predicting the end energy for heating in an early design stage and explains why after the energy crises of the seventies many European countries imposed thermal insulation requirements. A good thermal insulation of course has additional advantages such as upgrading thermal comfort and the environmental quality indoors.

How are thermal transmittances chosen? The sketch design should:
- Fix an annual end energy use for heating target.
- Calculate related specific transmission losses $U_m A_T$ with Formula (3.85) where A_T is the surface enveloping the protected volume (the sum of all spaces where thermal comfort is demanded, excluded garages, storage basements and unused attics).
- Equate the specific transmission losses with the sum of the weighted product of the thermal transmittance and area of roofs, opaque façade walls, lowest floors, windows and separations with adjacent unheated rooms. Add the maximum thermal transmittances imposed by law as boundary condition. Account for thermal bridging by dividing $U_m A_T$ by 1.05:

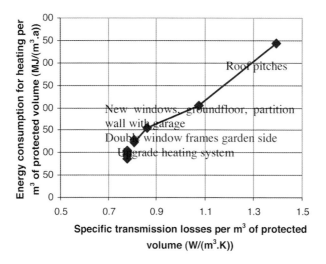

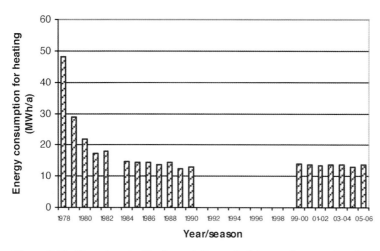

Figure 3.25. Stepwise retrofit of an existing end of the row house and end energy for heating used annually in relation to the specific transmission losses (both per unit of protected volume).

3.3 Energy efficiency

$$\frac{U_m A_T}{1.05} = U_{roof} A_{roof} + U_{facade} A_{facade} + \sum a_{fl} U_{fl} A_{fl} + U_w A_w + a U_{h,nh} A_{h,nh}$$

with: $U_{roof} \leq U_{roof,max}$ $U_{facade} \leq U_{facade,max}$ $a_{fl} U_{fl} \leq U_{fl,max} \left(R_{fl} \geq R_{fl,min} \right)$, etc.

- Fix all thermal transmittances minus one and calculate the one left. In doing so, distribute the insulation quality as uniformly as possible as this is economically preferred and most helpful in case of partial heating.

Thermal insulation anyhow gradually loses effectiveness. This is a direct consequence of the equation defining the clear thermal transmittance:

$$U = \frac{1}{R + d_{ins}/\lambda_{ins}} = \frac{\lambda_{ins}}{R \lambda_{ins} + d_{ins}} \tag{3.86}$$

Halving the value demands double insulation thickness. Very low clear wall thermal transmittances (< 0.1 W/(m² · K)) require thicknesses that generate important secondary costs, resulting in excessive space lost and complicated detailing. A legislation that mobilizes all other means to increase energy efficiency like the European EPDB (Directive 2002/91/EC) therefore makes sense.

Thermal inertia

In cool and cold climates net energy demand for heating is hardly influenced by thermal inertia, all other parameters being equal.

The higher the thermal inertia, the closer utilization efficiency nears 1 for moderate gains. But, the extra gain decreases with increasing capacity. Of course, in climates where short heating periods alternate with periods of no heating, inertia gains effectiveness compared to cool and cold climates, where the heating season lasts for 200 days. Even well insulated massive buildings cannot bridge such a long period.

The electric analogy of Figure 3.26 with a simple resistance-capacitance-resistance circuit $R_{i,x1} - C_{x1} - R_{x1,e}$ representing the opaque envelope helps in proving that statement, $R_{i,x1}$ being an equivalent thermal resistance between the node inside and the envelope's capacitance (C_{x1}) at temperature θ_{x1} and $R_{x1,e}$ being an equivalent thermal resistance between that capacitance and the outside.

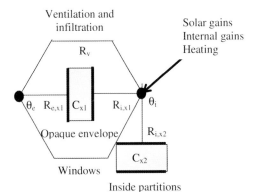

Figure 3.26. Electric analogy.

Partitions inside with a total exposed area A_i are coupled to the node inside by a capacitance-resistance circuit $C_{x2} - R_{i,x2}$ where $R_{i,x2}$ is an equivalent thermal resistance between indoors and a capacitance C_{x2} at a temperature of θ_{x2}. Ventilation and infiltration, given by $1/[0.34\,(n_V + n_{inf})\,V]$, pass in parallel with resistance R_v between the node inside and outdoors as does heat loss or heat gains through the windows. Solar gains, internal gains and the warmth by the heating system are injected in the node inside. Balance equations become (θ_i: inside temperature in °C, θ_e: outside temperature in °C, Φ_{SLW} solar gain in W, Φ_I internal gains in W and $\Phi_{heat,net}$ heating power in W):

Indoors:

$$\frac{(\theta_{x1} - \theta_i)(A_T - \sum A_w)}{R_{i,x1}} + \sum U_w\,A_w\,(\theta_e - \theta_i) + 2\frac{(\theta_{x2} - \theta_i)\,A_{x2}}{R_{i,x2}}$$

$$+\,0.34\,n\,V\,(\theta_e - \theta_i) + \Phi_{SLW} + \Phi_I + \Phi_{heat,net} = (\rho c)_{eq,a}\,V_a\,\frac{d\theta_i}{dt}$$

Envelope: *Partitions:*

$$\left(\frac{\theta_i - \theta_{x1}}{R_{i,x1}} + \frac{\theta_e - \theta_{x1}}{R_{e,x1}}\right) = C_{x1}\,\frac{d\theta_{x1}}{dt} \qquad \left(\frac{\theta_i - \theta_{x2}}{R_{i,x2}}\right) = C_{x2}\,\frac{d\theta_{x2}}{dt}$$

where A_w is the area and U_w the thermal transmittance of the windows, $(\rho c)_{eq,a}$ the equivalent volumetric specific heat capacity of the air inside (accounts for furniture and furnishings) and V_a the air volume in the building.

The difference between the temperature inside and the temperature θ_{x2} of the partition wall's capacitance should be minimal. The same holds for the temperature θ_{x1} of the opaque envelope's capacitance in massive buildings with outside insulation or cavity fill. That way, both temperatures and their derivatives equate with the inside temperature and its derivative ($d\theta_i/dt$), reducing the three equations to one (with the equivalent resistance $R_{e,x1}$ replaced by the mean thermal resistance of the whole envelope):

$$\underbrace{\left(\frac{A_T}{R_e} + 0.34\,n\,V\right)\theta_e + \Phi_{SLW} + \Phi_I + \Phi_{heat,net}}_{(1)} - \underbrace{\left(\frac{A_T}{R_e} + 0.34\,n\,V\right)\theta_i}_{H}$$

$$= \underbrace{\left[\rho_{eq,a}\,V + C_{x1}\,A_T + 2\,C_{x2}\,A_{x2} + (\rho c)_{eq,a}\right]}_{C_b}\frac{d\theta_i}{dt} \qquad (3.87)$$

With inside insulation the opaque envelope's capacitance hardly intervenes, or:

$$\underbrace{\left(\frac{A_T}{R_{x1}} + 0.34\,n\,V\right)\theta_e + \Phi_{SLW} + \Phi_I + \Phi_{heat,net}}_{(1)} - \underbrace{\left(\frac{A_T}{R_{x1}} + 0.34\,n\,V\right)\theta_i}_{H}$$

$$= \underbrace{\left[\rho_{eq,a}\,V + 2\,C_{x2}\,A_{x2} + (\rho c)_{eq,a}\right]}_{C_b}\frac{d\theta_i}{dt} \qquad (3.88)$$

3.3 Energy efficiency

Solving that first order differential equation for a step change of the term indicated with (1) at time zero gives:

$$\theta_i = \theta_{i\infty} + (\theta_{i\infty} - \theta_{i0})\exp\left(-\frac{H}{C_b}t\right) \quad (3.89)$$

where θ_{i0} is the temperature inside at time zero and $\theta_{i\infty}$ de final temperature inside. The inverse of the term C_b/H represents the time constant τ_b of the building in s. Looking to a periodic swing with period T of the outside temperature and the solar gains, amplitudes $\hat{\theta}_e$ and $\hat{\Phi}_{SLW}$, the complex inside temperature without heating $\alpha_{i,WH}$ becomes (internal gains constant):

$$\alpha_{i,WH} = \frac{\alpha_e}{\left(1+i\dfrac{2\pi\tau_b}{T}\right)} + \frac{\hat{\Phi}_{SLW}}{H\left(1+i\dfrac{2\pi\tau_b}{T}\right)} \quad (3.90)$$

where $[1 + I\,2\pi\tau_b/T]$ is the complex impedance of the building. For a period of one year, the net energy demand for heating equals:

Average:

$$\bar{Q}_{heat,net} = 31.5576\,H\left[\bar{\theta}_i - \left(\bar{\theta}_e + \frac{\Phi_{SLW}+\Phi_I}{H}\right)\right] \quad (MJ/a) \quad (3.91)$$

First harmonic:

$$\hat{Q}_{heat,net} = H\int_0^T (\alpha_i - \alpha_{i,WH})\,dt \quad (MJ/a) \quad (3.92)$$

In case of permanent heating without set-backs, thermal inertia dampens the temperature without heating and thus the net heating demand if the complex impedance for a one year period passes 1. If not, thermal inertia does not intervene. Moderately insulated, heavy buildings now have time constants touching 3 to 5 days. An excellent thermal insulation increases that value to 10 days. A well insulated light-weight building sees its time constant dropped to less than 1 day. Bad insulation reduces that 1 day to some 2 hours. In all 4 cases yet, the amplitude of the complex impedance remains close to 1 (see Table 3.20). Or, thermal inertia is not of importance.

Intermittent heating even turns inertia into a negative parameter. Warming up the assessable capacity after each set back period demands extra energy. The larger the inertia, the larger that extra is compared to a non-inert building. Of course, heat gain utilization efficiency increases.

Table 3.20. Impact of the thermal inertia on net heating demand.

Building	Capacitive		Well insulated		$\hat{Z}$
	Yes	No	Yes	No	
1	X			X	1.0012 / 1.0037
2	X		X		1.015
3		X		X	1.0
4		X	X		1.0

With night set-backs both effects neutralize each other. For longer set back periods, the extra net demand prevails. In case of extremely insulated buildings and optimal gains usage, capacity offers a benefit in spring and autumn when periods of well and no heating alternate. That positive affect at the beginning and end of the heating season, however, is marginal compared to the whole heating season.

Air-tightness

In cool and cold climates net energy demand for heating drops with better air-tightness of the envelope, all other parameters being equal.

Some become suspicious when they hear the word 'airtight'. What does ventilation mean in such cases? Put simply, one should no longer count on random infiltration through leaks and air permeable building parts but instead design a purpose designed ventilation system for the building. See Table 3.21. If not, an airtight envelope in fact may create a problematic indoor environment from a health and moisture tolerance perspective.

Table 3.21. Air-tightness and ventilation.

n_{50} h^{-1}	Climate	Way to ventilate
> 13	Moderate	Too air permeable
	Cold	Too air permeable
8–13	Moderate	Adventitious natural ventilation through infiltration
	Cold	Too air permeable
5–8	Moderate	Too airtight for adventitious natural ventilation through infiltration, too air permeable for a purpose designed natural or extract ventilation system
	Cold	Too air permeable
3–5	Moderate	Purpose designed natural or extract ventilation system
	Cold	Extract ventilation
1–3	Moderate	Balanced ventilation
	Cold	Balanced ventilation
< 1	Moderate	Balanced ventilation with heat recovery
	Cold	Balanced ventilation with heat recovery

Others still pretend walls should breathe, which air-tightness prevents. This fairy tale still survives, although it's the inverse. Lack of air-tightness causes so many drawbacks that the few advantages one may give are negated: (1) extra energy consumed, (2) draft problems, (3) degraded sound insulation, (4) increased moisture risks, (5) loss of thermal insulation quality and inertia, etc.

Air-tightness is measured using a blower door. Outside doors, windows and ventilation in- and outlets are closed, inside doors opened. The entrance door opening gets the blower door inserted which is equipped with a calibrated fan used for under- or overpressurizing the building (Figure 3.27). Per pressure step, the air flow (G_a in kg/s) and pressure difference (ΔP_a in Pa) with the outside is noted. The following relation is typically found:

$$G_a = C \, \Delta P_a^n$$

3.3 Energy efficiency

Figure 3.27. Blower door.

where C is the air leakage coefficient and and n the air leakage exponent. The characteristic value to evaluate air-tightness is the ventilation rate in ACH at 50 Pa air pressure difference (n_{50}). A lower n_{50}-value means a better air-tightness. Assuming a 3 to 4 Pa mean air pressure difference with the outside once a building is inhabited, taking the air leakage area equally distributed over the in and outflowing part of the envelope and using an air leakage exponent 0.6 to 0.8, then the mean infiltration rate plus the adventitious ventilation rate by window airing and outside door use (n_U) obeys the following rule of the thumb:

$$\frac{n_{50}}{20} + n_U \leq n \leq \frac{n_{50}}{9} + n_U$$

Summary

Building related parameters are the tools the designer has at his disposal. They should be combined in a creative way to realize an as low as economically advisible net heating demand lasting as long as the building does.

3.3.6 Parameters defining the net energy demand for cooling in residential buildings

The same set of parameters defining the net demand for heating governs cooling. Their effect however is different, see Table 3.22. The European EPDB-directive insists active cooling in dwellings should be minimized as much as possible. That turns building design into a prime concern. The measures applicable are: restricted use of glazing (competes with more solar gains during the heating season), solar shading of the glazed surfaces looking east over south to west, excellent thermal capacity and night-time ventilation.

Table 3.22. Parameters influencing the annual net energy demand for cooling.

Set	Parameter	Impact on net cooling demand, all other parameters being equal
Outside climate (cooling season means)	Temperature	Decreases when lower
	Insolation	Increases when more
	Wind speed	Depends on the difference between outside and inside air temperature. If negative a decrease, if positive an increase
Building use (cooling season means)	Inside temperature	Decreases when higher
	Temperature adaptation and control	Decreases when the principle 'cooling a zone when used only' is better applied
	Ventilation	May increase or decrease when higher. Depends on the difference between outside and inside air temperature. If negative a decrease, if positive an increase. More night ventilaton gives a decrease in regions with fresh night-time temperatures on condition the building is capacitive
	Internal gains	Decreases when lower. Is an incentive for energy efficient lighting, daylighting, dimming and presence control
Building design	Protected volume	Increases when larger
	Compactness	Increases with higher compactness
	Floor organisation	Increases when floor organisation augments solar gains
	Transparent surfaces (typically windows)	Increases when the envelope contains larger transparent parts that face the sunny sides and have high solar and low thermal transmittance
	Solar shading	Decreases when more efficient solar shading systems are used
	Thermal insulation	May increase or decrease with lower mean thermal transmittance of the envelope; depends on the difference between the mean outside sol-air temperature and the inside temperature. If negative, an increase, if positive, a decrease
	Thermal inertia	Decreases when higher. The effect assessable thermal capacity has is more pronounced when high ventilation rates are maintained during cool nights
	Envelope air-tightness	May increase or decrease when better. Depends on the difference between outside and inside temperature. If negative, a decrease. If positive, an increase

3.3 Energy efficiency

In the cool climate of Northwestern Europe, a better thermal insulation has a mostly negative impact on active net cooling demand. Anyhow, as long as the net heating demand prevails, excellent thermal insulation pays off.

3.3.7 Gross energy demand, end energy use

Here the HVAC-system is the first actor, though building and user still intervene. The gross demand equals the net demand, plus the unused distribution, control and emission losses. The ratio between the net demand and that sum is called the system efficiency.

End use stands for the transformation of any energy source into heat. For that to happen, boilers, heat pumps and chillers are used. The production efficiency then equals the ratio between the heat, injected in the distribution network and the energy used by the producing unit, see Table 3.23.

Table 3.23. Parameters of influence on the annual end energy use.

Set	Parameter	Impact on end energy use, all other parameters being equal
Heating and cooling	System efficiency	Decreases when closer to 1
	Boilers, production efficiency	Decreases when closer to 1
	Heat pumps, SPF	Decreases when higher
	Chillers, SPF	Decreases when higher
Ventilation	Extract	Decreases when the extract air is used as enthalpy source for a domestic hot water heat pump
	Balanced	Decreases when heat recovery is applied, on condition that the building is very air-tight, the installation perfectly mounted and the inhabitants instructed how to use the building

3.3.8 Residential buildings ranked in terms of energy efficiency

Talking about energy efficient buildings demands clear definitions. For that purpose, seven distinct efficiency ranks are considered: insulated buildings (level 1), energy efficient buildings (level 2), low energy buildings (level 3), passive buildings (level 4), zero energy buildings (level 5), energy plus buildings (level 6) and energy autarkic buildings (level 7).

3.3.8.1 Insulated buildings

Based on the assumptions that thermal insulation has the largest effect in terms of energy efficiency, that a U-value is a calculable quantity and that insulating a building does not limit architectural creativity, the focus is on a well insulated envelope. The only requirement imposed concerns the average thermal transmittance of the envelope and maximum thermal transmittances for the different envelope parts.

3.3.8.2 Energy efficient buildings

The objective is good thermal insulation, correct ventilation and an optimal use of solar and internal gains. The evaluation demands a standardized net energy demand calculation, for example based on EN 12831. The type of requirements imposed are: a maximum net energy demand for heating per m^2 floor area, a maximum net energy demand for heating per m^3 of protected volume or a maximum net energy demand for heating per m^2 of envelope area.

3.3.8.3 Low energy buildings

Normalised energy consumption for heating, cooling, air conditioning, domestic hot water and lighting is taken into account. Calculations are based on standardized methods as imposed by the EPDB-legislation in the EU-countries. The epithet 'low energy' is granted to residential buildings that consume less than 60 MJ/($m^3 \cdot$ a) primary energy for heating, have a good indoor climate in summer without mechanical cooling and show primary consumption for domestic hot water and household that is some 50% of the country's average. In moderate climates, such performances are achievable with an average thermal transmittance of the envelope reaching 0.3 to 0.6 W/($m^2 \cdot$ K) at the maximum depending on compactness (see Figure 3.28), air leakage not exceeding 3 h^{-1} at 50 Pa pressure difference and a well designed natural ventilation system with fan-support in bathroom and toilets.

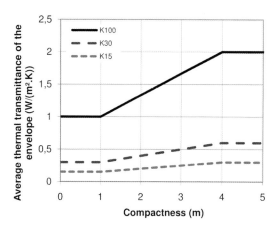

Figure 3.28. Average thermal transmittance of the envelope as a function of compactness.

3.3.8.4 Passive buildings

Comparable to low energy buildings. The epithet passive however is only granted to residential buildings that have a net energy demand for heating below 18 MJ/($m^3 \cdot$ a), have a good indoor climate in summer without mechanical cooling and whose overall primary energy consumption does not exceed 144 MJ/($m^3 \cdot$ a). To do that, air leakage is limited to 0.6 h^{-1} at an air pressure difference of 50 Pa, while an average thermal transmittance of the envelope touching 0.15 to 0.3 W/($m^2 \cdot$ K) depending on compactness is demanded (see Figure 3.28) and balanced ventilation with above 75% efficient heat recovery is a must.

3.3.8.5 Zero energy buildings

These are buildings which produce as much (primary) energy as is used for heating, cooling, domestic hot water and (sometimes) household. The objective demands lowering primary energy consumed as much as possible and producing the equivalent amount of electricity by PV. Analysis shows that even passive buildings need a respectable area of PV to achieve that objective. Also nobody considers the fact that if no measures at the electricity production, distribution and storage side are taken (the so-called smart grid initiative), zero energy on a large scale may result in a problematic unbalance between electricity produced and consumed.

3.3.8.6 Energy plus buildings

Are a next step. Energy-plus buildings produce more (primary) energy than used for heating, cooling, domestic hot water and (sometimes) household. Applying energy-plus on a large scale may even create more unbalance between electricity consumed and produced than zero energy does.

3.3.8.7 Energy autarkic buildings

Such buildings no longer depend on energy from fossil fuels and electricity. They produce their own energy from renewable sources such as wood chips, solar and wind. Requirements are solar collectors, PV-panels, wind turbines, heat storage and electricity storage.

3.4 Durability

3.4.1 In general

Durability is a building fabric requirement. Absolute durability does not exist. Even granite weathers. However relative durability, in terms of correct functioning of the fabric and its parts during a projected lifetime does. That projection depends on cultural and historical backgrounds, economics, local building tradition, building usage and which envelope part is looked for. People for example expect roofing felt to function properly for 10 to 15 years. A public building is projected to last longer than an industrial building. Timber framed construction is seen as having a shorter service life than massive construction. Finishes are trendier and therefore have shorter lives than load bearing parts.

Necessary conditions for a durable fabric are: (1) correct design from a structural and building physics point of view; (2) right material choices, that means materials that withstand the loads, characteristic for the position in the assembly and the nature of the building part; (3) good workmanship; (4) correct maintenance. The last concerns as well timely replacement of layers and parts with shorter service life as cleaning, repairing and refinishing the inside and outside surfaces.

When analyzing durability, terms like behaviour, environmental load, limit state, ageing, damage and risk pertain. Behaviour considers the evolution in time of the properties of a material or building part. Environmental load combines all external conditions that influence and change the response and behaviour of a material or building part (structural loading, temperature, moisture, electromagnetic radiation, etc.). Limit state means such damage that the material or building part no longer functions properly (too large deflection, frost damage, rot, collapse …). Ageing reflects the degradation of materials and building parts caused by inevitable physical, chemical and biological processes. Damage points to a too fast degradation as a consequence of design flaws, wrong material choices, poor workmanship or inadequate maintenance. Risk linked to damage is defined as:

$$\text{Risk} = p \cdot \text{Size} \tag{3.93}$$

where p is probability a given damage may happen whereas 'size' is the gravity of the negative consequences if happening. High risk thus includes rare damage with severe consequences as well as common damage with moderate consequences. The gravity of the damage depends on human and societal reaction. Those causing deaths are perceived as extremely severe. A deficiency triggering durability or functional integrity instead is generally judged less severe. While structural misfits may cause collapse and deaths, building physical shortcomings merely result in less usability, less comfort, worse indoor air quality, more energy used and higher environmental impact. Collapse probability from errors against physical integrity of a construction is rare, through not zero as Figure 3.29 shows. Timber beams supporting the roof of an indoor swimming pool rotted at their supports due to interstitial condensation behind the façade's inside insulation. Finally, some beams collapsed.

Figure 3.29. Indoor swimming pool, roof. Timber beam collapsed.

The severity of a consequence is often judged differently by inhabitants and society. Inhabitants hate bad odour, draft, rain penetration and mould. Higher energy consumption or more CO_2 emitted are less of a concern. Instead society bothers about higher energy uses and more emission of global warming gasses.

3.4.2 Loads

A distinction should be made between permanent and accidental loads. Permanent loads are:

- Structural loads, including own weight, dead load, live load, wind and inclination. They are static and dynamic, induce stress, strain and displacements. Time-related consequences include creep, relaxation and fatigue
- Thermal loads. These cause stress, strain, deformation, displacements and cracking. In combination with moisture, temperature oscillations around 0 °C may induce frost damage in brittle materials

3.4 Durability

- Hygric loads. The most damaging factor. They cause stress and strain, deformation, displacement and cracking. Moisture also degrades thermal and structural quality of materials and operates as a necessary boundary condition for many forms of biological attack. For the different types of material, the following moisture-related degradations are typical:

Material	Degradation
All, except metals, glass	Increased thermal and electrical conductivity, hygric displacement
Wood-based	Decrease in strength and stiffness, irreversible swelling, decay
Stony	Frost damage, salt damage, chemical attack
Synthetic	Hydrolysis
Metallic	Corrosion

- Electromagnetic load. Visible light dulls colours and enforces biological attack. UV may change the chemical composition of certain materials etc.
- Age. May slowly degrade the properties of certain materials without specific environmental loads acting as driving force.

Accidental loads are:

- Structural load. Earth quakes, hurricanes, explosions
- Fire. Exemplary case of an accidental load so devastating that the law imposes buildings should be designed and retrofitted taking into account probability 1.

All degrading actions are chemical, physical or biological in nature. To give some examples: chemistry governs the damage caused by moisture in combination with air pollution (SO_2, NO_x, Cl, NH_4). Diffusion, a physical phenomenon, causes out-gassing of blowing agents and alters the properties of some insulation materials that way. Damage caused by bacteria, algae, moss, moulds, fungi, birds, and rodents is biological in nature.

Ageing and damage are mostly the result of combined loading, which is why the following pages discuss damage patterns rather than loads

3.4.3 Damage patterns

3.4.3.1 Decrease in thermal quality

Thermal conductivity of water is more than twenty-fold that of air: 0.6 W/(m · K) versus 0.025 W/(m · K)). If for any reason a part of the air is chased by water in an open-porous material, more heat will be conducted. Latent heat transmitted by evaporation/condensation in the pores reinforces the effect whereas air permeation may add extra enthalpy flow.

How important the decrease in thermal quality is, depends on the air flow rate, the water content and which layer in a building assembly becomes wet. Moist layers at the warm side of thermal insulation are very negative, as well as wetting of the insulation itself, air looping around the insulation, wind passing behind the insulation and inside air passing between the insulation and the layers at its outside. For a roof in cellular concrete, 20 cm thick, and a fully filled cavity wall with a cavity depth of 6 cm, the thermal transmittances are:

Cellular concrete roof	U-value W/(m² · K)	Cavity wall	U-value W/(m² · K)
Air dry	0.8	Air dry	0.64
Built-in moisture	1.5	Veneer wetted by rain	0.65
Humid by interstitial condensation	1.3	Cavity fill saturated with water	1.80
Completely wet by rain	1.7		
Saturated with water	2.3		

Whereas the wet cellular concrete roof may be quoted as damaged, a wet veneer wall does not deserve that classification except if the cavity fill becomes wet.

3.4.3.2 Decrease in strength and stiffness

Strength and stiffness of most open-porous building materials decreases reversibly with higher moisture content. An irreversible decrease is noted for glued materials and materials such as particle board, plywood, fibre board and OSB. Reversible means that load removal restores the original situation or that stress and strain change sign when the load reverses. Irreversible indicates that part of the deformations become permanent. The cause of the irreversible reaction is hydrolysis of the glue as hydroxide ions cut the glue polymers into monomers. That causes the boards to swell and lose their cohesion. Figure 3.30 gives the decrease in bending strength of ureumformaldehyde (UF), melamine (Mel) and fenol formaldehyde (FF) bounded

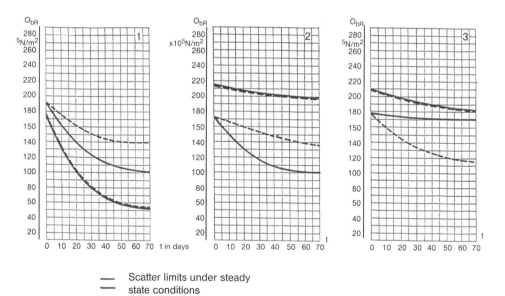

Figure 3.30. Bending strength of particle board under long lasting interstitial condensation. 1 = ureum formaldehyde boards, 2 = melamine boards, 3 = fenol formaldehyde boards.

particle board as a consequence of humidification caused by interstitial condensation. These and analogous data resulted in quality requirements for such boards depending on expected moisture load.

3.4.3.3 Stress, strain, deformation and cracking

Structural loads are not considered here, although the stress and strain they cause add to the thermal and hygric stress and strain. Again, reversible and irreversible effects intervene.

Thermal

Reversible deformation depends on the specific heat strain (α), defined as:

$$\alpha = \frac{1}{L}\frac{dL}{d\theta} \tag{3.94}$$

The value is highest for synthetic materials:

Material	α ($\cdot 10^{-6}$) K^{-1}
Wood-based	4 tot 30
Stony	5 tot 12
Synthetic	25 tot 200
Metallic	12 tot 29

If used as construction material in combination with other material, extra attention should go to detailing so the synthetic parts can move independently.

Temperature swings referenced by a temperature at construction establish the maximum expansion and contraction. The reference value adopted in practice is the annual mean outside. If movement is unhindered, reversible expansion and contraction are given by:

$$\Delta L = L_o \left[\exp(\alpha\, \Delta\theta) - 1 \right] \approx \alpha\, L_o\, \Delta\theta$$

with L_o the length at reference temperature.

However the probability of having unhindered deformation is small. Usually building assemblies and their layers are sandwiched between other parts or belong to the load bearing structure. The layers are also subjected to temperature gradients. As a result potential deformations cause stresses. For a fix-ended elastic layer, that stress in MPa equals $\alpha\, E\, \Delta\theta$, E being the modulus of elasticity in MPa. Stresses induced by temperature change increase when materials are stiffer and have a larger specific heat strain. Exceeding local tensile strength, results in micro-cracks. These evolve to macro-cracks depending on the nature of the load and fatigue response of the material. Thermal cracking probability diminishes at a decreasing ratio between the product $\alpha\, E$ and tensile strength. For layers where a high cracking resistance is important (outside stucco, paints, roofing felts), a low product $\alpha\, E$ should be coupled to high tensile strength. If that is not possible, thermal load must be minimized.

Irreversible thermal deformations are seen in plastic foams. Successive temperature shocks may lead to permanent, mostly anisotropic swelling or shrinkage. The reason is the structure of the foam: an aggregate of soap bubbles deformed by gravity into ellipsoids during hardening. Later

Figure 3.31. Damage to the low-sloped roof of a timber kiln, caused by an irreversible deformation of the PUR-insulation.

on higher temperatures lower the modulus of elasticity of the bubble walls while increasing gas pressure. That tends to restore the spherical form, resulting in shrinkage along the long axis and expansion along the short axis of the ellipsoid. In case the pores contain moisture, vapour pressure reinforces that irreversible deformation. The result may be severe damage, as Figure 3.31 shows for a low-sloped roof above a timber kiln.

Hygric

Hygric strain (ε) governs the reversible humidity-related deformation of open-porous materials. Its value is given by

$$\varepsilon(\phi \text{ of } w) = \frac{dL}{L} \tag{3.95}$$

Defining a specific hygric strain hardly makes sense. In fact, hygric deformations are not linear. The derivatives $d\varepsilon(\phi)/d\phi$ en $d\varepsilon(w)/dw$ are largest at low relative humidity or low moisture content and tend to zero for a relative humidity 100% which means capillary saturation. See Figure 3.32.

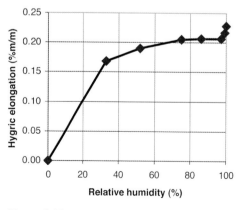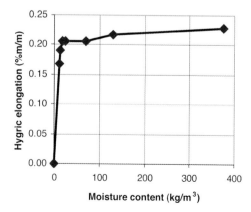

Figure 3.32. Cellular concrete, hygric expansion.

3.4 Durability

Relative humidity change compared to the value directly after processing, mostly 100%, fixes total reversible expansion or contraction. After application relative humidity slowly approaches an equilibrium which depends on where the material is used. The lower that equilibrium value, the more a material will shrink. Wood-based, lime-based and cement-based materials are very sensitive to hygric shrinkage. Once in use, the swings around the equilibrium relative humidity induce successive expansion and contraction. If unhindered, they can be expressed as:

$$\Delta L = L_o \left[\varepsilon(\phi_1) - \varepsilon(\phi_o) \right]$$

where L_o is the length at relative humidity ϕ_o. Again, unhindered movement probability is low. If hindered, part of the deformation transposes into stress, for a fix-ended elastic layer equal to $E\left[\varepsilon(\phi_1) - \varepsilon(\phi_o)\right] = E\,(d\varepsilon/d\phi)_m\,\Delta\phi$ in MPa, with $(d\varepsilon/d\phi)_m$ the mean derivative for the deformation in the relative humidity interval $\Delta\phi$. When the product $E(d\varepsilon/d\phi)_m$ and relative humidity swings increase, stress increases. Micro cracks appear when tensile strength is exceeded. With a decreasing ratio between the product $E(d\varepsilon/d\phi)_m$ and tensile strength the probability cracking will happen is lower. For layers where a high cracking resistance is important (outside stucco, paints, roofing felts), a low product $E(d\varepsilon/d\phi)_m$ should be coupled to high tensile strength.

Irreversible hygric deformation mainly figures as a chemical process or is linked to hydrolysis of the binder causing materials glued while pressurized to expand. In the first case, the irreversible part accompanies reversible initial shrinkage, in the second case swelling develops proportional to the decrease in strength and stiffness. In that sense, swelling is a measure for the severity of past moisture loading.

Limit state

Whether combined hygric and thermal load will cause cracking is difficult to predict. Both counteract one another. A temperature increase usually evokes lower relative humidity and vice versa. Some facts are nevertheless known. Timber normally spalls when moisture content drops below fibre saturation. Controlled drying may keep spalling minimal. Anyhow, at delivery, the moisture content must be as close as possible to the equilibrium once the building is in use.

Cement-based materials are sensitive to early cracking by combined chemical and hygric shrinkage, though at the moment of largest shrinkage final strength and stiffness is not reached yet. That lowers the stress. In parts with large thickness such as massive walls, columns and beams, thermal movement dominates once initial shrinkage is worked out. The main reason is the much larger penetration depth of temperature waves compared to relative humidity waves with equal period. In thin layers instead, such as stucco finishes, hygric movement wins. Thickness in fact is so small that temperature waves with periods less than one day hardly succeed in creating gradients across the layer while relative humidity waves with equal period do. That is why stucco shrinks under insolation and expands when cooling down!

It is not easy to say when cracks become damage. In general, the function of the cracked layer or building part, its structural integrity, rain tightness, air tightness or its sound attenuation should not be jeopardized. Cracks could also accelerate chemical attack. In some cases, appearance becomes degraded. Formulating clear performance requirements is difficult. For stucco, the rule could be: stress is limited to the extent a crack initiation is unlikely.

3.4.3.4 Biological degradation

The spectrum covered by biological degradation is very broad. Wood-based materials are challenged by bacteria, moulds, fungi and bugs. Fibrous and plastic foam insulation may suffer from rodents and birds. Stony materials get damaged by algae, moss and trees.

Wood-based materials

Bacteria

Aerobe and anaerobe bacteria help rot wood-based materials.

Moulds

See health issues. Wood-based materials with moisture ratio above 20 %kg/kg kept at 20 °C figure as preferred substrate for moulds of the deuteromycetes family. Their low metabolism keeps damage limited.

Fungi

Fungi are giant moulds. Those shown in Figure 3.33 belong to the basidomycetes family, which causes wood decay. Germination starts when spores settle on wood-based materials with moisture ratio far above 20 %kg/kg and temperatures between 8 and 35 °C. The hyphens pave their way into the wood while the fruiting body develops on the wood. Fungi have a high metabolism. Braun rot basidomycetes feed on cellulose while white rot basidomycetes attack the lignin skeleton. In doing so, both secrete water during digestion, creating their own optimal moisture ratio. The attacked wood looses strength, stiffness and cohesion.

 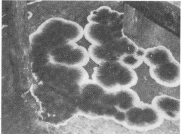 

Figure 3.33. Some fungi of the basidomycetes family.

Fungi may colonize large surfaces. The hyphens even penetrate mortar joints and porous bricks in search for feeding substrates. The damage inflicted is enormous: timber floors decay; doors are destruct, wooden finishes demolished. All that may stay concealed for quite a long time, as painted finishes are left intact.

Fungal attack is easily avoided by keeping wood dry, which means by avoiding moisture ratios above 20 %kg/kg. Combating fungal attack however is complicated. Decayed wood must be replaced by a more durable or preserved species, the intact parts have to be treated with a conserving liquid, brick work must be injected with poisonous liquid, moisture sources have to be removed. And all dismantled wood must be burned.

Bugs

Anobia, lycti and longicorns are common wood bugs in cool regions (see Figure 3.34). In warm climates, termites take over. Wasps also attack wood.

Anobia prefer wood infected by fungi. Some however look for dry, older wood. Lycti colonize porous timber species with pores containing yeast and a vitamin needed by their larva for growth. Longicorns like pine or sapwood stored in a warm, humid environment. Termites live in colonies below grade. A colony consists of workers, soldiers and females. They feed on

3.4 Durability

Figure 3.34. Left the longicorn, in the middle its larva, right the anobia.

decayed or dead organic material. Timber used in construction is dead and thus a preferred target if moist and contacting grade. The damage termites inflict is enormous. Estimates of lost capital in the US are up to $1,500,000,000 a year (in dollars of the early nineties). In Europe termites are active in most regions below 48° north. Wasps pick up cellulose to build nests.

A low enough moisture ratio keeps bugs out of wood. As 'low enough' is rather vague some countries require treated wood for use in construction. Also species-specific measures may be needed. To avoid termites, timber should not contact grade while the bottom plates in timber framed walls require protection by an impenetrable barrier.

Fibrous and plastic foam insulation

Rodents love nesting in fibrous and plastic foam insulation (Figure 3.35). Birds in turn use mineral fibres for building their nests (Figure 3.36). Avoidance demands a shielding of such insulation layers. Also modification of the binder in fibrous insulation so it loses its attractiveness as nesting material is an option.

 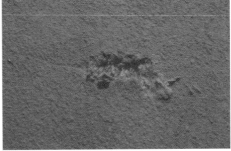

Figure 3.35. Rodent's nest in glass wool insulation.

Figure 3.36. Mineral fibre pecked off by birds.

Stony materials

Algae and moss demand light, nutrients, moisture and a stable temperature as is illustrated by the moss growing on slated roofs. Light is not a problem. The nutrients come from the organic dust accumulating between the slates. The water sucked up by capillarity between slates suffices as moisture source and, stable temperatures are encountered on pitches looking north-west over north to north-east (northern hemisphere). EIFS-systems typically suffer from algae. The moisture there comes from surface condensation by under-cooling at night while nutrients are found in the dust deposited on the stucco.

Whether the limit state deserves the notion 'damage', depends on the kind of problems greenery causes. Moss may widen the overlap between slates thereby facilitating rain infiltration. Plants in or on masonry walls erode the joints with their roots. Also a chemical attack is possible as moss, algae and some plants secrete organic acids.

3.4.3.5 Frost damage

Moist brittle open-porous stony materials may crack, spall and pulverize when subjected to successive frost/thaw cycles. The reason is that the water in the pores expands by 10% when freezing. In wide pores ice forms below 0 °C. In finer pores freezing needs much lower temperatures. Only at temperatures below $-x$ °C all water will be transformed into ice, x depending on the smallest pores present. Low moisture content excludes frost damage. Risk augments as moisture content increases. Zero risk demands restriction of moisture uptake by correct detailing, water retarding surface finishes or use of frost resisting materials.

In judging frost damage risk, two elements intervene: expected moisture content and material quality. The latter translates in the material's critical moisture content for frost (w_{crf}), so that the difference with the actual moisture content (w), called the frost number F, governs risk:

$$F = w_{cr,f} - w \tag{3.96}$$

A frost number below zero means frost damage is inevitable. Frost resistance is thereby a relative concept. A material with known critical moisture content for frost may show resistance in one application and deteriorate by frost attack in another.

As ice expands by 10% compared to liquid water, frost damage risk becomes acute once moisture fills 90% of the pores. The limit value for the critical moisture content for frost then becomes: $w_{crf} \leq 0.9\, w_m$, with w_m saturation moisture content. Several materials however degrade at lower moisture contents. Even below the limit value, a 10% expansion could induce stresses beyond the tensile strength of the material. The inverse is also true. At high tensile strength or such deformability the expansion may be buffered without damaging effects. Some materials are exclusively micro-porous in nature with icing starting at temperatures far below zero, meaning that frost damage is excluded in cool climates. The critical moisture content or critical saturation degree for frost thus depends on the pore structure, the tensile strength and the deformability of a material. See Table 3.24.

Expected moisture content differs between applications. A non-painted veneer wall never sees its moisture content exceeding capillary moisture content. Cavity fill retards drying but doesn't push moisture content beyond capillary. Painted veneer walls instead may absorb rain water through micro-cracks in the paint, while experiencing retarded drying. That upsets the moisture balance, pushing moisture content beyond capillary. Moisture content in tiles covering insulated roof pitches nears saturation in winter as under-cooling retards drying, while rain and surface condensation maintain wetting. Practice learns that unused chimneys also become very wet. These examples show that rain buffering façade materials may have moisture contents up to and beyond capillary. In case their critical moisture content for frost is below capillary, frost number turns negative and frost damage becomes very likely.

Besides material quality and exposure, detailing and the way frost attacks make a difference. Two-sided freezing gives different internal stress distributions than one-sided freezing. Stresses

3.4 Durability

Table 3.24. Critical saturation degree for frost of some materials.

Material	Critical saturation degree for frost $S_{cr,f}$ %
Natural stone	65–96
Normal concrete	90
Concrete with air bubbles	81
Lightweight concrete	60
Cellular concrete	40
Bricks	87
Sand-limestone	65
Cement mortar	90
Lime mortar	65

are larger in massive than in thin layers. Frost/thaw is also cyclic in nature. In cool climates, roof tiles have to withstand 5 to 70 cycles per winter. Cyclic loading induces fatigue, with visual damage appearing after a number of cycles that differs between materials.

Evaluation

The indirect G_c-method

Inspired by the frost number F, the G_c-method evaluates frost resistance of natural stone, tiles and bricks. The capillary moisture content, the saturation moisture content, the capillary water absorption coefficient (A) and the coefficient of secondary water uptake (A') are measured. Frost damage risk increases when capillary moisture content nears saturation. Instead, if the ratio between the capillary water absorption coefficient and capillary moisture content increases, as with macro-porous materials, risk decreases, as was proven by analyzing the correlation between porous structure, direct frost testing and observation on site. A mathematical expression for G_c was sought by correlating the results of direct frost tests under different degrees of exposure (higher moisture content in the materials, more severe freezing conditions) to the ratio between capillary and saturation moisture content at one side and capillary water absorption coefficient and moisture uptake at saturation ($w_m h$, with h height of the sample) at the other side:

Bricks and natural stone	Tiles
Suction curve with first and secondary water uptake	
$G_c = -14.53 - 240 \dfrac{A}{w_m h} + 20.3 \dfrac{w_c}{w_m}$	$G_c = -21.04 - 504 \dfrac{A}{w_m h} + 23\,913 \dfrac{A'}{w_m h} + 22.6 \dfrac{w_c}{w_m}$
Suction curve without secondary water uptake	
$G_c = -6.35 + 16\,676 \dfrac{A}{w_m h}$	$G_c = -6.35 + 16\,676 \dfrac{A}{w_m h}$

G_c was used to determine material classes:

G_c	Frost resistance	Class
< –2.5	Material frost resisting even under the most severe exposure	D
–2.5 ≤ G_c < –0.95	Usage of the material subjected to increasinggly severe limitations	C
–0.95 ≤ G_c < 0		B
0 ≤ G_c < 4.5		A
≥ 4.5	Material not frost resisting	O

Class and application were then linked for bricks and stone, see Table 3.25.

Table 3.25. Bricks and stone, link between class and application.

Application, detailing and orientation Building part	Cavity wall		Heated inside?	Protected thanks to detailing?			
	Unfilled	Filled		Yes		No	
				S+W	N+E	S+W	N+E
Above grade masonry	X		Yes	B	B	C	B
	X		No	C	C	D	C
		X	Yes	C	C	D	C
		X	No	C	C	D	C
At grade masonry	X		Yes	B	B	C	C
	X		No	C	C	D	D
		X	Yes	C	C	D	D
		X	No	C	C	D	D
Masonry parts extending above the roof	X	X				D	C
Horizontal masonry parts	X	X		D	C	D	D
Chimneys	X	X		C	C	D	D
Masonry in contact with water						D	D

The method is statistically sound: at a probability of 87% the class measured is correct, at 97% the error does not exceed 1 class above or below. The measurement of G_c has been standardized.

Direct methods

Direct methods were used to divide bricks and stones into classes. The extreme approach was to put them in a drained, sand-filled container with one surface weather-exposed. If after four winters no visual damage was noted, the materials were class D. If visually damaged, they were class O to C. Further classification included:

Class C Brick-laying the samples on a horizontal concrete floor and exposing them to the weather. If after four winters they did not show damage although degrading in the sand-filled containers, they are class C

3.4 Durability

Class B   Storing the samples on racks and exposing the racks to the weather. If after four winters they did not show damage although degrading on the concrete floor and in the sand-filled container, they are class B

Class A   Brick-laying the samples in a south-west oriented veneer of an unfilled cavity wall exposed to the weather. If after four winters they did not show damage although degrading on the rack, the concrete floor and in the sand-filled container, they are class A

Class O   Also the veneer wall test gives damage

Traditional frost/thaw tests consist of several 1 day cycles with the wet samples freezing in the air or in a sand-filled container over (24-y) hours, followed by y hours thawing in water. Variables considered are initial moisture content (saturation under vacuum, saturation at 1 atmosphere, capillary saturated), number of cycles, hours (y) of frost and freezing/thawing temperatures. Most important is the initial moisture content, Class D materials don't get damaged when saturated. Class C materials show damage then, but not when they are capillary moist. Class B, A and O materials get hardly, moderately or heavily damaged when they are capillary moist. The number of cycles should be 25 at least to see the effects of fatigue. The number of hours (y) must be large enough to get the samples completely frozen. A freezing temperatures of –10 to –25 °C is preferred, while thawing is best done in water at 20 °C.

Remedies

For limited frost damage eliminating the cause may be effective, for example by adding an extra line of defence against rain penetration. The only solution for severe damage is replacement of the degraded parts.

3.4.3.6 Salt attack

Salts and salt sources

Common are carbonates, sulphates, nitrates and chlorides of calcium (Ca), sodium (Na), potassium (K) and magnesium (Mg). Phosphates and ammonium nitrates are very occasionally found, while bricks may contain vanadium and manganese salts.

Some salts are material-specific:

Bricks   Clay may contain ferro-sulphates. During firing these react with oxygen forming SO_3. That in turn reacts with the alkali in the clay giving alkali sulphates (Na_2SO_4, $CaSO_4$, K_2SO_4, $MgSO_4$). If firing is done at temperatures beyond 950 °C these sulphates, except $CaSO_4$, react with the silicates in the bricks. Otherwise they deposit in the pores. If firing does not succeed in neutralizing all ferro-salts these continue to react after brick-laying.

Stone   May be loaded with salts when quarried.

Concrete Is by definition alkali-rich ($Ca(OH)_2$, KOH, NaOH).

Binders Portland cement contains $Ca(OH)_2$, KOH en NaOH. These react with CO_2 in the air, forming alkali-carbonates (Na_2CO_3, $CaCO_3$, K_2CO_3, $MgCO_3$). In the presence of moisture, SO_2 transforms these carbonates into sulphates.

Fresh lime mortar is a mixture of $Ca(OH)_2$ and sand. During curing calcium carbonate ($CaCO_3$) is formed.

Brick/mortar	NaOH and KOH in the mortar reacts with lime sulphate ($CaSO_4$) in the brick to form Na_2SO_4 en K_2SO_4, with $Ca(OH)_2$ as by-product, which in turn reacts with CO_2 in the air to form $CaCO_3$.

Others come from the environment:

By ground water	Calcium sulphates, sodium nitrates, potassium sulphates, potassium nitrates and ammonium nitrates (Na_2SO_4, K_2SO_4, $NaNO_3$, KNO_3, NH_4NO_3)
By air	Sodium chlorides in coastal areas, salt deposits in industrial areas. Air pollution releases CO_2, SO_2, NO_x and compounds of sodium carbonates, potassium carbonates, calcium carbonates and magnesium carbonates plus sulphates and nitrates of the same reagents
As thaw salts	Calcium chlorides ($CaCl_2$)
From cesspools	Nitrates

All dissolve in water (Table 3.26).

Table 3.26. Solubility of salts in water.

Salt	Solubility In g per 100 ml of water		
	0 °C	20 °C	100 °C
$CaCO_3$		0.0015	0.019
$CaSO_4 \cdot 2 H_2O$		0.24	0.22
Na_2SO_4		4.8	
$Na_2SO_4 \cdot 10 H_2O$		11.0	92.0
NaCl		36.0	39.0
K_2CO_3	105.5		156.0
K_2SO_4			24.1

Together with its capillary moisture content solubility determines how much salt the liquid phase in an open-porous material may contain. As the table indicates, solubility diminishes at lower temperatures, meaning that crystallization will be more intense then. Carbonates create few problems while chlorides, nitrates and sulphates are very degrading.

Salt transport

Dissolved salts displace together with the free pore water by advection. At the same time diffusion moves salt ions from spots with higher to spots with lower concentration. Part of these moving ions bind to the pore walls. Where capillary flow turns into diffusion and the decrease in water content over-saturates the liquid, salts crystallize. Successive dissolving and crystallizing happens when moisture content swings between hygroscopic and above hygroscopic, a thing rain absorption followed by drying causes. Built-in moisture is a temporary salt drive, while rising dampness is a permanent drive.

Once crystallized, hydration of the salt crystals occurs each time the relative humidity in the pores passes the equilibrium value for the saturated salt solution present.

3.4 Durability

Consequences of salt presence

Higher hygroscopicity

Salts change hygroscopicity. Once above the equilibrium relative humidity of the saturated solution, hygroscopicity increases compared to the salt-free material. The increase develops proportional to the salt concentration. Very active from that point of view are $MgCl_2$, K_2CO_3, $Mg(NO3)_2$, NaCl, KCl and K_2SO_4 with an equilibrium relative humidity at saturation equal to 33%, 45%, 52%, 75%, 86% en 97%. In case a material is loaded with a cocktail of salts, then the whole hygroscopic curve moves upwards, while the salts crystals present simultaneously hydrate. The other way around, salt-loaded materials dry only when the relative humidity drops below the equilibrium value of the saturated solution. In regions with cool but humid climate that fact moistens salt-saturated walls each summer.

Damage by crystallization and hydration

Crystallized salts and the remaining saturated solution occupy a larger volume than the over-saturated solution. If the pores lack the space to absorb that increase, crystallization pressure becomes:

$$P_{crystal} = \frac{RT}{v_m} \ln\left(\frac{c}{c_{sat}}\right) \qquad (3.97)$$

In that formula R is the universal gas constant (831.4 J/(kg · K)), v_m the molar volume of the salt and c/c_{sat} the ratio between the salt concentration in the over-saturated and in the saturated solution. Table 3.27 gives an idea of pressures expected. NaCl clearly ranks highest. However, thanks to its solubility, it is not the most eroding salt. That's Na_2SO_4 with its low solubility and high crystallization pressure!

Table 3.27. Crystallization pressures.

Salt	Molar volume	Crystallization pressure MPa			
		$c/c_{sat} = 2$		$c/c_{sat} = 10$	
	Ml	0 °C	50 °C	0 °C	50 °C
$CaCO_4 \cdot 1/2\ H_2O$	46	33.5	39.8	112.0	132.5
$CaSO_4 \cdot 2\ H_2O$	55	28.2	33.4	93.8	111.0
$MgSO_4 \cdot 1\ H_2O$	57	27.2	32.4	91.0	107.9
$MgSO_4 \cdot 6\ H_2O$	130	11.8	14.1	39.5	49.5
$MgSO_4 \cdot 7\ H_2O$	147	10.5	12.5	35.0	41.5
Na_2SO_4	53	29.2	34.5	97.0	115.0
$Na_2SO_4 \cdot 10\ H_2O$	220	7.2	8.3	23.4	27.7
NaCl	28	55.4	65.4	184.5	219.0
$NaCO_3 \cdot 1\ H_2O$	55	28.0	33.3	93.5	110.9
$NaCO_3 \cdot 7\ H_2O$	154	10.0	11.9	33.4	36.5
$NaCO_3 \cdot 10\ H_2O$	199	7.8	9.2	25.9	30.8

Depending on temperature and relative humidity, chemical equilibrium between the crystals and water vapour in the air may change from less to more bounded water molecules. That adds extra volume and heightens the internal pressure with:

$$P_{hydrat} = \frac{RT}{v_{cw}} \ln\left(\frac{p}{p_h}\right) \tag{3.98}$$

where v_{cw} is the volume of 1 mol of water, p the actual vapour pressure and p_h vapour pressure at hydration for the temperature T present (in K). See Table 3.28 for $CaSO_4$.

Table 3.28. Hydration pressure for $CaSO_4$.

Relative humidity	Hydration pressure MPa		
	0 °C	20 °C	50 °C
100	219	176	93
70	160	115	25
50	107	58	0

Examples of temperature and relative humidity induced reversible hydration reactions are:

$$CaSO_4 \cdot 2\,H_2O \underset{}{\overset{120-180\,°C}{\rightleftharpoons}} CaSO_4 \cdot 1/2\,H_2O\ (Gips)$$

$$NaSO_4 \cdot 10\,H_2O \underset{}{\overset{32.4\,°C}{\rightleftharpoons}} NaSO_4$$

$$Ca(NO)_3 \cdot 4\,H_2O \underset{}{\overset{30\,°C}{\rightleftharpoons}} Ca(NO)_3 \cdot 3\,H_2O \underset{}{\overset{100\,°C}{\rightleftharpoons}} Ca(NO)_3$$

$$CaCl_2 \cdot 6\,H_2O \rightleftharpoons CaCl_2 \cdot 2\,H_2O \underset{}{\overset{260\,°C}{\rightleftharpoons}} CaCl_2$$

$$Na_2CO_3 \cdot 10\,H_2O \underset{}{\overset{32\,°C}{\rightleftharpoons}} Na_2CO_3 \cdot 7\,H_2O \underset{}{\overset{35.4\,°C}{\rightleftharpoons}} Na_2CO_3 \cdot 1\,H_2O$$

Again $Na_2SO_4 \cdot 10\,H_2O$ is the most corrosive salt.

Other impacts
- Deposited salt crystals change the porous system, narrowing larger pores, and filling finer ones.
- Salt solutions have a higher density, a higher viscosity and different surface tension than pure water. That changes the moisture properties of a salt solution compared to water.
- Osmotic suction salt solutions induce changes the chemical potential influences.
- Dissolved salts may react with the matrix, see the formation of Candlot salt in mortar, causing significant expansion and pulverization:

$$Al_2O_3 \cdot 2\,CaO \cdot 12\,H_2O + 3\,CaSO_4 \cdot 2\,H_2O + 15\,H_2O$$
$$\rightarrow \underbrace{Al_2O_3 \cdot CaO \cdot 3\,CaSO_4 \cdot 32\,H_2O}_{\text{Candlot salt}} + Ca(OH)_2$$

Types of salt degradation and remedies

Type 1   Efflorescence of sodium sulphate (Na_2SO_4) and potassium carbonate (KCO_3). Degrades the appearance. Removing the efflorescence is unnecessary. Rain washes it away. However one should wait to paint until efflorescence has disappeared.

Type 2   Crypto-efflorescence of sodium sulphate (Na_2SO_4) and potassium carbonate (K_2CO_3). May cause severe degradation. No effective remedy exists.

Type 3   Carbonisation of $Ca(OH)_2$ on wall surfaces. Gives a tough lime deposit which degrades the look. Removal is possible by brushing, rubbing or scraping off, and then washing the surface with a 10% solution of HCl and rinsing it with water.

Type 4   Calcium sulphate attack. Results in severe damage caused by swelling and pulverisation of the joints. Only one remedy is applicable: tearing down the attacked wall parts and replacing by new masonry.

3.4.3.7 Chemical attack

Chemical attack changes the composition of the material's matrix. The reagents, which are part of the environment or material-specific, use moisture as a vehicle and catalyst. The term environment includes the air loaded with contaminants such as SO_2, chlorides, CO_2 and NO_x, acid rain, sewage water, sea water and others. Examples of chemical attack are: gypsum formation on limestone, steel bar corrosion by carbonation and chloral attack in reinforced concrete, sulphate and alkali-granulate reactions in concrete.

Gypsum formation in limestone

Calcite ($CaCO_3$) reacts with sulphur acid (H_2SO_4) to form gypsum ($CaSO_4$) on the surface and in the surface pores of limestone:

$$H_2SO_4 + CaCO_3 \rightarrow CaSO_4 + CO_2 + H_2O$$

Sulphur acid (H_2SO_4) is the outcome of oxidation of sulphur dioxide (SO_2) in the air, followed by a reaction with water:

$$SO_2 + O \rightarrow SO_3 \qquad SO_3 + H_2O \rightarrow H_2SO_4$$

Gypsum forms a fine-porous layer on the calcite that, intact, could protect the stone from further degradation. The reaction however causes a volumetric expansion, which induces stress and possibly cracking. But, even if not, the difference in thermal and hygric movement between gypsum and calcite will do it. Cracks allow the limestone to suck acid water. That way, not only gypsum formation goes on but the calcite also slowly dissolves, while the absorbed water may carry salts and cause frost damage. The end is complete weathering of the lime stone.

Could limestone last for ever without SO_2 in the air? No. Even carbon dioxide (CO_2) attacks the material as it reacts with water to form carbon acid (H_2CO_3) which in turn transforms calcite into bicarbonate:

$$CaCO_3 + H_2CO_3 \rightarrow Ca(HCO_3)_2$$

Bicarbonate dissolves in water, causing decalcification and weathering of the limestone.

Gypsum formation is best combated by eliminating moisture uptake. For that reason limestone often gets a water-repellent restorative treatment. The problem then is that the surface layer has different mechanical, thermal and hygric properties than the limestone beneath, while a water-repellent treatment cannot avoid cracks to remain capillary active. So water might get accumulated behind the treated surface. With salts present, crypto-efflorescence becomes the rule. Frost damage might develop, etc. ...

Steel bar corrosion in reinforced concrete

The calcium hydroxide ($Ca(OH)_2$) in fresh poured concrete creates an alkaline environment for the steel bars with pH between 12.5 and 13. That high value stabilizes bar corrosion. However, as soon as the concrete binds, carbon dioxide CO_2 starts diffusing from the surroundings into the pores where, in the presence of moisture, a reaction with the calcium hydroxide starts forming calcite ($CaCO_3$):

$$Ca(OH)_2 + CO_2 \to CaCO_3$$

Too much moisture produces bicarbonate, which dissolves in water. When the concrete dries, the solution moves to the surface where the bicarbonate crystallizes and washes away with the rain. The result is decalcifying of the concrete and a slow decrease of the pH around the steel bars which destabilizes their corroded surface layer. Diffusion of oxygen and the hygroscopic moisture present in the pores then suffice to reactivate corrosion. Ferro-oxide occupies a larger volume than steel. Stress building-up that way finally destroys the concrete cover. Cracks accelerate the process, commonly called concrete corrosion.

Chlorides show an even higher destructive effect. Some binding accelerators contain chlorines (NaCl of $CaCl_2$). Chlorines are used as thawing salts. Near coastlines sodium chloride is deposited by wind. Underwater concrete in the sea faces a saline environment... Chlorine reacts with cement to form calcium chloride salt compounds. When enough chlorine is present, not only the pH lowers, but the more voluminous chloride salt compounds, among them Friedel salt, help in spalling the concrete:

$$Al_2O_3 \cdot 3\,CaO \cdot 12\,H_2O + 3\,CaCl_2 \cdot 6\,H_2O \to \underbrace{Al_2O_3 \cdot 3\,CaO \cdot 3\,CaCl_2 \cdot 30\,H_2O}_{\text{Friedel salt}}$$

Chlorine ions reaching the reinforcement bars in the wet concrete, first attack the corroded surface layer, then cause pitting corrosion, eventually completely destroy the bars.

How to avoid bar corrosion? The concrete should be well compacted. Depending on the degree of exposure bar cover must be thick enough. Thicker cover retards diffusive flows from reaching the bars and enlarges the concrete mass subjected to decalcifying. If necessary, the concrete could receive a watertight or sorption-retarding finish. Another way would be to keep the concrete saturated with water. That is difficult to realize and might create problems such as frost attack.

Sulphate and alkali-granulate reactions

When enough sulphates diffuse into concrete, they will react with the cement to form Candlot salt with swelling, stress and cracking as a result. The term 'alkali-granulate reaction (AGR)' in turn indicates that in wet concrete the alkali hydroxides present in the mortar (NaOH, KOH, $Ca(OH)_2$) may react with the carbonate and silicate granulates used:

3.4 Durability

$$Na_2O + H_2O \rightarrow NaOH$$

$$2\, NaOH + SiO_2 + n\, H_2O \rightarrow Na_2SiO_3 \cdot n\, H_2O$$

$$Na_2SiO_3 \cdot n\, H_2O + Ca(OH)_2 + H_2O \rightarrow CaSiO_3 \cdot m\, H_2O + 2\, NaOH$$

The $CaSiO_3 \cdot m\, H_2O$ gel is more voluminous than the original granulates, causing the concrete to swell and fracture when the stress generated exceeds tensile strength. Unlike carbonation, chlorine and sulphate attack, which are diffusion-induced and thus acting from the surface inward the concrete, the alkali-granulate reaction happens in the concrete with cracking developing from inside. The result is a dramatic loss in tensile strength, although a three-dimensional reinforcement might transform the swelling into a kind of post stressing.

Avoiding AGR demands elimination of one of the three causes: (1) excessive wetness; (2) use of reaction sensitive granulates; (3) application of cement and water with high free alkali hydroxide content.

3.4.3.8 Corrosion

Moisture and oxygen play a decisive role in corrosion phenomena. Together, they produce the OH-ions at the cathode, while moisture acts as an ion-carrier between the cathode and the anode. Salts in the liquid facilitate the process.

Some observations are important for practice. Untreated steel starts corroding when the relative humidity exceeds 60%. For that reason, steel has to be treated even when used indoors in moderate climates. Ferro-oxide dissolves in water. Steel will thus go-on corroding until completely oxidized. For most non-ferro metals the oxides form a protective layer. However protection might be lost when water contacting the non-ferro contains ions (SO_3, Cl and others.). Ion-loaded surface condensation may induce pitting corrosion in most metals. In combining metals one should respect the voltage series of Table 3.29. Water draining from a higher to a lower voltage metal will corrode the lower one. The opposite case poses no problems.

Table 3.29. Metals: voltage series compared to a hydrogen electrode.

Metal	Chemical symbol	Voltage V
Magnesium	Mg	−1.87
Aluminum	Al	−1.45
Zinc	Zn	−0.76
Iron	Fe	−0.43
Nickel	Ni	−0.25
Tin	Sn	−0.15
Lead	Pb	−0.13
Copper	Cu	0.35
Silver	Ag	0.80
Gold	Au	1.50

3.5 Life cycle costs

3.5.1 In general

When it comes to building the main concern for the principal is the investment. Of equal importance, however less obvious because distributed over time, are the benefits generated, though not all can be translated into financial value. In terms of physical quality, the costs are the extra investments for better performance while the benefits are the advantages generated by the upgraded performance. Investing in more thermal insulation for example results in better thermal comfort, perhaps a fabric demanding less maintenance, a cheaper HVAC-system and less energy consumed. Of these, better comfort is hardly a financial factor. In terms of sound insulation, the costs include investment in tieless cavity walls between dwellings, heavy floors with floating screeds in apartment buildings, sound insulating glass and tight windows, while the benefits are more privacy and less stress. None are directly financial in nature, unless indirectly, for example in office buildings where a better overall comfort increases productivity.

3.5.2 Total and net life cycle cost

Total life cycle costs, also called total present value, are given by:

$$\text{TLCC} = I_{\text{b+HVAC}} + \sum_{j=x,y,z}\left[\frac{I_{\text{b+HVAC}}(1+r_o)^j}{(1+a)^j}\right] \\ + \sum_{i=1}^{n}\left[\frac{K_E(1+r_E)^i}{(1+a)^i}\right] + \sum_{i=1}^{n}\left[\frac{K_M(1+r_M)^i}{(1+a)^i}\right] + \sum_{i=1}^{n}\left[\frac{K_P(1+r_P)^i}{(1+a)^i}\right] + V_{\text{end}} \quad (3.99)$$

where $I_{\text{b+HVAC}}$ is the initial investment, r_o the inflation corrected mean annual increment in investment cost on a scale from 0 to 1 when reinvesting later-on, K_E the initial annual energy cost, r_E the inflation corrected mean annual increment in energy cost on a scale from 0 to 1, K_M the initial annual maintenance, r_M the inflation corrected mean annual increment in maintenance cost on a scale from 0 to 1, K_P the annual productivity in terms of man-year cost, r_P the inflation corrected mean annual increment in monetary value of increased productivity on a scale from 0 to 1, V_{end} the salvage value of the building at the end of the present value period and a the inflation corrected present value factor on a scale from 0 to 1.

The net life cycle costs, also called net present value, looks like:

$$\text{NLCC} = -I_{\text{b+HVAC}} - \sum_{j=x,y,z}\left[\frac{I_{\text{b+HVAC}}(1+r_o)^j}{(1+a)^j}\right] \\ + \sum_{i=1}^{n}\left[\frac{\Delta K_E(1+r_E)^i}{(1+a)^i}\right] + \sum_{i=1}^{n}\left[\frac{\Delta K_M(1+r_M)^i}{(1+a)^i}\right] + \sum_{i=1}^{n}\left[\frac{\Delta K_P(1+r_P)^i}{(1+a)^i}\right] + \Delta V_{\text{end}} \quad (3.100)$$

where Δ means the change in annual energy costs, annual maintenance costs, annual productivity in terms man-year costs and salvage value.

3.5 Life cycle costs

Random parameters in both formulas are the inflation corrected mean annual increments. They should be deduced from price evolution and inflation in the past. The present value factor accounts among other things for the perceived risk to invest in energy saving measures. The time span to be considered depends on the type of building, while the salvage value is impacted by several factors ranging from the fame of the designer to building location.

TLCC and NLCC allow searching for an optimal combination of influencing variables. For that to be true, TLCC should be minimal, NLCC maximal meaning that the partial derivatives to all independent variables $x_1 \ldots x_n$ should be zero:

$$\frac{\partial TLCC}{\partial x_1} = 0 \ldots \frac{\partial TLCC}{\partial x_n} = 0 \qquad \frac{\partial TLCC}{\partial x_n} = 0 \ldots \frac{\partial NLCC}{\partial x_n} = 0$$

Each derivative could be a non-linear function of all independent variables. Solution of the set of equations will thus generate several independent variable combinations, of which only one figures as 'the' optimum. In practice numerical techniques such as Monte-Carlo, genetic algorithms or a solution with successive stepwise change of each of the variables are used to find that optimum.

3.5.3 Examples

3.5.3.1 Optimal insulation thickness

Investment

Two elements intervene in the investment: material used (I_{mat}) and mounting (I_{mont}). Material costs are commonly assumed to increase linearly with insulation thickness (d):

$$I_{mat} = a_0 + a_3 d \tag{3.101}$$

where a_0 is a zero thickness cost that includes debiting of the production unit, worker's fees and the costs additional measures for guaranteeing correct performance (vapour barriers, air barriers, wind barriers, etc.). Mounting costs depend among others on the number of layers needed (n) and secondary costs once the thickness passes a limit value d_0:

$$I_{mont} = a_1 n + \max\left[0, \ a_2 (d - d_0)\right] \tag{3.102}$$

Summing up gives:

$$I_{insul} = a_0 + a_1 n + \max\left[0, \ a_2 (d - d_0)\right] + a_3 d \tag{3.103}$$

Benefits

Quantifiable benefits are a cheaper heating system and a lower annual end energy use for heating.

Heating system

Assume central heating with radiators. Thanks to a better insulation less power and thus a smaller boiler, smaller expansion vessel, smaller pump, thinner pipes and less radiator surface are needed. As the market only offers a discontinuous series of boilers, pipe diameters and radiators, the investment attributable to the 1 m² of envelope considered will be a discontinuous function of its thermal resistance. One nevertheless states that the investment includes a constant term (b_0) and a term reciprocal to the power per m² (b_1) needed:

$$I_{system,insul} = b_0 + b_1 \left(\frac{\theta_i - \theta_{eb}}{0.17 + R_0 + d/\lambda} \right) \quad (3.104)$$

In that formula R_0 is the thermal resistance of the part without thermal insulation.

End energy consumed for heating

Insulating a wall mainly impacts the transmission losses:

$$E_{use,heat} = \frac{0.0864 \, G^{\theta_i}_{d,\theta_e}}{\bar{\eta} \, B_1 \left(0.17 + R_0 + d/\lambda \right)} \quad (l, m^3 \text{ or kg per annum})$$

where $G^{\theta_i}_{d,\theta_e}$ are the degree days for transmission, $\bar{\eta}$ the building related heating season mean efficiency of the central heating system in reference to the lower heating value B_1 of the energy source used (in MJ/(unit)). The annual cost then becomes:

$$K_{energy,ann} = P_E \, E_{use,heat} \quad (3.105)$$

with P_E the price of 1 litre of oil, 1 m³ of gas, 1 kg of coal or 1 kWh$_E$.

Optimum insulation thickness

Assuming that insulation does not generate extra maintenance costs and has an equal service life as the heating system and the building, TLCC equals:

$$\begin{aligned} TLCC = &\, a_0 + a_1 \, n + \max\left[0, \, a_2 (d - d_0) \right] + a_3 \, d + b_0 + b_1 \left(\frac{\theta_i - \theta_{eb}}{0.17 + R_0 + d/\lambda} \right) \\ &+ \left[\frac{0.0864 \, G^{\theta_i}_{d,\theta_e}}{\bar{\eta} \, B_1 \left(0.17 + R_0 + d/\lambda \right)} \right] P_E \sum_{j=1}^{n} \left[\frac{(1+r_E)^j}{(1+a_n)^j} \right] \end{aligned} \quad (3.106)$$

with the insulation thickness d as independent parameter in that equation. The derivative zero gives:

$$\underbrace{\left\{ \frac{d\left[a_1 n + \max\left[0, a_2(d-d_0) \right] \right]}{d(d)} + a_3 \right\}}_{A_{insul}} - \frac{1}{(0.17 + R_0 + d/\lambda)^2}$$

$$\cdot \left\{ \underbrace{b_1(\theta_i - \theta_{eb})}_{A_{heat}} + \underbrace{\left(\frac{0.0864 \, G^{\theta_i}_{d,\theta_e}}{\bar{\eta}_{tot,gebouw} \, B_1} \right) P_E \sum_{j=1}^{n}\left[\frac{(1+r_E)^j}{(1+a_n)^j} \right]}_{A_{energy}} \right\} = 0$$

resulting in the following optimal thickness (d_{opt}) and optimal thermal transmittance (U_{opt}):

3.5 Life cycle costs

$$d_{opt} = \sqrt{\frac{A_{heat} + A_{energy}}{\lambda A_{insul}}} - \lambda(0.17 + R_0) \qquad U_{opti} = \sqrt{\frac{A_{insul}\lambda}{A_{heat} + A_{energy}}} \qquad (3.107)$$

That result is easy to see. The term A_{heat} stands for the investment in the heating system and the term A_{energy} for the total present value of the energy costs, both for a thermal resistance of the envelope part equal to 1 m²K/W. If both terms increase, so does the optimal thickness whereas the optimal thermal transmittance decreases. The same happens for a lower investment in insulation. (A_{insul}), which is smallest if only one layer is needed (dn/d(d) = 1). A higher thermal conductivity (λ) of the insulation material finally gives lower optimal thickness and higher thermal transmittance.

Of course, all quantities in the formula are variable. Happily the impact of the possible variations is tempered by the square root, which allows formulating quite general conclusions. In case of new construction for example the optimal thermal transmittance typically coincides with the value that does not generate extra construction costs. This is the case for exterior insulation, low-slope roof insulation and pitched roof insulation, in the last case as long as the thickness does not exceed rafter height. A two layer solution almost always passes the optimum. This is important for manufacturers as it should motivate them to produce in larger thicknesses.

3.5.3.2 Whole building optimum from an energy perspective

Methodology

A whole building approach complicates things. Variables now are thermal insulation, airtightness of the envelope, window area and type, glazing used, ventilation system and heating system implemented. The investments in thermal insulation differ between four or more building parts – low-slope roofs, pitched roofs, facades and floors – and invoke secondary costs. Thicker low-slope roof insulation may demand higher parapets and adjusted flashing. Larger wall thicknesses for an equal net floor area result in extra roof surface. Below grade wall thicknesses have to be adapted, while detailing of lintels, window edges and sills change and cavity ties need to be longer. Double glazing requires stiffer and deeper window frames, etc. The glass systems as well as the insulation thicknesses offered form a series of discontuous types and thicknesses. Energy consumed depends on transmission losses, ventilation and infiltration losses, usable solar and internal gains and heating system efficiency. Rebound must be accounted for and several different heating and ventilation systems can be used.

Only the numerical techniques mentioned above allow taking this all into account. The search for the optimum is a two-step operation: first calculating the investments in the building fabric and the effects on net energy demand, after that adding the investment in building services and then locating the primary energy use that generates lowest total or highest net life cycle costs.

Example

Five non-insulated dwellings with high air leakage and purpose designed natural ventilation are used as a reference. All had an 80–60 °C radiator heating controlled by a central thermostat, with a high efficiency gas boiler working at constant temperature and an AC-pump on the return in series with a closed expansion vessel. The boiler also produced domestic hot water. Figure 3.37 draws one of the dwellings. The fabric is typical for a country with brick construction: cavity walls pargetted inside; single-glazed wooden windows; reinforced concrete decks pargetted at the ceiling side with cement screed and pavement; heavy-weight low-slope roofs with concrete

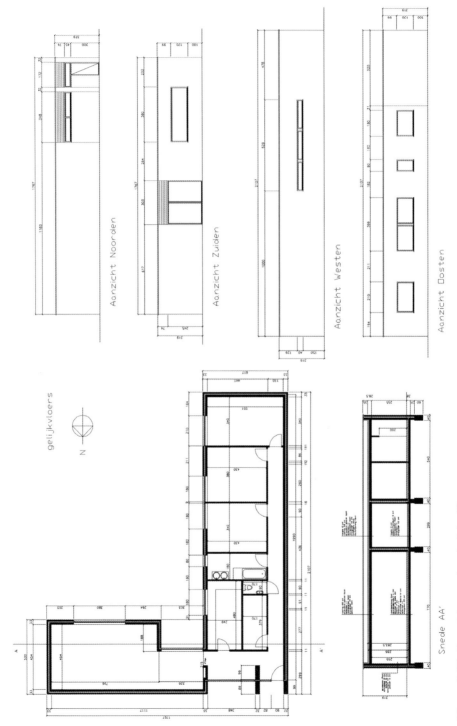

Figure 3.37. One of the two detached dwellings.

3.5 Life cycle costs

Table 3.30. Non-insulated reference, thermal transmittances.

Part	U-value W/(m² · K)
Floor on grade (EN ISO 13370)	0.54 to 0.78
Façade	1.4
Low-slope roof	1.2
Sloped roof	1.7

Table 3.31. The five dwellings.

Dwelling	Type	Heated volume m³	Envelope area A_T m²	Windows, % of A_T	Mean U-value W/(m² · K)	$Q_{prim,ref,T}$ MJ/a
1	Detached, one story, low-slope roof	472	555	5.2	1.31	126 200
2	Detached, two stories, sloped roof	625	712	11.4	1.88	159 350
3	End of the row house, one story, inhabited loft, sloped roof	421	317	9.1	1.64	93 480
4	Terraced house, two stories, inhabited loft, sloped roof	435	220	12.7	1.75	83 800
5	Corner apartment on the upper floor, low-slope roof	417	256	9.8	1.94	86 000

deck pargetted at the inside, lightweight screed and bituminous felt; sloped roofs with wooden rafters, underlay, battens and laths and a tiled deck. The thermal transmittances are listed in Table 3.30. Table 3.31 summarizes the characteristics of the five dwellings. Correct detailing neutralized overall thermal bridging.

Variation in envelope leakage was limited to $n_{50} = 3$ h^{-1} and, in case a balanced ventilation system with 80% efficient heat recovery was used, to $n_{50} = 1$ h^{-1}. That balanced system figured as only ventilation upgrade. The alternate for the central heating system was a 60–40 °C low temperature radiator heating, fed by a condensing boiler with supply temperature controlled by a PI sensor outside, DC-pump and decentralized control by thermostatic valves on each radiator.

Variables considered were: insulation thickness in the low-sloped roofs, sloped roofs, cavity walls and ground floor, glass type, air tightness level, ventilation system and heating system applied. Minimum and maximum insulation thickness and stepwise increase are labelled in Table 3.32. The insulation used was polyurethane (PU) with a thermal conductivity of 0.023 W/(m · K). Table 3.33 lists the glass applied.

Table 3.32. Minimum and maximum insulation thickness, step applied.

Envelope part	Minimum thickness mm	Maximum thickness mm	Step mm
Flat roof	30	300	10
Sloped roof	30	300	10
Cavity wall	30	200	10
Floor on grade	30	200	10

Table 3.33. Glass types.

Glass unit	Central thermal transmittance W/(m² · K)	Solar transmittance
Double	2.91	0.72–0.75
Double, low-e, argon filled	1.13	0.61
Triple, low-e, argon filled	0.80	0.50

Building and insulation had the same service life. Double glazing or better was expected to have 17% of its total surface replaced after 20 years. The boiler served for 20 years. The prices considered were from 2004. Financial information came from the industry, official catalogues, periodicals that publish price lists every year and the main energy supplier. Investments also included secondary costs. Heating power was calculated according to EN ISO 12 831 for –8 °C design temperature. The systems that were chosen delivered that power. Their cost thereby became a decreasing function of the mean thermal transmittance of the envelope as Figure 3.38 shows for one of the dwellings.

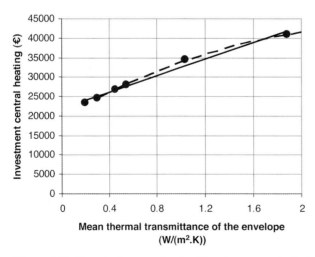

Figure 3.38. Detached two stories dwelling, investment in the hydronic central heating system (€ of 2004) as function of the mean thermal transmittance of the envelope.

3.5 Life cycle costs

The heating season mean inside temperature assumed was 18 °C, a value that accounts for some rebound. The reference year for Brussels, typical a cool climate, figured as outdoor condition. Primary energy consumption was calculated according to EN ISO 13790, considering each dwelling as a one-zone building. Dedicated ventilation and internal gains were calculated as:

$$\dot{V}_{a,dedic} = \left[0.2 + 0.5 \exp(-V/500)\right] V \quad (m^3/h) \tag{3.108}$$

$$Q_{I,m} = (220 + 0.67\, V)\, t_m \quad (MJ/month) \tag{3.109}$$

where t_m is the number of mega-seconds in a month. Primary energy consumption was evaluated by each dwelling's E-level:

$$E = 100\, Q_{prim,T} / Q_{prim,ref,T} \quad (\%) \tag{3.110}$$

where $Q_{prim,T}$ is annual primary energy consumption for heating, domestic hot water and auxiliaries and $Q_{prim,ref,T}$ is an annual reference defined as:

$$Q_{prim,ref,T} = 115\, A_T + 70\, V + 157.5\, \dot{V}_{dedic} \quad (MJ/a) \tag{3.111}$$

with $\dot{V}_{dedic}$ the dedicated ventilation in m³/h. Per dwelling 15 cases were evaluated, see Table 3.34. Per case, the calculations resulted in graphs showing the extra investment in insulation, glazing, air-tightening, heating and ventilation compared to the reference and giving the extra investment and total life cycle costs (TLCC or TPV) as a function of the E-level. As Figure 3.39 shows, the results form a series of clouds with the reference as point right above.

Table 3.34. cases per dwelling.

	Glazing	Heating	Ventilation	n_{50}, h^{-1}
1	Double (DG)	HE-boiler (HEB)	Purpose designed natural (NV)	12
4				3
7		C-boiler (CB)	Purpose designed natural (NV)	12
13				3
5		C-boiler (CB)	Balanced, heat recovery (BVHR)	1
2	Low-e, argon filled double	HE-boiler (HEB)	Purpose designed natural (NV)	12
5				3
8		C-boiler (CB)	Purpose designed natural (NV)	12
11				3
14		C-boiler (CB)	Balanced, heat recovery (BVHR)	1
3	Low-e, argon filled triple	HE-boiler (HEB)	Purpose designed natural (NV)	12
6				3
9		C-boiler (CB)	Purpose designed natural (NV)	12
12				3
15		C-boiler (CB)	Balanced, heat recovery (BVHR)	1

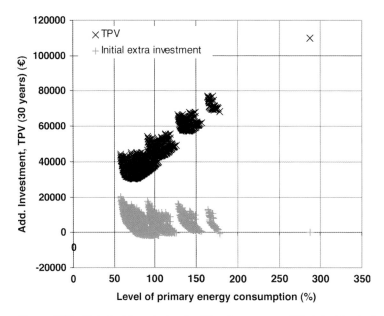

Figure 3.39. Detached two stories dwelling, low-e, argon filled double glazing, natural ventilation, $n_{50} = 12$ h^{-1}, present value factor 5%, increment in energy costs 3.2%, increment in investment costs 2%.

Table 3.35. Optimal choices for the dwellings.

Dwelling, optimum		E_{opt} %	Insulation thicknesses (PUR), cm			
			Low-sloped roof	Pitched roof	Wall	Floor
Present value factor 6.5%. Annual increment in energy cost 0.8%, in investment costs 0%						
1	Low-e, argon DG, $n_{50} = 3$ h^{-1}, HEB, NV	59.4	7		7	4
2		60.6	6	8	8	6
3		52.8		8	6	4
4		55.4	6	8	6	4
5		57.1	10		7	
Present value factor 3.5%. Annual increment in energy cost 3.2%, in investment costs 2%						
1	Low-e, argon DG, $n_{50} = 3$ h^{-1}, HEB, NV	50.8	10		10	6
2		57.8	8	8	10	6
3		48.6		10	8	5
4	Low-e, argon 3DG!	52.3	10	8	8	6
5	Same as 1, 2 and 3	55.0	10		10	

The clouds show an absolute minimum: the economic optimum, according to Table 3.35 with values E49 to E61. As energy for household and lighting is not included, the optima found are far from zero-energy: total annual primary consumption compensated by an equal amount of renewable energy produced on site. In the end of the row house 3 for example, total primary consumption may reach 21 000 kWh/a at the optimum. Annual compensation by installing photovoltaics demands 60 to 170 m^2 depending on the type of cells. Related investment exceeds the investment for the optimum. Also passive dwellings are to the left of the optimum.

3.6 Sustainability

3.6.1 In general

During the last decade, environmental concerns steadily grew. There are reasons for that. The two energy crises of the nineteen seventies showed how dependent the industrialized regions are on finite oil reserves in less stable parts of the world. In the same era, NASA saw the thinning of the higher ozone layer. Measurements on Mount Kauna in Hawaii showed an increase in CO_2 concentration in the atmosphere from 280 ppm before the industrial revolution to 380 ppm now. The United Nations Panel on Global Warming found human activity to be a most likely cause of that. The immense resource demand in the industrialized world further gave underuse elsewhere. Locally, waste, water pollution, air pollution and pressure on the open space colour the picture.

The notion 'sustainable development' was first used in 1987 in the Brundlandt report 'Our Common Future' by the World Council for Economic Development. It defines the concepts in the following terms *'Each generation must use all natural resources in a way the needs and well-being of future generations are not put at risk'*. The transposition to the built environment was first mentioned in a document published by the Civil Engineering Research Foundation (CERF) in 1996. In 1998, the International Council for Research and Innovation in Building and Construction (CIB) produced a report on sustainable development and the future of construction, followed by Agenda 21 and sustainable construction in 1999. The links between construction and actual and future needs and well-being are manifold. Constructing buildings and infrastructure consumes amazing amounts of non-renewable materials and occupies space for a long time. Conditioning the indoor environment and allowing for the activities for which buildings are designed results in tremendous energy consumption, without counting the energy used for transportation due to the distances between living, working and shopping areas.

Some interpret 'sustainability' in a strict way, looking to the planetary conditions only. Translated into sustainable construction, water usage, energy consumption, material use and waste then become the main items. A product of that vision is life cycle analysis. However, human needs and well-being requires buildings with a priority placed on functionality, accessibility for the disabled, thermal comfort, acoustical comfort, visual comfort and indoor air quality. That broadens the scope from a strict life cycle approach to what today are called high performance buildings.

3.6.2 Life cycle analysis

3.6.2.1 Definition

Life Cycle Analysis (LCA) looks to the building and its impact on the environment from cradle to grave, dividing the life cycle in four phases: (1) preparation, (2) construction, (3) service life, (4) demolition. Preparation considers building design and the production of all materials and elements needed to construct the building. Construction starts with the excavation and ends when the building is delivered. Then service life starts. The probability a building survives for 50 years passes 40%. Sooner or later the building is demolished. The waste produced can be dumped or recycled, in that way starting a new cycle.

In each phase, LCA considers inflows and outflows (Table 3.36).

The inflows are space, basic materials, fluids and energy, the outflows building materials and components, the building itself, solid, fluid and gaseous waste and noise. The fact that knowledge, creativity, human effort and money do not figure in that list follows from the system borders defined when applying LCA. Not all in and outflows have an environmental impact. Impact or a lack thereof is evaluated using a series of criteria:

Table 3.36. Life cycle analysis at the building level: in and outflows.

Phase	Inflow	Outflow
1. Preparation	Recycled materials Basic materials Water and other fluids Energy	Building materials and components Noise Solid waste Fluid waste Gaseous waste Environmental damage **BUILDING DESIGN**
2. Construction	*Building design* Space Building materials and components Water Energy	Hindering traffic Noise Construction waste Liquid waste Gaseous waste **BUILDING**
3. Service life	*Building* Goods Cleaning products Finishing materials Building materials Energie Water	Noise Solid waste Fluid waste Gaseous waste Construction waste Harming emissions inside and outside the building
4. Demolition	*Building* Energy Water	Noise Construction waste Solid waste Fluid waste Gaseous waste

(1) basic material exhaust, (2) energy consumption, (3) global warming potential, (4) ozone depletion potential, (5) ecotoxicity, (6) waste produced, (7) toxic waste produced, (8) acidification, (9) nitrification, (10) human toxicity, (11) photochemical smog formation, (12) consequences for the inside environment, etc. In most cases only some of these effects are analyzed when looking to opportunities for optimization.

3.6.2.2 Impacts

Global warming potential

Burning fossil fuels produces CO_2, ≈ 1.75 kg per m^3 gas and ≈ 2.8 kg per liter oil or kg coal. Once in the atmosphere CO_2, just as water vapour, CH_4, NO_2, SF_6 and all CFC's, absorbs the terrestrial and solar infrared radiation and re-emits both, half of it back to the earth. There, the infrared maintains the moderate temperatures needed for life. However, when the concentration of global warming gases increases, the re-emitted infrared also augments, resulting in a slow increase in terrestrial mean temperature: the global warming effect.

Growing fossil fuel usage worldwide ranks first as a cause. So, to moderate or even stop further global warming, total energy demand should be lowered drastically and a turn to more CO_2 neutral or CO_2-free energy supply should be made. For buildings, less energy means optimal insulation and excellent air-tightness of the envelope, intelligent use of glass and solar shading, passive measures against overheating, high efficiency HVAC, more renewable energy, a move to heat pumps, CHP, and heat and cold storage, among other measures.

Depletion of the ozone layer

CFC's were the main cause here. Until the end of the twentieth century their use as refrigerant, blowing agent in insulation materials (PUR and XPS) and fire fighting agent (halon) was widespread. The Montreal protocol of 1990 imposed a complete ban of the so-called hard CFC's (Freon 11, 12, 13, 111, 112, 113, 114, 115) and halons for the year 2000 (1997 in the EU), followed by a complete ban of the soft CFC's for 2012. That forced manufacturers to look for other refrigerants, blowing agents and fire fighting agents.

Acid rain

Coal and fossil fuels contain small percentages of sulphur. When burned, the reaction with the oxygen in the air produces SO_2 that in a moist atmosphere binds the water molecules to give sulphur-like acid first and then sulphur acid. Resulting acid rain attacks limestone and lime-containing building materials giving accelerated aging of landmark buildings. More strict legislation on sulphur content in fossil fuels has improved the situation.

3.6.2.3 Total energy use

Energy use by buildings extends beyond what is needed for heating, cooling, air conditioning, lighting, domestic hot water and function. It also includes building material and building component manufacturing, the transport to the building site, the construction and retrofit itself, demolition, etc. That additional package is called the embodied energy. When evaluating whole energy use embodied energy must be accounted for. The sum of both opens a window for optimization, provided one of the terms is non-linear. The result may be an energy-based optimum insulation thickness, the same way an economic optimum thickness exists. The elements needed are: the energy used to produce, transport, mount and demolish 1 m^2 of insulation material with a thickness of d meter, i.e. the embodied energy related to the LCA-phases 1, 2 and 4 and in a cool climate the energy consumed for heating during LCA phase 3. Embodied energy is given by:

$$E_{embodied,insul} = e_0 + e_1 d \qquad (3.112)$$

where e_0 is a constant, representing the thickness-independent part of the embodied energy and e_1 is the embodied energy per m³ of material. The energy consumed during phase 3 mounts to:

$$E_{use,heat} = \underbrace{\left[\frac{0.084\, G_{d,\theta_e}^{\theta_i}}{\eta\, B_1 (0.17 + R_0 + d/\lambda)}\right]}_{E_{use,heat,an}} T \qquad (3.113)$$

with T service life in years. The derivative for d of the sum of both formulas zero gives as optimal thickness:

$$d = \sqrt{\frac{T\, E_{use,heat,an}}{\lambda\, e_1}} - \lambda (0.17 + R_0) \qquad (3.114)$$

The equation shows that the energy-based optimal insulation thickness drops with higher conductivity of the insulation material, more embodied energy per per m³, shorter service life, lower annual energy consumption for heating and higher thermal resistance of the part without insulation.

The methodology is again transposable to the level of the whole building. Doing the exercise confirms that contrary to passive solar measures such as very high thermal capacity, large sun-oriented glazed surfaces, atriums, Trombe walls, providing excellent thermal insulation is a very smart decision in cool climates.

3.6.2.4 Total environmental load

Distinction again should be made between the load during phases 1, 2, 4 and phase 3. As with life cycle costs and energy, optimal choices are traceable which give a minimal load when looking from cradle to grave.

3.6.2.5 Recycling

When recycling is applied, the demolished building, insulating and finishing materials return to specialized facilities or manufacturers for low or high-valued re-use: low-valued when a building material is reused for landfill and levelling, high-valued when transformed into a new material or product. New concrete with crushed old concrete fill, mineral fibre insulation containing a percentage of melted old fibre, re-used aluminium, all are examples of the latter. Plastic waste may be used as fuel. That way one recycles the caloric value. Taxing has helped a lot in accelerating the move from dumping to recycling.

3.6.3 High performance buildings

The term 'high performance' couples sustainability to a much broader field of performances. A 'fitness for purpose' is a basic approach with as main elements: usability, energy consumption, environmental impact and durability.

The UK was the first to adapt a measuring instrument called BREEAM98, set up by BRE (Building Research Establishment). The checks include management, health and well-being, transport, water, materials, land use, ecology and pollution. A building gets a score per check, after which the sum situates the design in one of the following four classes: pass, good, very good, excellent. More than 100,000 buildings in the UK have been evaluated that way.

In the US the Green Building Council (USGBC), a voluntary organisation, developed a definition and measuring instrument called LEED. In the 2009 editions for new construction, existing buildings, specific building typologies, sectors and project scopes LEED focuses on the same aspects as BREEAM, though ordered differently. The five categories considered are sustainable sites, water efficiency, energy and atmosphere, materials and resources and indoor environmental quality. Additional aspects are innovation in green design and regional priority. In total, 100 points can be earned and four classes are defined:

Class	Points
Platinum	> 80
Gold	60–79
Silver	50–59
Certified	40–49

In Germany buildings will get a 'gesamtnote', with a higher scoring corresponding to a lower number (1 better than 2, 2 better than 3, etc.). Also France and the Netherlands are developing scoring systems. All systems listed have two characteristics in common. Though market-oriented, they are neutral and controls after construction are not included.

The author had the opportunity to develop a measuring instrument for government buildings. Three categories of performance requirements were considered: (1) Quality of use and well-being, (2) Energy, (3) Environment and durability. The first contains the services a building should deliver. The second focuses on the energy needed to realize these services. The third considers specific environmental issues and looks to durability, a main element of sustainability. Each category consists of a group of weighted performances.

Category 1: Quality of use and well-being	Category 2: Energy	Category 3: Environment and durability
Accessibility	Building level (net demand, gross demand, end use, primary use)	Location and services
Comfort (thermal, acoustics, visual)	Zone and building part level (U-value, damping)	Moisture response
Air quality and ventilation		Water usage
		Material usage

That weight rewards the importance of a performance related to the impact it addresses. Per performance, five quality levels are defined with scores of 0, 1, 2, 3 and 4. Zero means legal requirements are fulfilled whereas four indicates very high quality. The weights are distributed so that each category has the same impact on the final score. If a 4 is earned for all performances, then each category totals 80 points.

As mean score per category one has:

Category 1	Category 2	Category 3
$\bar{S}_1 = \dfrac{\sum_{i=1}^{n} g_{1,i} S_{1,i}}{\sum_{i=1}^{n} g_{1,i}}$	$\bar{S}_2 = \dfrac{\sum_{i=1}^{n} g_{2,i} S_{2,i}}{\sum_{i=1}^{n} g_{2,i}}$	$\bar{S}_3 = \dfrac{\sum_{i=1}^{n} g_{3,i} S_{3,i}}{\sum_{i=1}^{n} g_{3,i}}$

where $S_{j,i}$ is the score for performance i in category j and $g_{j,i}$ is its weight. The global score becomes:

$$S = \frac{(\bar{S}_1 + \bar{S}_2 + \bar{S}_3)}{3} \qquad (3.115)$$

The scores per category and the global score define the number of stars a building receives:

Scores		Stars	Meaning
Average per category	Global score		
No one zero	$0 < S \leq 1$	☆	Acceptable
No one below 1	$1 < S \leq 2$	☆☆	Good
No one below 2	$2 < S \leq 3$	☆☆☆	Very good
No one below 3	$3 < S \leq 4$	☆☆☆☆	Excellent

Designs that do not comply with legal requirements are excluded from star classification. Governmental buildings that do better than legally required can earn up to four stars. These will be visibly present at the building's main entrance. Defining the stars the way shown guarantees that a three star building scores very well in all three categories. That should be the goal when defining high performance buildings: a well balanced equilibrium between requirements.

Where possible, a definitive score is only assigned after testing if the building complies with the performance level intended. The design stage therefore ends with interim scores. Stars also get a date. In fact, each time legal performance levels increase, score levels have to be reviewed. The consequence might be a loss of stars. This would be couter-productive, which is why the date will be added.

3.6 Sustainability

The list of performances looks as follows:

Category	Performance	Weight
1: Quality of life and well being	Accessibility for disabled	4
	Thermal comfort, global, cold months	1
	Thermal comfort, global, warm months	1
	Thermal comfort, draft	1
	Thermal comfort, air temperature difference between head and feet	1
	Acoustical comfort: background noise	2
	Acoustical comfort: sound insulation against airborne and contact noise	1
	Vibration control	2
	Visual comfort: illuminance	2
	Visual comfort: daylight factor	1
	Indoor air: CO_2-concentration	1
	Indoor air: perceived quality	1
	Purpose designed ventilation: flows	2/1
	Purpose designed ventilation: peak airing	0/1
2: Energy	*Building level*	
	Net energy demand: insulation quality	3.5
	Net energy demand: air tightness	1
	Net energy demand: solar screening	2
	Net energy demand: night ventilation?	1
	Gross energy demand: heating, control	1/3
	Gross energy demand: heat emission	1/3 | 0
	Gross energy demand: pipe and duct insulation, heating	1/3 | 2/3
	Gross energy demand: cooling, control	0.5
	Gross energy demand: pipe and duct insulation, cooling	0.5
	End energy consumption: heat production	0.5
	End energy consumption: cold production	0.5
	Primary energy consumption: renewable energy	2
	Primary energy consumption: level of primary energy use (E)	5
	Zone and building part	
	Thermal transmittances	2
	Temperature damping	0.5
3: Environment and durability	Location	3
	Facilities	3
	Service life(durability)	5
	Water usage	3
	Material usage	6

Several performances outside the list are not scored, mainly because legal requirements are so severe that additional gradation in levels makes no sense.

3.7 References

[3.1] De Grave, A. (1957). *Bouwfysica 1*. Uitgeverij SIC, Brussel (in Dutch).

[3.2] Fanger, P. O. (1972). *Thermal comfort*. Mc Graw-Hill Book Company, New York.

[3.3] WTCB (1972). *Onderzoek naar de vorstvastheid van bouwmaterialen en bepaling van de eisen voor keuring der materialen*. Studie- en Researchrapport 15 (in Dutch).

[3.4] Meadows, D., Randers, J., Behrens, W. (1973). *Rapport van de club van Rome*. Uitgeverij het Spectrum (in Dutch).

[3.5] Vos, B. H., Tammes, E. (1973). *Thermal insulation, seen from an economic viewpoint*. Bouwcentrum Rotterdam.

[3.6] NBN B 62-100 (1974). *Hygrotermische eigenschappen der gebouwen: thermische isolatie, winteromstandigheden*. BIN (in Dutch).

[3.7] Fagerlund, G. (1974). *Critical Moisture Contents at Freezing of Porous Materials*. Symposium CIB-RILEM on Moisture Problems in Buildings, Rotterdam.

[3.8] Stichting Bouwresearch (1974). *Voorstudie Energiebesparing*. Rapport BII-46 (in Dutch).

[3.9] Olesen, B. W. (1975). *Thermal Comfort Requirements for Floors*. Technical University of Denmark.

[3.10] Caluwaerts, P. (1976–1985). *De warmtehuishouding in gebouwen 's winters en 's zomers*. Cursus Thermische isolatie en vochtproblemen in gebouwen, TI-KVIV (in Dutch).

[3.11] Benzinger, T. H. (1978). *The physiological basis for thermal comfort*. First International Symposium on Indoor Climate, Copenhagen.

[3.12] Anon. (1979). *Temperature without heating*. Proceedings of a CE-symposium, Brussels.

[3.13] RD-Energie (1980–1985). *Rationeel energiegebruik in de gebouwen*. Integratieoefeningen (in Dutch).

[3.14] Hens, H., Standaert, P. (1980). *Energiebesparing in de huishoudelijke sector, gezien door een bouwfysische bril*. Studiedag 'Feiten en mythen in zake energiebesparing in huishoudelijke en gelijkgestelde sectoren', Studiedag TI-KVIV (in Dutch).

[3.15] Gagge, A. P. (1980). *The new effective temperature ET**. An index of human adaption to warm environments, Environmental Physiology, Elsevier Amsterdam.

[3.16] Wouters, P., Hens, H. (1982). *De verbetering van bestaande gebouwen*. Studiedag ATIC over energiebesparing in flatgebouwen (in Dutch).

[3.17] Wouters, P., Hens, H. (1981). *Isolatie en haar impact op het bouwproces*. Ronde Tafel 'Isolatie' (in Dutch).

[3.18] TU-Delft (1981). *Afdeling der Bouwkunde*. Research in Energy Savings, Eindrapport.

[3.19] Vandermarcke, B. (1982). *Thermisch comfort: basis voor het ontwerpen van een zwembad, eindwerk*. KU-Leuven (in Dutch).

[3.20] Hens, H. (1982). *Bouwfysica 2: Warmte en Vocht, praktische problemen en toepassingen*. Vol. 1, Acco Leuven (in Dutch).

[3.21] Muzzin, G. (1982). *Uitbloeiïngen op baksteen*. WTCB-Tijdschrift 4, pp. 2–17 (in Dutch).

3.7 References

[3.22] Norm ISO 7730 (1984). *Moderate thermal environments*. Determination of the PMV and PPD indexes and specification of the conditions of thermal comfort.

[3.23] Vaes, F. (1984). *Hygrotermisch gedrag van lichte geventileerde daken*. Eindrapport onderzoek TCHN-KUL-IWONL (in Dutch).

[3.24] Nationaal Programma RD-Energie (1984). Ontwerp *en thermische uitrusting van gebouwen, deel 1*. DPWB (in Dutch).

[3.25] Nationaal Programma RD-Energie (1984). *Ontwerp en thermische uitrusting van gebouwen, deel 2*. DPWB (in Dutch).

[3.26] Hens, H. (1985). *Studie van een woning met laag energiegebruik*. Rapport RD-Energie (in Dutch).

[3.27] Opfergeld, D. (1985). *Guide de la gestion énergétique des établissements scolaires*. RD-Energie, Rapport E3/III.1.8 (In French).

[3.28] Ministerie van Economische Zaken, Administratie voor Energie, Energiebalansen 1984–1985–1986 (in Dutch).

[3.29] TI-KVIV, Kursus Termische Isolatie en Vochtproblemen in Gebouwen: Teksten (1985) (in Dutch).

[3.30] Sachverständigengremium 'Gesundes Bauen und Wohnen' (1986). *Gesundes Bauen und Wohnen: Antworten auf aktuelle Fragen*. Bonn (in German).

[3.31] PATO (1986). Sectie Bouwkunde, Syllabus van de leergang: Leidt energiebesparing tot vochtproblemen (in Dutch).

[3.32] Werner, H. (1986). *Berechnung des Heizenergiebedarfs von Gebäuden nach der Methode ISO-DP 9164*. HLH 37, Heft 11, p. 541–545 (in German).

[3.33] NBN B62-003 (1986). *Berekening van de warmteverliezen van gebouwen* (in Dutch and in French).

[3.34] Huber, J. W., Baillie, A. P., Griffiths, I. D. (1987). *Thermal comfort as a predictive tool in home environments*. CIB-W77 meeting, Holzkirchen.

[3.35] WCED (1987). The Brundtland Report.

[3.36] Leerstoel Prof. Ir. R. Van Cauteren (1987). Restauratie in de civiele techniek: tekstboek (in Dutch).

[3.37] NBN B62-002 (1987). Berekening van de warmtedoorgangscoëfficiënten van wanden (in Dutch and in French).

[3.38] Bogaerts, F. (1987). *Syntheserapport Energie audits*. Rapport RD-Energie (in Dutch).

[3.39] IEA-EXCO ECBCS, Annex 8 (1987). *Inhabitants' behaviour with regard to ventilation*. Final report.

[3.40] Fanger, O. (1988). *The Olf and the Decipol*. ASHRAE Journal, October, pp. 35–382.27.

[3.41] Fanger, O. (1988). *Hidden Olfs in sick buildings*. ASHRAE Journal, November, pp. 40–43.

[3.42] Fanger, O. (1988). *Introduction of the Olf and Decipol Units to Quantify Air Pollution Perceived by Humans Indoors and Outdoors*. Energy and Buildings 12, pp. 1–6.

[3.43] Fanger, O., Lauridsen, J., Bluyssen, P., Clausen, G. (1988). *Air Pollution in Offices and Assembly Halls*. Quantified by the olf Unit, Energy and Buildings, 12, pp. 7–19.

[3.44] Labs, K., Carmody, J., Sterling, R., Shen, L., Huang Yu, Joe, Parker, D. (1988). *Building Foundation Design Handbook*. Chapter 11.1 on Termite Control, ORNL.

[3.45] KU-Leuven, Laboratorium Bouwfysica (1982–1988). Rapporten over de Haram-al-Shariff te Jeruzalem, Het Oude Gerechtshof te Brugge, Arenberg kasteel te Heverlee e.a. (in Dutch, and English).

[3.46] De Bie, C., Deschrijver, F. (1988). Aantasting van Natuursteen: gevallenstudie Jeruzalem, eindwerk KU-Leuven (in Dutch).

[3.47] Verbruggen, A. (1988). Investeringsanalyse, Rationeel Energiegebruik in Kantoorgebouwen, Wilrijk (in Dutch).

[3.48] Poffijn, A. (1989). *Binnenluchtverontreiniging door Radon.* TI-KVIV studiedag Gezond Bouwen, fabel of realitei (in Dutch).

[3.49] Wouters, P. (1989). *De mogelijkheden en beperkingen van ventileren met betrekking tot gezond wonen'.* TI-KVIV, studiedag Gezond Bouwen, fabel of realiteit (in Dutch).

[3.50] Uyttenbroeck, J. (1989). *Thermisch comfort.* TI-KVIV studiedag Gezond Bouwen, fabel of realiteit (in Dutch).

[3.51] Norgard, J. (1989). *Low Electricity Appliances-Options for the Future, in 'Electricity, Efficient End-Use and New Generation Technologies and Their Planning Applications'.* Lund University Press.

[3.52] Mayer, E. (1990). *Untersuchung der physikalischen Ursachen von Zugluft.* Gesundheitsingenieur, 1, pp. 17–30 (in German).

[3.53] Mayer, E. (1990). *Thermische Behaglichkeit, Einfluß der Luftbewegung auf das Arbeiten in Reinraum.* Reinraumtechnik 4, pp. 30–34 (in German).

[3.54] Raatschen, W. (1990). *Demand Controlled Ventilating System, State of the Art Review.* Report IEA, Annex 18.

[3.55] Barron, E. (1990). *Ongewisse Warmte.* Natuur en Techniek 2, pp. 96 105 (in Dutch).

[3.56] Hens, H. (1991). *Bouwen en gezondheid: mythe of werkelijkheid.* Onze Alma Mater 1991/4, pp. 285–304 (in Dutch).

[3.57] Hens, H. (1991). *Analysis of Causes of Dampness, Influence of Salt Attack.* L'Umidita ascendente nelle murale: fenomenologia e sperimentazione, Bari.

[3.58] Environment International (1991). *Special Issue Healthy Buildings.* Pergamon Press, New York.

[3.59] IEA-Annex 14 (1991). *Condensation and Energy.* Final Report, volume 1: Source Book, ACCO Leuven.

[3.60] IEA-Annex 14 (1991). *Condensation and Energy.* Final Report, volume 3: Case Studies, ACCO Leuven.

[3.61] American Hotel and Motel Association (1991). Mold and Mildew in Hotels and Motels, the Survey.

[3.62] Cziesielski, E. (1991). *Energiegerechte Sanierung von Korrosionsschäden bei Stahlbetongebäuden.* Bauphysik 5, pp. 138–143 (in German).

[3.63] Gertis, K. (1991). *Verstärkter baulicher Wärmeschutz-ein Weg zur Vermeidung der bevorstehenden Klimaveränderungen?* Bauphysik 5, pp. 133–137 (In German).

[3.64] Spengler, J., Burge, H., Su, J. (1991). *Biological Agents and the Home Environment.* Bugs, Mold & Rot 1, Proceedings, p. 11–18.

[3.65] Gertis, K. (1991). *Verstärkter baulicher Wärmeschutz-ein Weg zur Vermeidung der bevorstehenden Klimaveränderung.* Bauphysik 13 (Heft 5), pp. 132–137 (in German).

3.7 References

[3.66] Van den Ham, E. R., Ackers, J. G. (1992). *Radon in het binnenmilieu*. Bouwfysica 1, pp. 21–25 (in Dutch).

[3.67] Van der Wal, J. F. (1992). *Belasting van de binnenlucht door emissie uit (bouw)materialen en producten*. Bouwfysica 1, pp. 12–21 (in Dutch).

[3.68] Radon, J., Werner, H. (1992). *Quantifizierung des Solar-Ausnutzungsgrades zur Berechnung des Heizenergiebedarfs von Gebäuden*. Bauphysik 14, Heft 1, p. 7–11 (in German).

[3.69] Fachvereinigung Mineralfaserindustrie e. V., Umgang mit MineralwolleDämstoffen (in German).

[3.70] CEN/TC 89 (1992). Working draft 'Thermal performance of buildings-Calculation of energy use for heating-Non Residential buildings'.

[3.71] Stevens, W. J. (1992). *Het Sick Building Syndrome, een Air Conditioning Ziekte*. WTCB-Beroepsvervolmaking: Luchtkwaliteit in Gebouwen (in Dutch).

[3.72] Rousseau, E. (1992). *Bouwproducten: asbest en andere probleemstoffen*. WTCB-Beroepsvervolmaking: Luchtkwaliteit in Gebouwen (in Dutch).

[3.73] Van Bronswijk, J. E. M. H. (1992). *Biologische agentia en bouwfysica*. Bouwfysica 1, pp. 7–12 (in Dutch).

[3.74] Winnepenninckx, E. (1992). *Het Vlaamse isolatie- en ventilatiedekreet, Hoofdstuk over Isoleren en Milieu*. Eindwerk IHAM (in Dutch).

[3.75] CEN/TC 89 (1993). Working draft 'Thermal performance of buildings-Calculation of energy use for heating-Residential buildings'.

[3.76] Johansson, T., Kelly, H., Reddy, A., Williams, R., Renewable Energy (1993). *Sources for Fuels and Electricity*. Island Press, 1160 pp.

[3.77] Hens, H. (1993). *Toegepaste*. Bouwfysica 1, Randvoorwaarden en Prestatie-eisen, ACCO, Leuven (in Dutch).

[3.78] Hens, H. (1993). *Thermal Retrofit of a Middle Class House: a Monitored Case*. Proceedings of the International Symposium on Energy Efficient Buildings, Stuttgart, 9–11 March.

[3.79] Hens, H., Ali Mohamed, F. (1993). *Thermal Quality and Energy Use in the Existing Housing Stock*. Proceedings of the International Symposium on Energy Efficient Buildings, Stuttgart, 9–11 March.

[3.80] Verbruggen, A. (Ed.) (1994). *Leren om te Keren*. Milieu en Natuurrapport Vlaanderen, Galant, Leuven, 823 p. (in Dutch).

[3.81] Adan, O. C. G. (1994). *On the fungal defacement of Interior Finishes*. Doctoraal Proefschrift, TU/e.

[3.82] Anon. (1994). *World energy outlook*. IEA, Paris, 305 pp.

[3.83] Vliet-SENSIVV (1995–1999). Onderzoek naar de bouwfysische kwaliteit van woningen, gebouwd na 1993. WTCB (in Dutch).

[3.84] IEA (1995). *Solar Heating and Cooling Programme*. Solar low energy houses of IEA Task 13, James & James Publishers, London.

[3.85] AIVC (1996). A Guide to Energy efficient Ventilation, 254 p.

[3.86] Cerf (1996). *Construction Industry, Research Prospectuses for the 21st Century*. Report #96-5016.T, Civil Engineering Research Foundation, Washington DC, 130 pp.

[3.87] Geselschaft für rationelle Energieverwendung (GRE) (1996). *Heizenergieeinsparung im Gebäudebestand*. BAUCOM, Böhl-Iggelheim, 99 pp. (in German).

[3.88] Hens, H. (1996). *Toegepaste Bouwfysica 3, gebouwen en installaties*. ACCO, Leuven, 329 p. (in Dutch).

[3.89] Lévêque, F. (1996). *Modélisation de la prévision de la consommation d'électricité du secteur résidentiel belge*. Memoire, UCL, 82 p. + annexes.

[3.90] Verbruggen, A. (Ed.) (1996). *Leren om te Keren, Milieu en Natuurrapport Vlaanderen*. Galant, Leuven, 585 p.

[3.91] DOE (1997). *Office of Energy Efficiency and Renewable Energy*. Scenarios of U.S. carbon reductions.

[3.92] Feist, W. (1997). *Lebenszyklus-Bilanzen im Vergleich: Niedrigenergiehaus, Passivhaus, Energieautarkes Haus*. Wksb, Heft 39, 42 jahrgang, Juli, pp. 53–57 (in German).

[3.93] Hens, H., Verdonck, B. (1997). *Wonen, verwarmen: energie en emissies*. CO_2-project Electrabel-SPE, 52 p. (in Dutch).

[3.95] Laboratorium Bouwfysica (1997). Databases STUD9497.xls, ANE9497.xls, ANAF9497.xls.

[3.95] Anon. (1997). *Enhancing the market deployment of energy technology, a survey of eight technologies*. IEA, Paris, 231 pp.

[3.96] ASHRAE (1998). *Field Studies of Thermal Comfort and Adaptation*. A Collection of Papers from the ASHRAE Winter Meeting in San Francisco.

[3.97] CIB (1998). *Sustainable Development and the Future of Construction*. Publication 225, Rotterdam.

[3.98] Erhorn, H. (1998). *Fördert oder schadet die Europäische Normung der Niedrigenergiebauweise in Deutschland*. Gesundheidsingenieur 119, heft 5, s. 236–239 (in German).

[3.99] Rudbeck, C. (1999). *Methods for Designing Building Envelope Components Prepared For Repair and Maintenance*. Report R-035, Technical University of Denmark.

[3.100] Hens, H. (1999). *Fungal Defacement in Buildings: A Performance Related Approach*. International Journal of HVAC&R Research, Vol. 5, nr 3, pp. 265–280.

[3.101] CIB (1999). *Agenda 21 on Sustainable Construction*. Publication 21, Rotterdam, 120 pp.

[3.102] prEN 12831 (2000). *Heating systems in buildings, method for calculation of the design heat load*. Final Draft, August, 77 p..

[3.103] Wouters, P. (2000). *Quality in Relation to Indoor Climate and Energy Efficiency. An analysis of trends, achievements and remaining challenges*. Doctoraal proefschrift, UCL.

[3.104] Hendriks, L., Hens, H. (2000). *Building Envelopes in a Holistic Perspective*. Final report IEA-ECBCS Annex 32, ACCO, Leuven, 102 pp. + ad.

[3.105] ASHRAE (2001). Handbook of Fundamentals.

[3.106] Mohamed, F. A., Hens, H. (2001). *Vermindering van de CO_2 uitstoot door ruimteverwarming in de residentiële sector: rendabiliteit van bouwkundige maatregelen*. Rapport Kennis CO_2-emissies, Electrabel/SPE (in Dutch).

[3.107] Sedlbauer, K. (2001). *Vorhersage von Schimmelpilzbildung auf und in Bauteilen*. Doktor-Ingenieur Abhandlung, Universität Stuttgart (in German).

[3.108] Desmyter, J., Potoms, G., Demars, P., Jacobs, J. (2001). *De alkali-silicaat reactie. Basisbegrippen en belang voor België*. WTCB Tijdschrift, 2^e trimester, pp. 3–16 (in Dutch).

[3.109] Ali Mohamed, F., Hens, H. (2001). *Vermindering van de CO_2 uitstoot door ruimteverwarming in de residentiële sector: rentabiliteit van bouwkundige maatregelen*. Eindrapport project 'Kennis van de CO_2-emissies', 85 p. (in Dutch).

3.7 References

[3.110] Verbeeck, G., Hens, H. (2002). *Energiezuinige renovaties: economisch optimum, rendabiliteit.* Rapport Kennis CO_2-emissies, Electrabel/SPE (in Dutch).

[3.111] KU-Leuven, Laboratorium Bouwfysica (1994–2003). Rapporten over het thermisch comfort in grote verkoopsruimten, trading zalen en woningen (in Dutch).

[3.112] CEN (2002). EN ISO 13790). *Thermal Performance of Buildings.* Calculation of Energy Use for Heating.

[3.113] Anon. (2002). Directive 2002/91/EC of the European Parliament and of the council of 16 december 2002 on the energy performance of buildings.

[3.114] Heron (2002). Special Issue on ASR, Vol. 47, no 2.

[3.115] Shum, M. (2002). *An Overview of the health effects due to mold exposure.* Proceedings of the Indoor Air 2002 Conference, Vol. 3, pp. 17–22, Montery, California.

[3.116] Weiss, J. S., O'Neill, M. K. (2002). *Health effects from stachybotris exposure in indoor air, a critical review.* Proceedings of the Indoor Air 2002 Conference, Vol. 3, pp. 23–28, Montery, California.

[3.117] Fanger, P. O. (2002). *Prediction of thermal sensation in non-airconditioned buildings in warm climates.* Proceedings of the Indoor Air 2002 Conference, Vol. 3, pp. 23–28, Montery, California.

[3.118] Du Plessis, C. (2002). Agenda 21 for Sustainable Construction in Developing Countries, published for CIB and UNEP by CSIR Building and Construction Technology, Pretoria, 82 pp.

[3.119] CEN (2003). EN 12831. Heating Systems in Buildings, Method for Calculation of the Design Load.

[3.120] Hens, H. (2003). *Bouwen, gebouwgebruik en milieu, geen problemen of toch?* Wetenschap op nieuwe wegen, Lessen van de 21^e eeuw, Raymaekers en Van Riel, ed. Universitaire Pers, Leuven, p. 219–243 (in Dutch).

[3.121] NBN EN 12831 (2003). *Heating systems in buildings. Method for calculation of the design heat load*, 76 pp. (in English)).

[3.122] Grumman, D. (Ed.) (2003). *ASHRAE GreenGuide.* Atlanta, 163 pp.

[3.123] Anon. (2004). *Decreet houdende eisen en handhavingsmaatregelen op het vlak van de energieprestaties en het binnenklimaat voor gebouwen en tot invoering van een energieprestatiecertificaat.* Belgisch staatsblad, 30/07, ed.3 (in Dutch).

[3.124] Hens, H. (2004). *Cost efficiency of PUR/PIR insulation.* Report 2004/14, written on demand of the European PU-industry, Laboratory of Building Physics, 76 pp.

[3.125] Anon. (2005). *Calculation Method for the Characteristic Annual Primary Energy Consumption in Residential Buildings.* Add I to the Flemish EPB-legislation, Belgisch Staatsblad/Moniteur Belge, June, 15 (in Dutch and French).

[3.126] Seppänen, O., Fisk, W. (2005). *Indoor climate and productivity.* Proceedings of the International Conference on the Energy performance of Buildings, AIVC Brussels, 21–23 September.

[3.127] Flannery, T. (2005). *The Weather Makers: The History and Future Impact of Climate Change.* Text Publishing, Melbourne.

[3.128] Vlaamse regering (2005). Besluit tot vaststelling van de eisen op het vlak van de energieprestaties en het binnenklimaat in gebouwen met bijlagen (in Dutch).

[3.129] REHVA (2005). *Clima 2005 8th World Congress.* Proceedings, Lausanne (CD-ROM).

[3.130] D'haeseleer, W (redactie) (2005). *Energie vandaag en morgen, beschouwingen over energievoorziening en -gebruik.* ACCO, Leuven, 292 pp. (in Dutch).

[3.131] Geurts, H., Van Dorland, R. (2005). *Klimaatverandering*. Teleac KNMI, Kosmos-Z&K uitgevers, Utrecht/Antwerpen, 128 pp. (in Dutch).

[3.132] Hens, H. (2006). *Duurzaam bouwen*. Francqui leerstoel VUB, 80 pp. (in Dutch).

[3.133] Flannery, T. (2006). *De weermakers*. Uitgeverij Atlas, Amsterdam, 324 pp. (in Dutch).

[3.134] Verbeeck, G., Hens, H. (2006). *Development of extremely low energy dwellings through life cycle optimization*. Research in Building Physics and Building Engineering (P. Fazio, H. Ge, J. Rao, G. Desmarais Ed.), Taylor & Francis, London, p. 579–586.

[3.135] Johannesson, G. (2006). *Building energy – a design tool meeting the requirements for energy performance standards at early design – validation*. Research in Building Physics and Building Engineering (P. Fazio, H. Ge, J. Rao, G. Desmarais Ed.), Taylor & Francis, London, p. 627–634.

[3.136] Van der Veken, J., Hens, H., Peeters, L., Helsen, L., D'haeseleer, W. (2006). *Economy, energy and ecology based comparison of heating systems in dwellings*. Research in Building Physics and Building Engineering (P. Fazio, H. Ge, J. Rao, G. Desmarais Ed.), Taylor & Francis, London, p. 661–668.

[3.137] Federaal Planbureau (2006). *Het klimaatbeleid na 2012, analyse van de scenario's voor emissiereductie tegen 2020 en 2050*. Studie in opdracht van de federale minister van leefmilieu, 246 pp. (in Dutch).

[3.138] Commission Energy 2030 (2006). *Belgium's Energy Challenges Towards 2030*. Report commissioned by the federal minister of energy.

[3.139] ASHRAE (2007). *Handbook of HVAC Applications*. Chapter 36, Atlanta.

[3.140] Verbeeck, G. (2007). *Optimisation of extremely low energy residential buildings*. PhD-thesis, K. U. Leuven.

[3.141] Verbeeck, G., Hens, H. (2007). *Life cycle inventory of extremely low energy dwellings*. Proceedings Plea 2007, Singapore, 22–24 november.

[3.142] Anon. (2007). *Assessment of office buildings: towards a sustainable housing of the Flemish civil services*. D/2008/3241/080, 126 p. (in Dutch).

[3.143] Hens, H., Verbeeck, G., Van der Veken, J., De Meulenaer, V., Janssens, A., Van Londersele, E., Willems, L., Peeters, L., D'haeseleer, W., Vermeyen, P. (2007). *Ontwikkeling via levenscyclusanalyse van extreem lage energie- en pollutiewoningen (EL^2EP)*. Eindrapport GBOU-project (in Dutch).

[3.144] ASHRAE (2008). *Advanced Energy Design Guide for K-12 School Buildings*. ISBN 1-931862-41-9, 171 p.

[3.145] Gertis, K., Hauser, G., Sedlbauer, K., Sobek, W. (2008). *Was bedeutet "Platin"? Zur Entwicklung von Nachhaltigkeitsbewertungsverfahren*. Bauphysik 30 (4), pp. 244–256 (in German).

[3.146] Hens, H., Rose, W. B. (2008). *The Erlanger House at the University of Illinois-A Performance-based Evaluation*. Proceedings of the Building Physics Symposium, Leuven, 29–31 October, pp. 227–235.

[3.147] NIBS (2008). *High Performance Buildings*. Assessment to the US Congress and US Department of Energy in Response to Section 914 of the Energy Policy Act of 2005 (Public Law 109-058), 27 p.

[3.148] USGBC (2009). LEED 2009 for New Construction and Major Renovations, 88 p.

4 Heat-air-moisture performances at the envelope level

4.1 Introduction

Chapter 3 looked at some main performances at the building level. This Chapter 4 steps one level down, looking to the building envelope with the heat-air-moisture performances as exemplary case. For the opaque parts, these are: (1) air-tightness, (2) thermal transmittance, (3) thermal transient response, (4) moisture tolerance and (5) hygrothermal load. In that quintet, air-tightness figures as the throughline for those that follow. If it is lacking, thermal transmittance decouples from insulation quality, transient response degrades and moisture tolerance becomes more risky. For the transparent parts, mastering solar gains replaces thermal transient response. For the floors belonging to the envelope, the contact coefficient of the floor finish should not be overlooked.

4.2 Air-tightness

4.2.1 Flow patterns

Not only the other hygrothermal performances but also the acoustical performances and performances at the building level such as thermal comfort, primary energy consumption and fire safety are impacted by lack of air-tightness.

When judging air-tightness as an envelope performance seven flow patterns may interact:

Pattern	Cause	Consequences
Outflow (In → Out)	Envelope part not airtight Difference in temperature between the inside and the outside Envelope parts at leeside Overpressure inside	Thermal transmittance no longer reflecting insulation quality High interstitial condensation risk, larger deposits Faster drying to the outside Uncontrolled adventitious ventilation indoors
Inflow (In ← Out)	Envelope part not air-tight Difference in temperature between the inside and the outside Envelope parts at the windside Under pressure inside	Thermal transmittance no longer reflecting insulation quality Worse transient thermal response Increased mould and surface condensation risk Faster drying, mainly to the inside Drop in sound insulation for airborne noise outside Uncontrolled adventitious ventilation indoors

Applied Building Physics: Boundary Conditions, Building Performance and Material Properties. Hugo Hens
Copyright © 2011 Wilhelm Ernst & Sohn, Berlin
ISBN: 978-3-433-02962-6

Pattern	Cause	Consequences
Cavity ventilation	Cavity at the outer side of thermal insulation with air inlets and outlets in the cladding or an air permeable cladding Wind pressure differences along the outside surface Temperature difference between the cavity and outdoors Inlet and outlet at different heights	Small increase in thermal transmittance Typically considered as beneficial for moisture tolerance, though condensation by undercooking at the cavity side of the cladding more likely Drop in sound insulation for airborne noise
Wind washing	Cavity at the inner side of the thermal insulation disclosed for outside air, cavity filled with air-permeable insulation material Wind pressure differences along the outside surface, temperature difference between the cavity and outside	Large increase in thermal transmittance Worse transient thermal response Increased risk on mould and surface condensation inside Drop in sound insulation for airborne noise
Inside air ventilation	Cavity at the inner side of thermal insulation disclosed for inside air Temperature and height differences along the inside surface Air pressure differences along the inside surface	Small increase in thermal transmittance Drop in sound insulation for airborne noise
Inside air washing	Cavity at the outer side of thermal insulation disclosed for inside air Temperature and height differences along the inside surface Air pressure differences along the inside surface	Large increase in thermal transmittance High interstitial condensation risk, larger deposits Drop in sound insulation for airborne noise
Air looping	Air cavity at both sides of the thermal insulation, leaks at different heights in the insulation layer or, air permeable insulation. Temperature difference across the insulation	Large increase in thermal transmittance Somewhat higher interstitial condensation risk

4.2 Air-tightness

Limiting air in and outflow to the utmost demands inclusion of an air barrier in the envelope. If mounted inside, such a barrier also minimizes inside air washing. At the outside, it acts as wind-barrier, controlling wind washing while allowing outside air ventilation in a cavity between it and the cladding. For tempering both indoor air and wind washing to a maximum, one should combine an air barrier inside with a wind barrier outside of the thermal insulation. In case the insulation layer itself is perfectly airtight, wind washing, inside air washing and air looping are excluded. Eliminating the cavity between an airtight insulation layer and the outside cladding excludes outside air ventilation, inside air washing and air looping. With no cavity at the backside of an airtight insulation layer, wind washing, inside air ventilation and air looping are avoided.

4.2.2 Performance requirements

4.2.2.1 Air in and outflow

The answer to the question of how air-tight an envelope should be is: perfectly. In practice, however, this is fiction. Even if imposed and even if the design should guarantee perfection, limits in building ability will induce imperfections that relegate the 'perfect' requirement to the realm of fairy tales. Therefore, another approach is advisable. Air leakage short-circuits the diffusion resistance between the inside and interfaces in the assembly where condensation is probable. Whether this will result in unacceptable moisture deposits there depends on the overall composition of the envelope part and the climate in and outdoors with vapour and air pressure excess inside as main players. The air-tightness requirements should therefore be coupled to the indoor climate class and the air pressure differentials expected. That gives the following upper limit for indoor climate class 1, 2 and 3 buildings: (1) no concentrated leakages in terms of cracks, perforations, open joints, etc., (2) area averaged air permeance coefficient $\leq 10^{-5}$ kg/(m$^2 \cdot$ s $\cdot$ Pa^b)

4.2.2.2 Inside air washing, wind washing and air looping

Assume we call equivalent thermal transmittance the area- and time-averaged heat flow rate across the assembly, whatever may be the cause, divided by the difference in inside and outside reference temperature. That quantity could also be written as the thermal transmittance, multiplied with a so-called Nusselt number, Inside air washing, wind washing and air looping now should not increase the equivalent thermal transmittance compared to the thermal transmittance with a percentage beyond x. If for example x is set 10%, than Nusselt may not pass 1.1.

Example: partially filled cavity wall, wind washing

Consider a cavity wall with 9 cm thick brick veneer, 3 cm wide cavity, partial fill with 10 cm PUR, inside leaf in 14 cm thick light-weight fired clay blocks ($R = 0.88$ m² · K/W) and airtight pargetting inside (Figure 4.1). The wall is 2.7 m high. Top and bottom of the veneer wall contain two weep holes per meter run. Bad workmanship however causes the cavity fill to stop above the lower and below the upper weap holes while the fill is mounted so carelessly that the layer stands somewhere between the inner leaf and the veneer wall.

How does wind washing affect the thermal transmittance?

Figure 4.1

Wind washing redistributes the outside air flow between the cavity behind the veneer wall (suffix 1) and the air layer behind the insulation (suffix 2), proportional to the third power of their widths:

$$G_a = G_{a,1} + G_{a,2} \qquad G_{a1} = G_a \frac{d_1^3}{d_1^3 + d_2^3} \qquad G_{a2} = G_a \frac{d_2^3}{d_1^3 + d_2^3}$$

As the heat flow across the insulation will be small compared to the one across the veneer, temperature in the outer cavity will hardly differ from the equilibrium value without ventilation, giving as fair approximation for the temperature in the air layer behind the insulation:

$$\theta_2 = \theta_{2,\infty} + (\theta_e - \theta_{2,\infty}) \exp\left(-\frac{R_1 + R_2}{c_a \, G_{a,2} \, R_1 \, R_2} z\right)$$

where R_1 is the thermal resistance across the thermal insulation between the air layer behind the insulation and outside, R_2 is the thermal resistance across the inside leaf between the inside and that air layer behind and $\theta_{2,\infty}$ the temperature one should have in that air layer without wind washing. The effective thermal transmittance, the Nusselt number and thermal insulation efficiency then are:

$$U_{eq} = U \underbrace{\left\{1 - c_a \, G_{a,2} \frac{U \, R_1^2}{H} \left[\exp\left(-\frac{1}{U \, R_1 \, R_2 \, c_a \, G_{a,2}} H\right) - 1\right]\right\}}_{\text{Nu}}$$

$$\eta_{is} = 100 \frac{d_{ins,eq}}{d_{ins}} = 100 \frac{U}{U_{eq}} \left(\frac{1 - U_{eq} \, R_0}{1 - U \, R_0}\right)$$

where U is the thermal transmittance, equal to $1/(R_1 + R_2)$, H the height between the upper and lower weap holes in m and R_0 the thermal resistance of the assembly if the insulation was a layer with thickness zero.

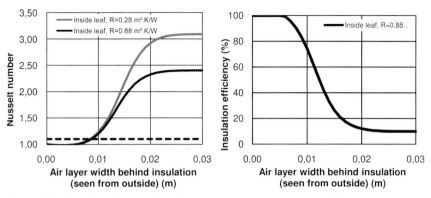

Figure 4.2. Partially filled cavity wall, wind washing at a mean free field wind speed of 4 m/s at 10 m height, Nusselt number and insulation efficiency as function of the air layer width behind the insulation. Nu passes 1.1 for a width beyond 8 mm.

Figure 4.2 shows how the insulation efficiency and the Nusselt number depend on the air layer width behind the insulation. Nusselt 1.1 already requires careful workmanship. In fact the insulation efficiency drops quickly once the air layer exceeds an average width of 8 mm.

4.3 Thermal transmittance (U)

4.3.1 Definitions

4.3.1.1 Envelope parts

When air flow is excluded whole thermal transmittance of an envelope part calculates as:

$$U = U_\beta + \frac{\sum_j (\psi_j L_j) + \sum_k \chi_k}{A} \quad (\text{W}/(\text{m}^2 \cdot \text{K})) \tag{4.1}$$

where U_0 is the clear wall thermal transmittance, ψ_j linear thermal transmittances (W/(m · K)) and L_j the length of all linear thermal bridges and χ local thermal transmittance (W/K) of the local thermal bridges within the part with area A. That formula does not apply for floors on grade, floors above basements, floors above crawl spaces and transparent parts.

4.3.1.2 Envelope

The building envelope contains a whole of building parts, linear and local thermal bridges, a lowest floor and transparent parts, all coupled in parallel. The average envelope thermal transmittance then looks like:

$$U_m = \frac{U_{fl} A_{fl} + \sum_{\text{opaque}} U A + \sum_j (\psi_j L_j) + \sum_k \chi_k + \sum_w U_w A_w}{A_T} \quad (\text{W}/(\text{m}^2 \cdot \text{K})) \tag{4.2}$$

where U_{fl} is the mean thermal transmittance and A_{fl} the area of the lowest floor, U the thermal transmittance and A the area of all other opaque building elements and A_w the area and U_w the thermal transmittance of all transparent parts.

4.3.2 Basis for performance requirements

4.3.2.1 Envelope parts

Values should be low enough to keep mould risk in outside edges and corners below 5%. In cool climates that demands thermal transmittances U_0 below 0.46 W/(m² · K). At the same time, the optimum in terms of life cycle costs should be aimed for, giving a range from 0.2 to 0.6 W/(m² · K). Of course one could also pose minimal total energy consumption or minimal total pollution as a target, a track leading to thermal transmittances below 0.15 W/(m² · K).

4.3.2.2 Envelope

The only way to get optimum values from a life cycle cost perspective is by applying a whole building approach, as explained in Chapter 3.

4.3.3 Examples of performance requirements

Already before the EU Energy Performance Directive of 2003 went into force, many European countries imposed legal requirements to the thermal transmittances of opaque and transparent building parts (U_{max}). Some also limited the envelope's thermal transmittance in relation to the compactness of the building. Even today, due to the long service life of a good thermal insulation, energy performance requirements remain complemented by insulation requirements.

4.3.3.1 Envelope parts

Table 4.1 gives maximum thermal transmittances for normally heated buildings as required in a few countries and regions.

4.3.3.2 Envelope

Imposing maximum thermal transmittances per building element has disadvantages. It does not dissuade the use of large glazed surfaces. Even with the well insulating glass systems of today, large surfaces do not offer much benefit in cool climates as insolation in winter is low. Also compactness is not observed. An alternative therefore is imposing upper limits to the end energy demand per unit of protected volume, as done by the EPR. A not so harsh approach consists of limiting that part of the net heating demand that is most easily controlled during design: the transmission losses. These are proportional to the product of the envelope area (A_T) and its thermal transmittance (U_m):

$$Q_{T,ann} \div U_m \, A_T \quad (MJ/a) \qquad (4.3)$$

Requiring the product $U_m A_T$ to be proportional to the protected volume can be expressed by:

$$U_m = \alpha V / A_T = \alpha C$$

with C compactness in m. So, in a [C, U_m]-coordinate system a straight line through the origin with slope α is found. The smaller that slope, the more severe are the limits to the transmission losses (Figure 4.3a). Keeping that line straight under all circumstances however is not possible.

4.3 Thermal transmittance (U)

Table 4.1. Maximum thermal transmittances.

Element	U_{max} (W/(m² · K))	
	New construction	**Retrofit**
Belgium (Flanders)		
Walls	0.4	0.4
Roofs	0.3	0.3
Floors above grade	0.4[1] (R_{min} = 1 m² · K/W)	1.2
Floors above basements and crawlspaces	0.4[1] (R_{min} = 1 m² · K/W)	0.9
Floors above outdoor spaces	0.6	0.6
Walls contacting the ground	R_{min} = 1 m² · K/W	0.9
Separation walls and floors between dwellings	1.0	1.0
Glass	1.6	1.6
Windows	2.5	2.5
Germany (normally heated buildings)		
Walls		0.24
Roofs		0.24
Low sloped roofs		0.20
Floors above grade		0.30
Floors above basements and crawlspaces		0.30
Floors above outdoor spaces		0.30
Walls contacting the ground		0.30
Glass		1.10
Windows		1.30
UK		
Walls	0.30/0.35	
Roofs	0.16	
Floors	0.25	
Windows	2.00	
Sweden	**Oil or gas heating**	**Electrical heating**
Walls	0.18	0.10
Roofs	0.13	0.08
Floors	0.15	0.10
Windows	1.30	1.10
Outer doors	1.30	1.10

At high compactness, the insulation requirements may become so weak that mould growth, surface condensation and comfort complaints become likely. At very low compactness, the insulation requirement may be of such severity that the investments explode, worse, buildability becomes questionable. Actually, the necessary usage of glazed surfaces turns that straight line through the origin anyhow into a fiction. In fact, one has:

$$U_m = U_{m,op} + (A_{T,w}/A_T)(U_w - U_{m,op}) \approx U_{m,op} + [A_{T,w}/V(U_w - U_{m,op})]C$$

i.e. a straight line of the form $a + b C$ with $U_{m,op}$ the average thermal transmittance of the opaque building elements, U_w the average thermal transmittance and $A_{T,w}$ total window area. As the protected volume V may be written as $A_{fl} h$ with h the floor height, the slope b seems

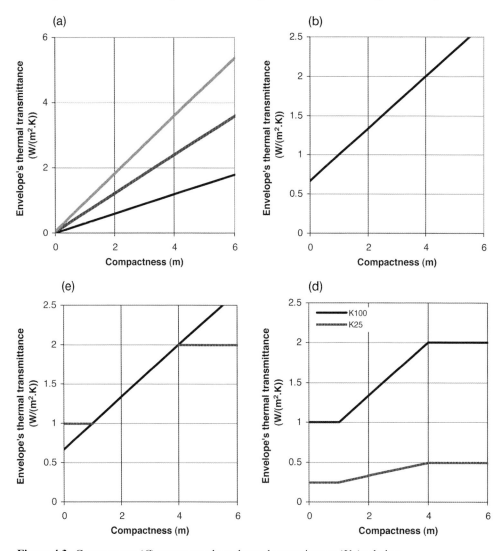

Figure 4.3. Compactness (C) versus envelope thermal transmittance (U_m) relation.

4.3 Thermal transmittance (U)

proportional to the ratio between glass and floor area and inversely proportional to the floor height h (Figure 4.3b). Fixing a and b delivers the basis for formulating an envelope thermal transmittance performance requirement. An example is found in the Belgian legislation which defines a reference line with a equal to 2/3 and b equal to 1/3. By that, the straight line intersects the compactness axis in the point $(-2, 0)$. Weakening the thermal transmittance requirement at low and upgrading them at high compactness is done by keeping a value 1 W/(m² · K) for a compactness below 1 m and 2 W/(m² · K) for a compactness above 4 m (Figure 4.3c). The broken line found that way is called 'level of thermal insulation K100'. Each building with the (C, U_m)-pair on that line obtains that level. Any other level is now defined by a broken line proportional to the K100 reference. As an equation:

$$\begin{array}{ll} C \leq 1\text{ m} & K = 100\, U_m \\ 1 < C < 4\text{ m} & K = \dfrac{100\, U_m}{2/3 + C/3} \quad (\text{W/(m}^2 \cdot \text{K)}) \\ C \geq 4\text{ m} & K = 50\, U_m \end{array} \quad (4.4)$$

See Figure 4.3d. Imposing a performance requirement on the envelope thermal transmittance is easy that way. Low energy for example demands more or less K25. The only thing still needed are rules to calculate compactness, envelope and building part surfaces and, thermal transmittance of all separate building parts.

Some countries define compactness the other way around: not V/A_T in m, but A_T/V in m^{-1}. The straight line than becomes a hyperbola. As an example Figure 4.4 gives the actual German envelope thermal transmittance requirements. The same corrections are applied as explained above: constant values, now below compactness 0.2 m^{-1} and above compactness 1.05 m^{-1}, hyperbolic in between:

$$\begin{array}{ll} C' \leq 0.2\text{ m}^{-1} & U_m = 1.05 \\ 0.2 < C' \leq 1.05\text{ m}^{-1} & U_m = 0.3 + \dfrac{0.15}{C'} \quad (\text{W/(m}^2 \cdot \text{K)}) \\ C' > 1.05\text{ m}^{-1} & U_m = 0.44 \end{array} \quad (4.5)$$

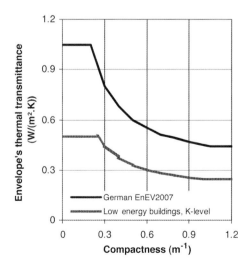

Figure 4.4. German envelope thermal transmittance requirements.

4.4 Transient thermal response

4.4.1 Properties of importance

4.4.1.1 Opaque envelope parts

In regions with a cool climate, transient thermal response of the enclosure is one of the parameters determining summer thermal comfort. In regions with a warmer climate, energy consumption for cooling is a main beneficiary. Characteristics determining the transient thermal response of an opaque one-dimensional building element are:

Harmonic load (period: 1 day)	Others
Temperature damping D_θ	Time constant τ
Dynamic thermal resistance D_q	
Admittance Ad	

An important advantage of the harmonic properties is that they are analytically calculable. Quantifying a time constant instead demands simplified models or a numerical approach.

Example of a simplified model:
the building element seen as a resistance-capacitance-resistance circuit

First the thermal capacity per layer ($C = \rho\, c\, d$ in J/K) is calculated and considered as a vector in the layer's centre. All layers together give a vector field whose resultant ($\Sigma\, C_j$) is situated in what is called the point of action. Assume x is the ordinate of that point along an x-axis with origin in the contact interface with environment 1. Thermal resistance between environment 1 and x is called R_{1x}, thermal resistance between environment 2 at the other side and x R_{2x}. The heat balance for the circuit $R_{1x} / \Sigma\, C_j / R_{2x}$ then becomes (Figure 4.5):

$$\frac{\theta_1 - \theta_x}{R_{1x}} + \frac{\theta_2 - \theta_x}{R_{2x}} = \left(\Sigma C_j\right) \frac{\theta_x}{dt}$$

where θ_x is the temperature in the point of action, θ_1 the uniform temperature in environment 1 and θ_2 the uniform temperature in environment 2. A step increase or decrease of temperature θ_1 or θ_2 at time zero gives as a solution:

$$\theta_x = \theta_{x,\infty} + \left(\theta_{x,0} - \theta_{x,\infty}\right) \exp\left(-\frac{t}{\overline{R} \Sigma C_j}\right)$$

where $\overline{R}$ is the harmonic mean of R_{x1} and R_{x2}:

$$\overline{R} = \frac{R_{1x} R_{2x}}{R_{1x} + R_{2x}}$$

The time constant is: $\tau = \overline{R} \Sigma C_j$

That formula shows the time constant increases with both total capacity and the harmonic mean of both thermal resistances. In that mean the smallest thermal resistance has the largest impact.

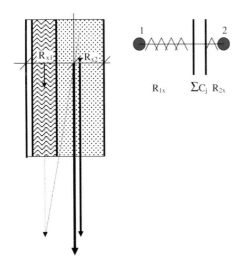

Figure 4.5. The <thermal resistance R_1/capacitance Σ C/thermal resistance R_2> analogue.

4.4.1.2 Transparent envelope parts

Transparent parts act as a hatch for short wave solar radiation and a source of indirect solar gains by convection and long wave radiation of absorbed short wave irradiation. Both fix the solar transmittance g of the part.

4.4.2 Performance requirements

4.4.2.1 Opaque envelope parts

Imposing limit values for the harmonic properties is less evident. The maximum thermal transmittance (U_{max}) fixes the lowest value the dynamic thermal resistance amplitude will touch, as following relation holds: $[D_q] > 1/U_{max}$. The lower U_{max}, the larger the minimal dynamic thermal resistance will be.

The maximum thermal transmittance also borders the lowest possible admittance amplitude, while the highest amplitude possible never passes the thermal surface film coefficient: $U_{max} \leq [Ad] \leq h_i$. That way, low U_{max}-values and high h_i-values open a large window of admittance values. A high admittance makes heat storage easier, which is why a performance requirement could be: $[Ad] \geq h_i/2$

Temperature damping amplitude may finally have values between 1 and infinity. The relevance of very high values, however, is relative. If for example the sol-air temperature outside swings between 10 and 80 °C on a daily basis, than the amplitude of the complex inside temperature will equal $35/[D_\theta]$, which translated into numbers gives:

$[D_\theta]$	$\hat{a}_i$ °C
1	35.0
2	17.5
4	8.8
8	4.4
16	2.3
32	1.1
74	0.6

A difference between night and day of 2.3 °C will hardly be decisive for thermal comfort. For that reason it suffices to impose a lower limit, for example: $[D_\theta] \geq 15$.

4.4.2.2 Transparent envelope parts

In cool climates solar transmittance should accommodate two conflicting requirements: in view of energy efficiency as close as possible to 1 during the heating season, in view of thermal comfort and energy efficiency if cooling is needed, as low as possible during the warm half-year, however without hindering daylighting. The best solution therefore follows from a combined end energy consumption/summer comfort analysis, using building energy software tools. A possible reference for summer comfort is the number of weighted temperature excesses

(WTE). If all other parameters are invariant, glass surface area and solar shading should be fixed in a way that number does not exceed 100. The WTE-hours are given by summing up the excess factors (EF) the hours the building is used:

$$|PMV| \leq 0.5 \quad EF = 0$$
$$|PMV| > 0.5 \quad EF = 0.47 + 0.22|PMV| + 1.3|PMV|^2 + 0.97|PMV|^3 - 0.39|PMV|^4 \quad (4.6)$$

where PMV is the predicted mean vote at an hourly basis (see Chapter 3, thermal comfort). Following array with $f_{gl} = A_{gl}/A_{fac}$ the glass to façade surface ratio allows a quick rating of the advisable solar transmittance (g) during the warm half-year:

Inside partitions	$g f_{gl}$	
	Low ACH	High ACH
Light	0.12	0.17
Heavy	0.14	0.25

4.4.3 Consequences for the building fabric

4.4.3.1 Opaque envelope parts

How are high admittances structurally achieved? The following simple model demonstrates the answer. Take an assembly composed of two layers, one light and insulating, thermal resistance R, and the other heavy and hardly insulating, capacitance C. The thermal surface film coefficients at both sides are h_1 respectively h_2. The heat balance becomes:

$$\frac{\theta_1 - \theta_x}{R + 1/h_1} + \frac{\theta_2 - \theta_x}{1/h_2} = C \frac{\theta_x}{dt} \quad (4.7)$$

where θ_x is the central temperature in the capacitance, θ_1 the temperature in the environment at the insulation side and θ_2 the temperature in the environment at the heavy layer side. Assume environment 1 is the outside ($h_1 = h_e$, θ_1). The outside temperature fluctuates harmonically, period T. In such a case also the inside temperature (θ_2) will vary harmonically with a same period T and an amplitude α_2 dampened and shifted in time compared to the temperature outdoors:

$$\theta_2 = \alpha_2 \exp\left(\frac{i 2 \pi t}{T}\right) \quad (4.8)$$

Temperature damping is defined now for a complex heat flow rate zero at the inside surface (surface 2). That presumes a heat flow rate of zero between the capacitance and indoors or: $\alpha_x = \alpha_2$. That way the heat balance (4.7) is reduced to (after elimination of the time function $\exp(i 2 \pi t/T)$):

$$\frac{\alpha_1 - \alpha_2}{R + 1/h_1} = \frac{i 2 \pi C}{T} \alpha_2$$

Complex temperature damping then is ($R_1 = R + 1/h_1$): $D_\theta^{1,2} = \alpha_1/\alpha_2 = 1 + i 2 \pi R_1 C/T$ with as amplitude:

$$\hat{D}_\theta^{1,2} = \sqrt{1 + \left(\frac{2\pi R_1 C}{T}\right)^2}. \tag{4.9}$$

That value increases quasi linearly with the part's time constant ($R_1 C$).

Now, the situation is reversed, with 2 being the outside and 1 the inside, or θ_2 the cause and θ_1 the consequence. The complex heat flow rate q'_1 now becomes zero, changing the heat balance into ($h_2 = h_e$):

$$h_2(\alpha_2 - \alpha_1) = \frac{i 2\pi C}{T}\alpha_2$$

and giving as complex temperature damping $D_\theta^{2,1} = \alpha_2/\alpha_1 = 1 + i 2\pi C/(T h_2)$, and an amplitude of:

$$\hat{D}_\theta^{2,1} = \sqrt{1 + \left(\frac{2\pi C}{T h_2}\right)^2} \tag{4.10}$$

Due to a surface thermal resistance value $1/h_2$ which is far below the sum of the thermal resistance $1/h_1 + R$ ($h_1 = h_2 = h_e$!), the temperature damping amplitude $\hat{D}_\theta^{1,2}$ is much larger now than $\hat{D}_\theta^{2,1}$. If 1 in the square root is neglected, then the ratio between the two ($\hat{D}_\theta^{1,2}/\hat{D}_\theta^{2,1}$) equals $1 + R h_e$. For an insulation with thermal resistance 1.5 m² · K/W that ratio equals 38.5 ($h_e = 25$ W/(m² · K)). Achieving a high temperature damping thus demands an opaque assembly composed of a heavy and an insulating layer with the insulating layer outside and the heavy layer inside. A damping amplitude 15 requires a time constant $R_1 C$ of 205 800 s or 2.38 days. Higher thermal resistances R_1 allow reducing the thickness of the capacitive layer for the same damping result. A lower thermal resistance instead demands a thicker capacitive layer for the same damping result. As an illustration:

R_1 m² · K/W	C J/(m² · K)	Assembly (the parts in italic not buildable in practice) ($\hat{D}_\theta = 15$)
1	205 800	4 cm thermal insulation at the outside
		9 cm concrete or 19 cm thick hollow fired clay bricks inside
2	102 900	8 cm thermal insulation at the outside
		5.5 cm concrete or 9 cm thick hollow fired clay bricks inside
4	51 450	16 cm thermal insulation at the outside
		2.75 cm concrete or 4.5 cm thick hollow fired clay bricks at the inside
8	25 725	32 cm thermal insulation at the outside
		1.375 cm concrete or 2.25 cm thick hollow fired clay bricks at the inside

A same discourse for the admittance shows highest values are attained with a capacitive layer inside, not screened by even a thin insulating layer or a low thermal surface film resistance inside ($1/h_i$).

4.4.3.2 Transparent envelope parts

Solar screening is the way to go, either by using sun absorbing or reflecting glazing systems, movable outside screens or fixed shading elements.

4.5 Moisture tolerance

4.5.1 In general

Several humidity sources must be considered when moisture tolerance is at stake:

Contact with water	Built-in moisture
	Rain
	Rising damp
	Pressured water
	Accidental (leaky pipes, etc)
Contact with water vapour	Hygroscopic moisture
	Surface condensation
	Interstitial condensation

Consequences of a failed moisture tolerance were reviewed in Chapter 3. Moisture is a quasi necessary actor when it comes to building damage, less usability and less well-being of the users.

4.5.2 Built-in moisture

4.5.2.1 Definition

The term built-in moisture embraces all excess humidity present in the fabric at the start of the building operation period. Many sources contribute: rain and snow during construction; using more water in mortar, concrete and gypsum plaster than chemically needed; bricks humidified to facilitate brick laying; timber arriving too humid on site. Cellular concrete contains up to 250 kg water per m^3 after production. Carbonisation generates water. The hygroscopic moisture content in apparently dry materials corrresponds to the relative humidity outside, not to that expected inside …

The amounts of built-in moisture in massive constructions are quite impressive. Public housing showed values of up to 5000 l per dwelling. Tests on low-sloped roof screeds just before water-tightening gave values of up to 120 kg/m^3.

Built-in moisture causes end energy use for heating to be higher during the first years of operation. Drying also increases the water vapour pressure and relative humidity indoors. That may temporarily cause mould and surface condensation. Happily both fade away after a couple of years. In heat-air-moisture transport, built in moisture fixes the initial conditions.

4.5.2.2 Requirements

Built-in moisture must have (1) the ability to dry without causing unacceptable interstitial condensation in thermally insulating or other moisture sensitive layers, whereas (2) the way the envelope part or fabric part is assembled should not unnecessarily retard drying. Point (2) does not mean a correct assembly will give short drying time. Duration depends on the amount of built-in moisture, the thickness and hygrothermal properties of the different layers

4.5 Moisture tolerance

(air permeability, thermal conductivity, volumetric heat capacity, diffusion resistance factor, moisture permeability, moisture retention curve) and the boundary conditions present (air temperature outside, insolation, under-cooling, precipitation, relative humidity in and outside, operative temperature inside and air pressure in and outside).

4.5.2.3 Consequences for the building fabric

Both requirements result in some specific rules: (1) never sandwich moist and dry layers together between vapour retarding foils, (2) insert a vapour retarder between a vapour permeable thermal insulation or a moisture sensitive layer and wet layers at their warm side, (3) wait with painting, wallpapering, flooring until the fabric is nearly air-dry.

In cool climates, ventilation and heating both accelerate drying. Consider for example a humid enclosure layer facing the inside. The first drying stage gives as vapour flow rate:

$$g_{v,d} = \beta \left(p_{sat,s} - p_i \right) = \frac{h_{c,i}}{N \lambda_a} \left(p_{sat,s} - p_i \right) \tag{4.11}$$

where β is the surface film coefficient for diffusion, h_i the thermal surface film coefficient for convection (both increasing with air speed), p_{sat} vapour saturation pressure at the inside surface and p vapour pressure inside.

Fast drying assumes a high thermal surface film coefficient for convection and a large difference between vapour pressure inside and saturation pressure at the surface. That demands heating the surface, augmenting the air speed and lowering the relative humidity inside. With a ventilation flow G_a and A as area of the drying surface, vapour pressure inside becomes:

$$p_i = p_e + \frac{g_{v,d} \, A}{6.21 \, 10^{-6} \, G_a} \tag{4.12}$$

Combining the Equations (4.11) and (4.12) gives for the vapour flow rate:

$$g_{v,d} = \frac{h_{c,i} \left(p_{sat,s} - p_e \right)}{N \lambda_a + h_{c,i} \, A / \left(6.21 \cdot 10^{-6} \, G_a \right)} \tag{4.13}$$

That equation could be quantified if vapour saturation pressure at the surface, in other words, surface temperature θ_s, is known. In case of internal partitioning, with the same environmental temperature at both sides, surface temperature follows from:

$$l_b \, g_{v,d} = \left(h_{c,i} + 5.4 \, e_L \right) \left(\theta_i - \theta_s \right) \tag{4.14}$$

with e_L long-wave emissivity of the surface and l_b heat of evaporation in J/kg. Entering Equation (4.13) in (4.14) results in a first relation between saturation vapour pressure and surface temperature. The equation of state offers a second one. That way the drying flow rate is known. Figure 4.6 gives the result for a newly pargetted room with volume $4 \times 4 \times 2.7 = 43.2$ m^3 and a wet surface of 56.8 m^2 (four walls and ceiling minus the window) for a January temperature of 2.7 °C. The parameters are: ventilation rate and inside temperature. Ventilation without heating looks hardly effective. Heating is even important. The benefit of additional ventilation also decreases with increasing ventilation rate.

However, pushing drying to higher rates, causes the second drying stage to occur at a higher transition moisture content:

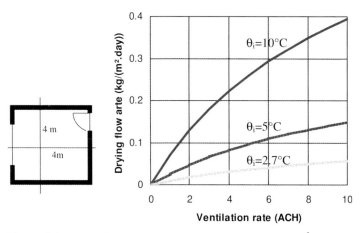

Figure 4.6. Drying flow rate in a newly pargetted $4 \times 4 \times 2.4$ m³ room.

$$w_{tr} = w_{cr} + \frac{g_{vd}\, d}{3\, D_w} \quad (4.15)$$

In that second stage the drying flow rate drops to:

$$g_{v,d} = \frac{p_{sat,x} - p_i}{N\, \lambda_a / h_{c,i} + \mu\, N\, x} = \frac{p_{sat,x} - p_i}{Z_x} \quad (4.16)$$

where $p_{sat,x}$ is the vapour saturation pressure at the retiring drying front, which at the moment considered is at a distance x of the inside surface, and μ is the vapour resistance factor between that front and the inside. Air speed has hardly any impact because drying becomes increasingly dependent on the material properties (μ). The inside temperature (fixes $p_{sat,x}$) and relative humidity instead remain as important as during the first drying stage. A premature finishing of the inside enlarges the diffusion resistance between the drying front and the inside, which further retards drying:

$$g_{v,d} = g_{v,d,o}\, \frac{Z_x}{Z_x + Z}$$

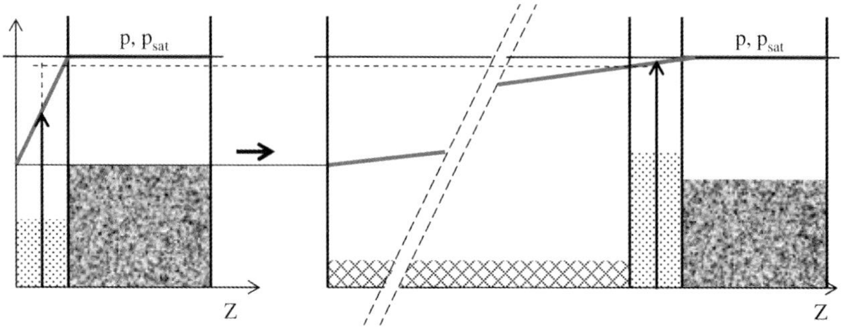

Figure 4.7. Humidification of the already dry thickness after applying a vapour retarding finishing too early.

4.5 Moisture tolerance

Simultaneously, moisture in the drying layer gets redistributed, humidifying the already dry thickness once again (Figure 4.7)!

4.5.3 Rain and rain penetration

4.5.3.1 Definition

When layers in the envelope, whose function or location require dryness, become wet by rain, rain penetration occurs. Take the thermal insulation and all layers at its inside. Capillary insulation materials may suck penetrating rain water. The absorbed rain in turn will evaporate and perhaps condense elsewhere in colder layers within the assembly. When the indoor finish gets wet, a drying surface is created that releases water vapour to the inside while consuming energy since drying absorbs the heat of evaporation and may cause evaporative cooling that way. When at the same time that drying increases the water vapour pressure inside, the possible consequences are mould and surface condensation elsewhere on the enclosure.

Additional heat demand or evaporative cooling?

Consider a room with volume V (m³) and an area A (m²) of identical exterior walls. The room is ventilated. In case all inside surfaces are air-dry, mean heating power needed becomes (Φ_H) (solar gains and internal gains not accounted for, adjacent rooms at same temperature):

$$\Phi_H = (U\, A + 0.34\, n\, V)(\theta_i - \theta_e)$$

If all inside surfaces are wet, the steady state heat balance changes to (per m²):

$$h_i(\theta_i - \theta_{si}) + \frac{\theta_e - \theta_{si}}{R_T + 1/h_e} = l_b\, g_{vd}$$

where g_{vd} is the drying flow rate, l_b the heat of evaporation and R_T the thermal resistance of the walls. At room level one gets as heat balance:

$$h_i\, A(\theta_i - \theta_{si}) + \Phi'_H + 0.34\, n\, V\, \theta_e = 0.34\, n\, V\, \theta_i$$

resulting in a new heating power demand ($U' = 1/(R_T + 1/h_e)$):

$$\Phi'_H = \left(\frac{h_i\, U'}{h_i + U'} A + \frac{l_b\, g_{vd}}{\theta_i - \theta_e} + 0.34\, n\, V\right)(\theta_i - \theta_e)$$

In case the thermal permeance U' is much smaller than the surface film coefficient h_i, that equation may be simplified to:

$$\Phi'_H = (U\, A + 0.34\, n\, V)(\theta_i - \theta_e) + l_b\, g_{vd}\, A$$

If the inside temperature remains unchanged, the extra heating needed consists of the heat evaporation, absorbed by $A\, g_{vd}$ kg of water per time unit. In case heating power is maintained equal to Φ_H, then the inside temperature drops to:

$$\theta_i = \theta_e + \frac{\Phi_H - l_b\, g_{vd}\, A}{U\, A + 0.34\, n\, V}$$

i.e. lower than the dry room. Evaporative cooling is a fact.

Rain penetration as defined above does not assume a single layer in an assembly may not become wet. Nobody, except manufacturers of non frost-resisting bricks or bricks containing unbounded salts, will worry about a brick veneer sucking rain. In cool climates, tiles turn wet in autumn. They stay wet the whole winter and only start drying in the spring. That should not trouble anyone.

4.5.3.2 Requirements

These follow from the definition given: (1) each exposed envelope part must be protected or assembled so that even under extreme rain conditions neither the thermal insulation nor the layers at its inside get humidified beyond the hygroscopic equilibrium, (2) claddings and veneers should absorb rain induced hygric and thermal loads without unacceptable functional or aesthetic degradation.

4.5.3.3 Modelling

Rain impingement

During wet weather, horizontal and inclined surfaces collect rain independent of wind speed. Vertical surfaces only get wet when rain and wind act together, giving wind driven rain. The quantity of wind driven rain hitting a vertical wall is calculated as:

$$g_{r,v} = (0.2\, C_r\, v_w\, \cos\vartheta)\, g_{r,h} \tag{4.17}$$

where ϑ is the angle between wind direction and the normal to the surface while C_r is the wind driven rain factor, a function of the location of the building, its surroundings, the spot on the façade, local detailing, etc. The product $0.2\, C_r\, v_w\, \cos\vartheta$ is called the catch ratio. Its magnitude follows from measurements or calculations that combine CFD for the wind field with droplet trajectory tracing (see Chapter 1, Figure 1.16). Catch ratios are shown to be highest at corners (see Figure 4.8 and Table 4.2) and up the façades that face the main wind direction.

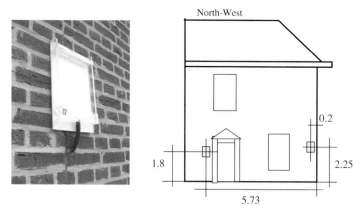

Figure 4.8. End of the row house in a residential neighbourhood, measuring the wind driven rain load.

4.5 Moisture tolerance

Table 4.2. End of the row house. Measured catch ratio north- and south-west at a height of 2.25 m, 0.2 m from the corner, and a height of 1.8 m, 5.73 m from the corner, catch ratio north-east.

Location	NW 5.73/1.8 m	NW 0.2/2.25 m	SW 0.36/2.25 m	SW 6.53/1.8 m	NE 1.72/2.25 m	NE 4.38/2.7 m
Catch ratio	0.016	0.26	0.23	0.047	0.008	0.014

Rain control

Geometrical shielding is very efficient through difficult for roofs but easily done above and in facades. If shielding is excluded then one should play with the triplet drainage, storage and transmission (Figure 4.9). Drainage happens along the exterior plane and between layers in free contact, on condition that other forces than suction transmit the rain water (gravity, wind). Storage indicates that capillary layers suck rain water and retain it.

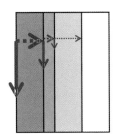

Full arrow: = drainage
Gray: = storage
Dotted arrow: = transmission

Figure 4.9. Rain control: drainage, storage and transmission.

Exterior surface, drainage and buffering

Wind-driven rain humidifies exterior layers in two-steps. First the sprinkled layer sucks the droplets. Then, once the exterior surface is capillary saturated, water droplets form. For a while these droplets are retained by adsorption before running off as a fingering water film. That results in water collection and capillary suction on surfaces below the sprinkled ones.

How long suction takes (t_f), depends on the wind driven rain intensity ($g_{r,v}$) and the water absorption coefficient of the exterior layer:

$$\frac{A}{2\,g_{r,v}^2} \leq t_f \leq \frac{\pi^2\,A}{16\,g_{r,v}^2} \qquad (4.18)$$

If the water absorption coefficient is large and wind driven rain intensity is low, it takes a while before the water film forms, allowing the exterior layer to act as buffering volume. If the water absorption coefficient is small, then film formation is fast, turning the cladding into a drainage plane. Bricks for example have typically large water absorption coefficients (0.2 to 1 kg/(m² · s$^{0.5}$)). A brick veneer therefore acts as a buffering volume, limiting moisture load on joints, façade details and others. Conversely the water absorption coefficient of concrete, concrete blocks, sand-lime stone, water-repellent stuccoes, paints and timber claddings is low (≤ 0.02 kg/(m² · s$^{0.5}$)). A façade finished that way will act as a drainage plane. Water-tight materials such as glass, plastics, metal and bituminous layers don't suck at all. They can generate important water loads below.

At first sight, run off along non-sucking sprinkled surfaces should see increased flow top down and constant flow once below the sprinkled zone. If, as is the case for high-rises, the sprinkled zone may be large, heavy wind driven rain could generate important run-offs. However that is not shown by observation. The explanation is that only the top of a high-rise catches much rain, while friction, obstruction by façade relief and evaporation impede and minimize run off.

Exterior surface, transmission

Capillarity mobilizes buffering. However, when sucking contact exists between successive layers, rain water can be transmitted by capillarity. In all other cases, gravity flow, pressure flow and kinetic energy is needed as intermediary. Gravity intervenes when run-off creates water puddles above facade reliefs or fills leaky joints, which empty through wide pores and cracks. Wind squeezes run-off through cracks wider than 0.5 mm. Kinetic energy is doing so when yawning fissures, cracks and joints are present. Once a crack or leak gets filled with water and when more run-off collects at its entrance than transmission removes, leakage flow may be estimated to equal:

$$g_w = -\frac{\rho_w b^3}{12\eta} \, grad \, P_w \qquad (4.19)$$

where η is the dynamic viscosity of water (0.00015 kg · s/m²), b the width of and grad P_w the pressure gradient along the fissure, crack or joint. In case of gravity, P_w equals the product $\rho_w g z$, with g the acceleration by gravity (9.81 m/s²) and z the height of the water column intervening.

4.5.3.4 Consequences for the envelope

When designing the envelope, rain control could be realized by (1) shielding, (2) exterior surface drainage (one step control), (3) buffering, (4) combining drainage with buffering and limited transmission (two step control).

Shielding

The principle is simple. What is higher shields what is lower. From the top, the first shield encountered is the cornice: protruding gutter and drain outside the façade. The same for sills and wall covers: protruding, a slope towards and past the facade's exterior surface, weather moulding below. The same goes for window profiles (Figure 4.10).

One step

The outside surface acts as drainage plane. To function properly the finish must be watertight, water-repellent or fine-porous. Watertight finishes neither buffer nor transmit water, except when leaking. Water repellent finishes have contact angles close to but below 90°. Water repellent stucco anyhow must be sufficiently thick to avoid the moisture front touching the substrate. How thick (d_{pl}), follows from $d_{pl} = B \sqrt{T}$ with B the water penetration coefficient and T rain duration. Thin stucco should thus have a lower water penetration coefficient than a thick one. As $B \approx A/w_c$ that demands a low water absorption coefficient and high capillary moisture content. Fine-porous finishes finally prohibit the substrate from sucking much water. However such a finish is quite vapour-retarding.

One drawback of a one step control is its damage sensitivity. Once perforated or cracked, rain penetration becomes unavoidable, the more the rain barrier also acts as a wind barrier.

4.5 Moisture tolerance

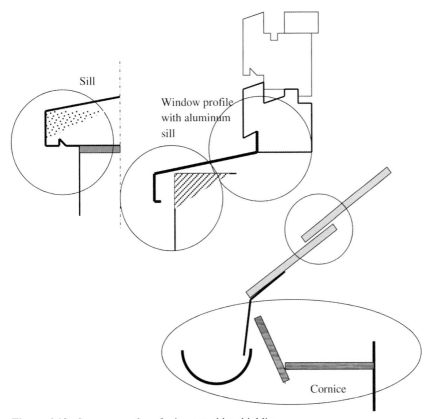

Figure 4.10. Some examples of rain control by shielding.

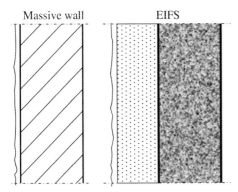

Figure 4.11. One step control, two examples.

Buffering

In buffering, drainage outside is combined with moisture storage in the wall. To function properly the wall should be thick enough to keep the moisture front far from the inside surface even after a long-lasting rain period. In cool climates 'long lasting' means some 30 hours of continuous rain while 'far' stands for not passing the wall's midline.

Example

Take a wall in cellular concrete. With a water absorption coefficient of 0.08 kg/(m² · s^0.5) and a capillary moisture content of 350 kg/m³, the thickness needed is :

$$d = 2\left(\frac{A}{w_c}\right)\sqrt{T} = 2\left(\frac{0.08}{350}\right)\sqrt{2.5 \times 24 \times 3600} = 0.21 \text{ m}$$

In brick-laid walls the joints act as short-circuits, which is why buffering brick walls are built 1 : 1/2 stone thick. That way a continuous mortar layer is created which acts as a rain stop (Figure 4.12).

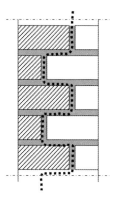

Figure 4.12. Rain control by buffering.

Two steps

A two step system contains two drainage planes, the outside surface and the backside of the rain barrier. Transmission of rain from the backside to the layers behind is halted by an air cavity or water repellent, highly air permeable insulation layer, that in combination with an airtight inner part. Rain and air/wind tightness are split that way. Many different rain barriers can be used:

- A buffering veneer wall with tray and weep holes to the outside down the second drainage plane. A cavity wall with brick veneer is an example of that (Figure 4.13).
- All kinds of cladding systems such as tiles, slates, façade elements with open joints, window profiles.
- Non porous tight exterior layers such as metallic cladding systems.

Wind and air-tightness of the inside layer is of paramount importance. That holds for an inside leaf of a cavity wall, roof pitches below the tiles, the joint seals inside and the inside weather strip inside between sash and frame of operable windows. In view of its three-dimensional nature the cavity behind the second drainage plane can never negate all wind pressure differences across the rain barrier. Detailing must account for that. In extreme cases, one should compartmentalize the cavity.

The major advantage of a two steps rain control is its damage insensitivity. A crack or perforation of the rain barrier does not negate its control function. A problem for professionals is understanding that at the inside the air plus wind barrier demands more care than the outside

4.5 Moisture tolerance

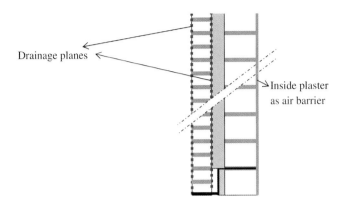

Figure 4.13. Filled cavity wall as two step rain control.

rain barrier. Two step joint seals especially suffer from that lack of knowlege. The best seal should sit inside. Most professionals put it outside.

4.5.4 Rising damp

4.5.4.1 Definition

The term 'rising damp' is used when walls get wet from below. For that to happen the whole wall or some layers must be capillary and contact phreatic water, below grade capillary moisture, below grade sink water or run-off rain water. The last will even cause rising damp at higher floors. It is also not just a massive wall problem. Cavity side run-off in a cavity wall may be sucked from below by the inside leaf.

4.5.4.2 Requirements

Rising damp must be avoided by all means as large surfaces get wet and stay wet. The drying surfaces indoors created that way have as consequences a higher heating consumption, possible evaporative cooling, increased vapour release, high relative humidity and accrued mould risk. Moisture content in the wall also reaches high values ($w_{cr} \leq w \leq w_c$) resulting in plaster, wall paper and paint damage. In many cases rising damp transports dissolved salts. Once in the wall, crystallization and hydration increases damage, while a salt-loaded wall becomes so hygroscopic that drying is not possible anymore.

4.5.4.3 Modelling

How high dampness will rise depends on the balance between capillary suction and evaporation. In contact with the phreatic surface or run-off, suction starts from a water plane whereas in contact with below grade capillary water it starts from a wet layer. In that case, the suction potential of wall and wet ground define what will happen. If larger for the wet ground, no rising damp will be seen. If instead it is larger for the wall, rising damp will develop, although at a lower rate and lower moisture content than for a wall contacting a water plane.

Temperature plus relative humidity in the surroundings and the vapour diffusion resistance of the wall finish determine evaporation. The tighter the finish, the less vapour will evaporate and the higher the dampness will rise in the wall for equal suction.

A simple model clarifies that balance between suction and evaporation. In a homogeneous wall, suction speed is given by:

$$v_m = r \left(\frac{\sigma \cos \Theta}{4 \eta}\right) \left(\frac{1}{h} - \frac{1}{h_{max}}\right) \qquad (4.20)$$

where h is the damp height and h_{max} the maximum damp height without evaporation, equal to $2 \sigma \cos \Theta / (r \rho_w g)$. The dynamic viscosity η of water, surface tension σ of water, contact angle Θ and mean radius r of the capillary pores in the material in turn determine the water penetration coefficient:

$$B = \sqrt{r \frac{\sigma \cos \Theta}{2 \eta}} \qquad (4.21)$$

That equation allows writing $r \sigma \cos \Theta / (4 \eta)$ as $B^2/2$. With $B = A/w_c$ and the sucked water flow G_m equal to $v_m w_c d$ where d is the tickness of the wall and w_c the capillary moisture content of the material, rising damp flow and maximum damp height without evaporation become:

$$G_m = \frac{A B d}{2} \left(\frac{1}{h} - \frac{1}{h_{max}}\right) \qquad h_{max} = \frac{5.5 \cdot 10^{-4}}{B^2} \qquad (4.22)$$

At equilibrium damp height, sucked flow G_m must equal the evaporated flow G_d, or:

$$G_m = G_d = h \left[\frac{p_{sat,1}(1-\phi_1)}{1/\beta_1 + Z_1} + \frac{p_{sat,2}(1-\phi_2)}{1/\beta_2 + Z_2}\right] \qquad (4.23)$$

with β_1 and β_2 the surface film coefficients for diffusion and Z_1 and Z_2 the diffusion resistances of the wall finish at both sides. Combining (4.22) with (4.23) gives:

$$h^2 - \left[\frac{A B d}{2 h_{max} (g_{d,1} + g_{d,2})}\right] h + \frac{A B d}{2 (g_{d,1} + g_{d,2})} = 0$$

a quadratic equation with positive root:

$$h = \frac{A B d}{4 h_{max} (g_{d,1} + g_{d,2})} \left[\sqrt{1 + \frac{4 h_{max}^2 (g_{d,1} + g_{d,2})}{A B d}} - 1\right] \qquad (4.24)$$

Moisture profile in the wall follows from:

$$\frac{d}{dz}\left(D_w \frac{dw}{dz}\right) = \frac{g_{d,1} + g_{d,2}}{d}$$

If moisture diffusivity (D_w) is assumed constant and the drying rate identical along the damp height (h), than the solution is ($z = 0: w = w_c; z = h: w = w_{kr}$):

$$w = \left(\frac{g_{d,1} + g_{d,2}}{2 d D_w}\right) z^2 - \left[\frac{w_c - w_{kr}}{h} + \frac{(g_{d,1} + g_{d,2}) h}{2 d D_w}\right] z + w_c \qquad (4.25)$$

4.5 Moisture tolerance

i.e. a parabola with highest moisture content at the contact plane and lowest moisture content at the damp height. For a moisture diffusivity, function of moisture content ($D_w = f(w)$), the profile changes from parabolic to more rectangular with a smaller difference over the damp height. Nonetheless deceasing moisture content along the damp height may be seen as a rising damp characteristic, although dissolved salts can obscure that picture. Figure 4.14 gives the damp height in an inside partition wall as function of the product of the water penetration coefficient and the water absorption coefficient (AB) for mean indoor conditions ($\theta_1 = 20\,°C$, $\phi_1 = 50\%$) and the wall basement contacting phreatic water.

With increasing value of the product AB (material less fine-porous, pore volume smaller) damp height first climbs to a maximum to drop thereafter. Very limited heights are typical for materials with ultra-fine pores only or coarse pores only. Figure 4.15 gives the damp height

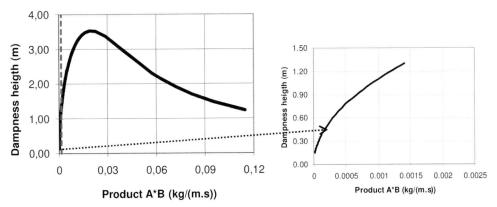

Figure 4.14. Unpainted inside wall without joints: damp heigth. Wall 30 cm thick, 20 °C, 50% RH, surface film coefficient for diffusion $2.6 \cdot 10^{-8}$ s/m. The right figure shows the part of the curve for bricks (dotted grey line at the left (A: capillary water sorption coefficient, B: capillary water penetration coefficient)).

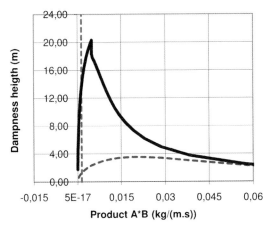

Figure 4.15. Painted inside wall without joints: maximum damp heigth. Identical boundary condition as Figure 4.14. The paint has a diffusion thickness of 1 m. Comparison with the unpainted wall (dotted grey line).

after the partition wall is painted (extra diffusion thickness 1 m). Hindered drying increases it with a factor of 6! With other words, although still done, hiding the damp surface behind a vapour retarding cladding is not a solution. After a few years, the moisture front will reappear, this time above the finish.

Departure from the simple model is seen when the wall contacts capillary moist ground, in brick-laid walls and when dissolved salts are transported into the wall. Capillary moist ground can be modelled by lowering the water absorption factor and the capillary moisture content of the wall. How much lower, demands testing. Nonetheless the two together clearly limit the damp height and keep moisture content in the wall below capillary.

In brick-laid walls mortar joints may act as capillary stops. If coarser porous than the stone, they will hardly suck moisture from the stones. In the opposite case, the stones will poorly pick-up moisture from the joints. Of course, joints may also act as continuous suction paths and turn wet without humidifying the stones. Nonetheless they typically retard moisture uptake and limit the wet height to a few layers of stones, though the plaster may act as short-circuit and get wet over considerable height. Salts finally facilitate rising damp and mitigate the joint effect. High salt concentration at the wall's surfaces and at the moisture front could lead to such high hygroscopicity that even after curing, high moisture content will persists.

4.5.4.4 Consequences for the building fabric

New construction

The measure is simple: provide a watertight foil in all walls just above grade or just above where run-off rain water is collected (as is the case in cavities). See Figure 4.16.

Figure 4.16. Cavity wall, inside leaf.
Watertight foil above run-off collection in the cavity.

Retrofit

In case problems with rising damp surface, measures to be taken are:

Avoid suction

Possible by inserting watertight foils or steel plates above grade in all damp walls, by filling the pores above grade by injection or infusion or turning the pore walls above grade into water repellent surfaces by injection or infusion. This first strategy is by far the most efficient. Failure is possible in case the damp walls are loaded with salts.

Change suction into repellence

In theory, electro kinesis accomplishes that. In fact, during capillary suction, an electrical potential exists over the water layer contacting the pore walls. If its direction can be inverted, suction should turn into repellence, pressing the water out of the pores. However inversion requires energy. When applying the passive electro-kinesis, an electric conductor is embedded above grade in the walls and coupled to the earth using electrodes. Until broken, corrosion of the contacts between conductor and electrodes delivers the energy needed. With active electro-kinesis, energy supply is guaranteed by maintaining a less then two volt electrical potential difference over the conductor. Helps with corrosion but consumes electricity!

Nonetheless none of the two guarantees drying, among others because solved salts makes the adsorbed water electrically conducting.

Activate drying

Best remedy is removing all finishes that retard drying. Drying pipes above grade has no effect as the air they contain turns 100% humid, limiting the drying area to the section the pipes have, with a flow depending on the surface film coefficient for diffusion at the aperture and the temperature and vapour pressure outside:

$$G_{v,d,pipe} = \beta \left(p_{sat,se} - p_e\right) \left(\frac{\pi d_{pipe}^2}{4}\right)$$

The worst thing todo is to cover the damp walls with a vapour retarding finish. Prohibited drying in that case allows dampness to climb higher.

Problem walls, among them former stable walls or walls touching former cesspools, should be investigated for salt presence first. Too many times problems salts cause are mistaken for rising damp and cured with the measures listed above. That doesn't help. The only solution is hiding the salt-laden walls behind a non capillary insulation or a brick veneer.

4.5.5 Water heads

4.5.5.1 Definition

The term water heads applies to moisture transport having aeraulic or hydraulic pressure differences as a driving force. Air pressure gradients are generated by stack, wind and fans. In general the differentials in that case are small (from 2 to ±1700 Pa). Hydraulic pressure gradients intervene below phreatic, in swimming pools, water reservoirs and water wells, or in spots below grade where sinking rain water accumulates. Pressure differences are typically large and quite stable. A 10 cm deep water pond already generates pressures of 1000 Pa, while 1 m of water head gives 10 000 Pa.

4.5.5.2 Requirements

Wetting, filtering through as well as discharge should not disturb the function of the spaces that have walls subjected to external pressures. Thermal insulation in any wall assembly should stay dry, while in walls containing a water pressure barrier, humidification should stop there.

4.5.5.3 Modelling

Pressure flow is calculated using Darcy's law for saturated water displacement. The resulting moisture content in open-porous materials exceeds capillary with substantial increase in frost damage probability as one of the risks.

4.5.5.4 Consequences for the building fabric

Two tracks can be followed: (1) removing the pressure difference, (2) caring for water-tightness of the walls.

Removing pressure differences

One possibility below phreatic or where sinking rain causes pressure heads is to use drains at the water side of the below grade construction (Figure 4.17). The system functions thanks to the high pressure head in the ground, the pressure head zero in the porous pipes and the high water conductivity around these pipes. A drain changes a horizontal phreatic surface into a profile comparable with the temperature course in the ground after a sudden drop in surface temperature. The moisture flow per meter run then equals:

$$G_w = \sqrt{\frac{w_b \, k_{w,b}}{\pi \, t}} \, \Delta P_w$$

where w_b is the saturation moisture content and $k_{w,b}$ the water conductivity of the ground. Drainage pipes function properly as long as the driving water head ΔP_w is large enough, the ground not too permeable and the pressure head in the pipes close to zero.

The water head between pipes and phreatic surface should not exceed 0.5 m or, should be temporary for example caused by rain water sinks.

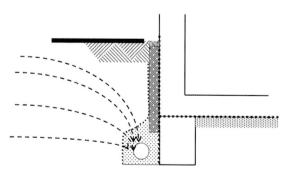

Figure 4.17. Drainage.

4.5 Moisture tolerance

Building element watertight or water retarding

Damming constructions should be watertight. If they contain an insulation layer, the watertight layer must be located at the water head side of that layer.

If usage of the below grade spaces allows, an alternative is to retard the water permeation across the enclosing walls to a level the equilibrium plane with vapour removal by diffusion does not reach the inside surface. For that to happen the following relation should hold (single-layered wall, steady state):

$$k_w \, \mu = \frac{p_{sat,x} - p_i}{\Delta P_w} \left(\frac{d-x}{N \, x} \right)$$

where μ is the vapour resistance factor of the material, x the location from inside of the equilibrium plane in the wall, ΔP_w the water head the wall is subjected to, $p_{sat,x}$ saturation vapour pressure in x and p_i vapour pressure indoors. As materials with low open porosity are vapour retarding (high vapour resistance factor of μ), the formula presumes low water permeability. That imposes severe restrictions in terms of total open porosity, pore diameters and pore tortuosity.

4.5.6 Hygroscopic moisture

4.5.6.1 Definition

'Hygroscopic' is used to denote moisture content in a material in equilibrium with relative humidity (ϕ) in the surroundings. The sorption curve of a material is S-shaped, with quite a strong increase at low relative humidity, a platform between 30 and 80% and a steep increase again at higher relative humidity. Also a hysteresis exists between sorption and desorption, with the desorption curve giving higher moisture contents for a same relative humidity (Figure 4.18).

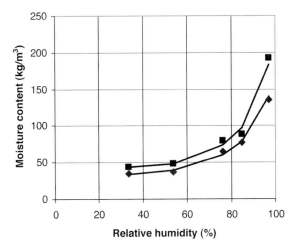

Figure 4.18. Sorption/desorption.

4.5.6.2 Requirements

The presence of hygroscopic moisture reflects an equilibrium state. However, a too high or too low relative humidity and related too high or too low hygroscopic moisture content causes problems. Timber, textiles and paper for example shrink substantially once the relative humidity drops below 30%, resulting in cracking of massive pieces, damage to paint, etc. Too high relative humidity in turn accelerates biological activity on material surfaces resulting in mould attack and dust mite overpopulation.

4.5.6.3 Modelling

Evaluating cracking risk demands combined heat-moisture-stress-strain models. If mould attack is the annoyance considered, fixing the temperature factor needed to avoid germination imposes a two step approach:

1. Calculate per month the mean saturation pressure that should not be exceeded anywhere inside on the envelope using the most likely vapour pressure indoors (p_i) as a starting point. Transpose that saturation pressure into a dew point temperature (θ_d):

$$p_{sat,si} = p_i/0.8 \qquad p_{sat,si} \rightarrow \theta_d \left(p_{sat,si} \right)$$

2. Take as acceptable lower limit of the monthly mean temperature factor inside anywhere on the envelope:

$$f_{h_i} \geq \frac{\theta_d \left(p_{sat,si} \right) - \theta_e}{\theta_i - \theta_e}$$

4.5.6.4 Consequences for the building fabric

The lowest temperature factor found fixes the value to be respected when designing the building envelope. Especially potential thermal bridges must be controlled using three-dimensional heat flow models with the lowest number expected at the spot considered as surface film coefficient indoors and the operative temperature at the centre of the room, 1.7 m above floor level as reference temperature inside. If everywhere on the inside surfaces a temperature factor above the lower limit is found, then it complies. If not, changes should be considered and a new control done.

4.5.7 Surface condensation

4.5.7.1 Definition

The term surface condensation is used to characterize water vapour condensing on the inside or outside surface of any envelope or building fabric part.

4.5.7.2 Requirements

These are not unambiguous. Surface condensation on single glass is annoying but without consequences for the glazing. Nonetheless soft wood window sashes moistened by condensate run-off may rot, while the condensate deposited on aluminium window profiles may get sucked by the reveals, causing mould to develop. Surface condensation on any kind of double glazing

4.5 Moisture tolerance

is a signal vapour release inside is quite high or ventilation is insufficient. At the same time, mould may start germinating on envelope parts that are somewhat warmer at their inside surface than the glass. Alternate humidification by surface condensation followed by drying may push the mean relative humidity at a surface beyond 80%, the mould limit. Long lasting surface condensation will also result in high moisture content in the finish with potentially annoying consequences.

The requirement often advanced is that on a daily basis one may not have accumulating condensation anywhere on inside surfaces. Some condensation is thus acceptable as long as even on days with the daily mean outside temperature equal to the design temperature for heating, humidification and drying are in balance.

4.5.7.3 Modelling

Modelling starts with gathering information on vapour pressure in the air on the side surface condensation is expected. If that value exceeds surface saturation pressure, condensate will be deposited, with a flow rate of:

$$g_c = \beta(p - p_{sat,s}) \approx 7.4 \cdot 10^{-9} \, h_c (p - p_{sat,s}) \tag{4.26}$$

where β is the surface film coefficient for diffusion and h_c the convective surface film coefficient. Heat transfer is:

$$q = g_c \, l_b \approx 2.5 \cdot 10^6 \, g_c$$

In case of a 1 m² large flat part, the heat balance at the condensing side becomes:

$$-2.5 \cdot 10^6 \, g_c = h_1 (\theta_1 - \theta_s) + \frac{\theta_2 - \theta_s}{1/U - 1/h_1}$$

resulting in the following surface temperature:

$$\theta_s = \frac{h_1 \theta_1 + U' \theta_2 + 2.5 \cdot 10^6 \, g_c}{U' + h_1} \quad \text{with} \quad \left(U' = \frac{1}{1/U - 1/h_1} \right) \tag{4.27}$$

In those formulas suffix 1 denotes the condensation side and suffix 2 the other side whereas θ_1 and θ_2 are the reference temperatures at both sides, operative indoors and sol-air outdoors. If one truly wishes to quantify a surface condensation problem, the system (4.26) (4.27) should be solved iteratively. If after a condensing period, the surface warms up and its vapour saturation pressure moves beyond the vapour pressure in the air close by, drying starts and continues until all condensing moisture is gone. Also drying is calculated combining the system of Equations (4.26) (4.27) with the following mass balance:

$$\int_0^t g_d \, dt \leq m_c \tag{4.28}$$

Surface condensation outdoors has as its cause under-cooling or sudden changes in weather, from cold to warm and humid. The results are icy roads, frost on automobiles and wetness or rime on roofs. In moderately rainy climates the wetness so deposited during a whole winter compares well with the amount delivered by rain.

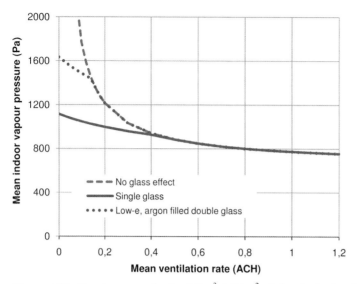

Figure 4.19. Sleeping room, $4 \times 4 \times 2.5$ m³, 2.52 m² of glass in the façade wall. Monthly mean inside vapour pressure in case 2 persons sleep there during 8 hours a day. Monthly mean outside temperature and vapour pressure 2.7 °C respectively 663 Pa, monthly mean inside temperature 13.9 °C.

Surface condensation indoors results from too high vapour pressures or too low surface temperature, the first being caused by abundant vapour release and/or insufficient ventilation, the second by poor thermal insulation and/or details leaving ample room for thermal bridging. Of course at the same time condensation dries the indoor air and stabilizes the indoor vapour pressure at a level that depends on the thermal resistance of the preferential condensing surface, typically the windows. See Figure 4.19.

4.5.7.4 Consequences for the building fabric

These are quite clear: insulate well and avoid thermal bridges with too low temperature factors when designing the building envelope.

4.5.8 Interstitial condensation

4.5.8.1 Definition

Interstitial condensation stands for water vapour accumulating as moisture inside a building part. For a long time, only water vapour released inside was considered a cause. However, also built-in moisture, hygroscopic moisture, rising damp and absorbed or leaking rain water may evaporate and be deposited as moisture elsewhere in a part each time a temperature gradient exists. Without such gradient, capillarity and hygroscopicity will redistribute the moisture present but condensation due to 100% relative humidity somewhere is excluded. Driving forces causing interstitial condensation are diffusion and convection, with convection usually taking the lead. Design flaws and/or workmanship errors have to be blamed when problems arise, although damage caused may stay hidden for long periods of time.

First modelling is discussed and then the requirements are listed.

4.5.8.2 Modelling

The models used have evolved a lot. In 1958–1959, Glaser advanced a simple method to control interstitial condensation in cold store walls. As most materials used in cold stores are non-capillary and non-hygroscopic, as the walls are airtight and as the temperature and vapour pressure difference are close to constant, steady state vapour diffusion was the only moisture transport mechanism involved. Thus, his method gave quite good results for that type of walls. That motivated other authors to extend the Glaser rationale (called dew point method in North America) to building envelopes, which was a step too far. None of the conditions fulfilled by cold store walls are typical for building envelopes. Among these conditions are transient loading, air-tightness that is not guaranteed, many capillary active and hygroscopic materials, substantial gravitational force, etc. Although the Glaser approach was charming in its simplicity, in the years that followed, it was upgraded by using more realistic boundary conditions and introducing a simple methodology for including capillarity as redistribution force. Others went for more complete transient heat-air-moisture models, transposing the governing PDE's into numerical computer software.

During the sixties and seventies the Glaser method figured as a reference when judging moisture tolerance of envelope parts. Interstitial condensation by diffusion thus became the number one moisture risk, which caused a vapour barrier mania. This of course turns the world upside down. In fact, the assumptions behind the method are too far reaching for use as a unique instrument to judge moisture tolerance of envelope parts except for one case: an airtight sandwich with non-hygroscopic inside finish, airtight insulation layer perfectly filling the space between finish and a non-hygroscopic, non-capillary outside cladding.

Further discussion is limited to the upgraded Glaser and the convection/diffusion method.

Glaser's method

Assumptions

Geometry

1. Building parts composed of plan-parallel layers.

Moisture flow

2. Moisture migrating as vapour only. Acceptable as an assumption at relative humidity below 80–90% but false in capillary materials above critical moisture content.
3. Diffusion as sole driving force. No convective transport. Valid as long as building assemblies and their layers are airtight. Testing showed this to be untrue.
4. Hygroscopicity is not considered, eliminating hygric inertia as factor, making the vapour balance steady state under all circumstances. Softened as an assumption by using monthly mean climate values.
5. Vapour resistance factor of any material constant. Does not fit with reality but makes calculation easy. That restriction may be softened by using different vapour resistance factors depending on where a layer sits: inside or outside of the thermal insulation.

Heat flow

6. Heat flow by equivalent conduction only. Excludes any form of enthalpy transport, included latent heat. For the last to be true, interstitial condensate flow rate must be minimal.

7. Thermal inertia not intervening. Heat flow is steady state, even under varying boundary conditions. Softened as an assumption by using monthly mean climate values.
8. Thermal conductivity of all materials a constant. Does not fit with reality but makes calculation easy.

Calculations

Method based on the equivalent outside temperature for condensation and drying

Boundary conditions

Assume slope, orientation and shortwave absoptivity of the outside surface known. That allows calculating the equivalent outside temperature for condensation and drying [θ_{ce}^*] as explained in Chapter 1.

Take the monthly mean outside vapour pressures for the location considered and fix inside temperature and inside vapour pressure. If no measured data are available, use:

Monthly mean indoor temperature (°C):

Building type	Annual mean	Annual amplitude
Dwellings, schools, office buildings	20	3
Hospitals	23	2
Natatoriums	30	2

Monthly mean indoor vapour pressure:
- Calculated starting from the monthly mean value outside, the monthly mean ventilation rate and the monthly mean vapour release indoors.
- If neither ventilation rate nor vapour release are known, take the upper boundary of the indoor climate class the building belongs to.
- The step from annual mean and annual amplitude to monthly value looks like: monthly value = annual mean + annual amplitude · $C(t)$, with $C(t)$ the time function used to calculate the monthly mean equivalent outside temperature for condensation and drying.

Dry building part
- Knowing thickness, thermal conductivity and vapour resistance factor for each layer, calculate the thermal and diffusion resistances.
- Redraft the wall in a [vapour pressure (p_{sat}, p) versus diffusion resistance (Z)]-axis system and calculate temperature and vapour saturation pressure gradient (p_{sat}) for the coldest month of the year. In most cases, fixing saturation pressure in each interface and linking successive values with straight lines suffices. Introduce the vapour pressure straight line. If not intersecting saturation pressure, no interstitial condensate is deposited. If intersecting, interstitial condensation is a fact during the coldest month.

If condensation is present, then:
- Trace the ingoing vapour pressure tangent to the saturation curve through the vapour pressure inside (interface $Z = Z_T$) and the outgoing vapour pressure tangent to the saturation curve through the vapour pressure outside (interface $Z = 0$). If the points of contact coincide, that is the condensation interface. If they do not coincide, a tangent scan from highest to lowest contact point allows deciding if additional condensation interfaces exist or if both delimit a condensation zone.

4.5 Moisture tolerance

- Calculate per condensation interface the difference in slope between the in and outgoing tangent. The number found gives the amount condensing:

    $m_c = 86\,400\, d_{mo}\, g_c$

    where g_c is the condensation flow rate in kg/(m² · s) and d_{mo} the number of days in the month.

- In case of one condensation interface, calculate the amount condensing or drying assuming saturation pressure in that interface per month. Add the values found. If condensing becomes drying and the amount drops below zero, set it at zero and use vapour pressure in the condensation interface for the months that follow until intersection returns and condensation restarts. When drying to zero occurs on an annual basis, long lasting moisture accumulation is excluded. When not, an annual accumulation will occur until a limit state is reached.

- Two or more condensing interfaces makes counting more difficult. The annual result is quickly controlled by calculating the algebraic sum of the twelve slopes of the in and outgoing tangent assuming vapour saturation pressure at their contact points. However assessing allowance of the winter maximums demands a much finer scan with condensation and drying balances per condensing interface.

- If the contact points of the in and outgoing tangent delimit a condensation zone, then the curved saturation line allows calculating moisture distribution in that zone.

Building part with wet layer

When a layer in an assembly is wet by built-in moisture, rain or rising damp, calculation becomes more complex. Vapour pressure in that layer is assumed to equal saturation. Evaporation happens in the interfaces with adjacent layers. In the [vapour pressure (p_{sat}, p)/diffusion resistance (Z)]-axis system, connect the interface inside of the wet layer with the vapour pressure inside and the interface outside of the wet layer with the vapour pressure outside with straight lines. If these intersect the vapour saturation curve, condensation of humidity from the wet layer happens in the inside or outside part of the assembly. The next steps reflect the dry wall case: replacing the straight line by the tangents through the interface of the wet layer intervening and the those through the in- or outside vapour pressure, points of contact fixing the condensation interface or condensation zone, etc.

Applying the European standard

Contrary to the methodology just explained, the European standard uses monthly mean outside air temperatures, which fits for surfaces protected from sun and under-cooling.

Diffusion/convection method

Assumptions

Geometry

No difference with Glaser.

Moisture flow

1. Moisture migrating as vapour only.
2. Exfiltration/infiltration and diffusion as driving forces.
3. Hygroscopicity not intervening.
4. Diffusion resistance factor a constant.

Heat flow

6. Heat flow the result of equivalent conduction and enthalpy flow.
7. Latent heat release/adsorption too small to be considered.
8. Thermal inertia not intervening.
9. Thermal conductivity a constant.

Calculations

As infiltration/exfiltration short-circuits hygric and thermal inertia, calculation is not performed year round on a monthly basis but for a representative cold week only. Such a week should be defined for each location. For the cool climate of Uccle it looks like:

Temperature	Relative humidity	Resulting radiation, hor. surface	Thermal surface film coefficient	Wind velocity (free field, 10 m)
°C	%	W/m^2	W/(m^2·K)	m/s
−2.5	95	−30	17	3.8

Slope, orientation, composition, radiation properties of the outside surface, thickness, air permeance, equivalent thermal conductivity and equivalent vapour resistance factor of all layers must be known. First the air flow rate through the part is quantified. For that to be possible, wind, stack and fan induced air pressure differences have to be known. Then thermal and diffusion resistance of the composing layers is calculated. Further evaluation is as follows:

Boundary conditions

Fix the weekly mean inside temperature and vapour pressure inside. If measured data are missing or best guesses are not available, take the values for the Glaser method.

Dry building part

Calculate temperature, saturation and vapour pressure in the building part. Vapour pressure now curves exponentially in the [vapour pressure/diffusion resistance]-axis system. If that exponential exceeds saturation pressure, interstitial condensation is a fact. If not, no condensate will be deposited.

If condensation occurs, then:

- With vapour pressure inside and vapour pressure outside as starting points, trace the in and outgoing tangent exponentials to the saturation curve. When the points of contact coincide, condensate is deposited in that one interface. If they don't coincide, a tangent scan starting at the inner point of contact and jumping from interface to interface till the outer point of contact will allow fixing all intermediate condensation interfaces or condensation zones.
- Calculate per condensation interface or over the condensation zone the flow deposited (G_c in kg/(m^2·week)).

Building part with wet layer

If a layer in an assembly is wet by built-in moisture, rain or rising damp, then calculation becomes more complex. The problem is solved using the same methodology as explained for the Glaser method.

4.5.8.3 Requirements

Allowance is expressed in terms of amounts of condensate permitted per square meter of interface, assuming the condensate between or against capillary layers is sucked. Examples of allowance are (d thickness in m, w_{cr} critical moisture content in kg/m³, r density in kg/m³):

Belgium	Allowed kg/m²
1. Annually accumulating deposit	Limit state
Permitted when deposited against capillary active stony materials or when problem-free drainage is possible	
– Stony materials that are frost resisting and have no vapour-tight finish at the outside	$m_c \leq w_{cr} d$
– Stony materials that are frost resisting and have a vapour-tight finish at the outside	$m_c \leq 0.05\, w_{cr} d$
– Stony materials that are not frost resisting	$m_c \leq 0.05\, w_{cr} d$
2. No annually accumulating deposit	Winter maximum
– Stony materials that are frost resisting and have a vapour-tight finish at the outside	$\leq 0.05\, w_{cr} d$
– Stony materials that are not frost resisting	$\leq 0.05\, w_{cr} d$
– Wood and moisture proof wood-based materials (plywood, particle board, fibre board, OSB)	$\leq 0.03\, \rho\, d$
– Non moisture proof wood based materials	≤ 0.05
– Non-capillary foils with slope (s) below 15°	$\leq 0.4 - 0.3\, s/15$
– Non-capillary foils with slope (s) beyond 15°	≤ 0.1
– Insulation materials showing hardly any moisture sensitivity (mineral fibre, EPS, XPS, PUR)	$\leq \max\left[\dfrac{12.5\,\lambda}{U_0(0.6-\lambda)}, 0.5\right]$
Switzerland	**Allowed kg/m²**
1. Annually accumulating deposit	
Not allowed	
2. No annually accumulating deposit	Winter maximum
– Stony materials	≤ 0.8
– Wood and moisture proof wood-based materials (plywood, particle board, fibre board, OSB)	$\leq 0.03\, \rho\, d$
– Non-moisture proof wood based materials	0
– Non-capillary foils with any slope	0.02
– Insulation materials showing hardly any moisture sensitivity (mineral fibre, EPS, XPS, PUR)	$\leq 10\, d$

4.5.8.4 Consequences for the building fabric

Interstitial condensation must be avoided primarily by a good air-tightness of any building part. Secondly, the assembly should be designed in a way the diffusion resistances decrease from the warm and more humid side to the colder and drier side. An alternative is providing a vapour retarding layer at the warm side of the thermal insulation. The neccessary diffusion resistance has to follow from calculation.

The second step introduces main differences between cool/cold climates and warm, humid climates. In cool/cold climates warm and more humid means the inside, while in warm, humid climates the widespread air conditioning inside turns the outside into 'warm and more humid'. That moves the location of a vapour retarder from the inside of the insulation to its outside. In rainy climates the second step may also conflict with rain tightness and mechanical quality of the outside cladding. In case of a vapour-tight finish outside, vapour retarders should not enclose thermal insulation together with moist layers. Further-on, one must take care with rain buffer layers at the outside of the insulation as sucked rain water may condense against the backside of the vapour retarder and in the insulation during warm, sunny weather.

4.5.8.5 Remark

In building assemblies, what is called interstitial condensation may stay limited to changes in hygroscopic moisture content. Interstitial condensation in the sense vapour deposits as liquid droplets is only possible against non-porous layers such as metal claddings, glass and synthetic finishes. Sucking layers have to be capillary wet before condensate forms at their surface. But even then, the coefficient of secondary moisture uptake may be high enough to absorb the deposit.

4.5.9 All sources combined

4.5.9.1 Modelling

In real world situations built-in moisture, rain, rising damp, hygroscopic moisture, surface condensation and interstitial condensation all act together. Thanks to the overall heat-air-moisture models available today, judging that combined moisture response is now possible The main elements in these models are:

Geometry

Always simplified. Some use one-dimensional dummies of real building parts; others go for a two-dimensional approach which allows judging idealized details Also three-dimensional calculations still consider idealized assemblies.

Air flow. Most models apply a hydraulic circuit analogy. Some use CFD. Properties needed are: air permeability of the porous materials used, air permeance of cavities, leaks, cracks, composite layers, etc.

Moisture flow

- Moisture migrates as vapour and liquid. Below 0 °C transformation into solid state starts. Some models handle vapour transport only. That way, they simulates reality quite well as long as moisture content remains within the hygroscopic range.
- Driving forces for vapour flow are equivalent diffusion and convection. Many models only consider equivalent diffusion, which makes them unusable for air leaky assemblies.
- Driving forces for liquid transport are capillary suction, gravity and pressure heads, though most models only consider capillary suction.
- Hygroscopicity causes hygric inertia, with air transport acting as efficient short-circuit.
- Material properties: density, moisture retention curve as function of suction, water vapour permeability as function of suction, moisture conductivity as function of suction.

Heat flow

- Heat flow combines conduction, convection, radiation and enthalpy displacement.
- Latent heat plays a role.
- Thermal inertia intervenes.
- Material properties: density, specific heat capacity as function of moisture content, equivalent thermal conductivity as function of moisture content and temperature, surface properties for radiation.

Boundary conditions

Hourly or even ten minute values of all parameters defining the outside and inside climate.

Contact conditions

Suction contact, diffusion contact, mixed contact, contact with draining water.

4.5.9.2 Requirements

The requirements relate to the impact combined heat-air-moisture has on end energy consumption and durability. For energy, the increase of the thermal transmittance is the drawback considered. Durability in turn looks to unwanted physical, chemical and biological degradation.

Energy

Insulated buildings

$$\Delta U_0 \leq 0.1 \, U_0$$

Low energy

$$\Delta U_0 \leq 0.05 \, U_0$$

Durability

Physically

- No cracking that harms aesthetics, compromises air and rain tightness and accelerates dirt accumulation.
- Moisture content in layers that freeze and thaw during frost periods below the critical value for frost ($w < w_{cr,f}$) or, what is the same, a frost factor F above zero ($F = S_{cr,f} - S > 0$). Frost resisting materials that freeze may stay capillary wet. Frost sensitive materials should remain drier.
- Relative humidity in humidity resistant wood-based materials not above 95%.
- Relative humidity in salt-laden stony materials low enough to avoid damaging hydration (how low depends on the salt), moisture content low or high enough to avoid successive salt dissolution and crystallization (how low and how high depends on the salt and its concentration).

Biologically

- Relative humidity (ϕ) in interfaces between materials below the mould germination value. The following is the relation between that value and the period considered (ϕ on a scale from 0 to 1):

$$\phi = \min\{1, \quad 0.8 \cdot [1.25 - 0.074 \ln(t)]\}, \quad t \text{ in days}$$

- Depending on presumed mould probability, relative humidity in wood is limited to:

Temperature	Mould probability		
	RH ≤ 75	RH 75–90	RH > 90
20 °C	0	0–0.5	0.5–1
0–8 °C	0	0–0.1	0.1–0.8
< 0 °C	0	0–0.02	0.02–0.5

- Moisture content in wood never above fibre saturation.
- Time-averaged relative humidity in floor coverings is limited to:

Material	RH %
Synthetic carpet with mould sensitive lower layer	80
Non alkali-resisting floor covers glued on a cement screed	
Six months periods	
Composite products	90
Homogeneous synthetic carpets	85
Shorter periods	
Composite products	95
Homogeneous synthetic carpets	90
Cork floor cover without synthetic protective layer	80
Cork floor cover with synthetic protective layer	85
Linoleum	85–90
Self-levelling screed	90–95
Synthetic carpet without mould sensitive layer, needle felt	99

Chemical

Avoid premature perforation of metallic coverings. For example, weight loss by corrosion of an unprotected aluminium sheet after 4 years of exposure to the outside climate is given by (Δm in g/m^2):

$$\Delta m = 0.85 + 0.029 \cdot \text{TOW} \cdot [SO_2] \cdot [O_3] + 80 \cdot g_R \cdot [H^+]$$

where TOW is the time fraction, the sheet is wet on a scale from 0 to 1, the terms between brackets [...] are the concentrations of the gasses in the outside air (µg/m^3) and g_R is the average precipitation in m per year. If an exponential increase is assumed, weight loss is 0 for $t = 0$, and Δm is equal to the calculated value after 4 years and half of it after 1 year, then total weight loss after x years is quantifiable and comparable with the initial weight.

4.5.9.3 Where actual models fail

In theory, building parts could be evaluated in their entire complexity. In reality there are many reasons why this is not possible.

Material properties

As explained, the continuity approach used by all models assumes that something like a reference elementary volume (REV) exists. Most building materials, however, are too non-homogeneous and randomly anisotropic to be represented by identical REV's. This makes any assumption about single-valued property relationships quite obvious. These relationships are also uncertain. To give an example, the main properties defining moisture response of an open-porous material are specific moisture capacity and moisture permeability. Specific moisture capacity is the derivative of a water retention curve measured spot-wise. Differentiating is done by fitting an analytical function to the separate data points. Such a fit is always approximate, while the derivative is even more sensitive to errors than the fit! At the same time, hysteresis affects the derived moisture capacity. Moisture permeability is calculated from a measured diffusivity. That diffusivity follows from a [position/time]-scan with γ-ray, x-ray or NMR of successive moisture profiles in a sample of the material during capillary suction or drying. Most profiles show a flat part at high moisture content, a steep front between higher and lower moisture content and a flat part again at low moisture content. The weak slope and scatter in measured values at low and high moisture content makes a correct calculation of the diffusivity troublesome. The steep part tells that diffusivity increases quickly as moisture content augments. A precise relation between both however is difficult to calculate, which is why a Boltzmann transform is used, which unifies all consecutive moisture profiles into one, although with much scatter and thus, uncertainty

Geometry

The real geometry of a building assembly, including all randomly present joints, cracks, voids, leaks, air spaces, etc. is never known, which is why models always consider a virtual dummy, even when three-dimensional. That way important flow components such as air infiltration and exfiltration, buoyancy induced air rotation, wind washing, water ingress by gravity and pressure heads, water run-off and drainage may be overlooked. Typical examples of uncertain geometry are: (1) a brick veneer where each joint stands for a random system of cracks and voids, (2) settled soft insulation, (3) casual air voids between stiff insulation and timber beams, (4) cracks in inside linings, etc.

Contact conditions

Most models consider perfect hydraulic contact between layers. In reality, four types of contact are possible:

(1) Perfect hydraulic contact characterized by continuity in suction and moisture flux. Although handled as the reference, that case is the exception.
(2) A thin airspace between layers. A much more probable situation than perfect contact! Only air and vapour are transferred, except if condensation or water ingress from outside succeeds in filling the thin space. Then, gravity may activate water run off and drainage. The airspace may also belong to a network of cracks, joints, voids and spaces that cross the assembly. This will activate air flow, sometimes with detrimental effects on moisture response and thermal integrity.

(3) Natural contact characterized by a random distribution of air voids and spots of perfect hydraulic contact. All inconveniences of (2) may be active here.
(4) Real contact with inter-penetration of materials. That typically happens when different materials are moulded together with bricklaying and rendering as two examples. In that case, the contact creates an additional layer with its own but unknown properties.

Boundary conditions

Wind driven rain

Wind driven rain has been a subject of research for several decades now. However, even today, with the use of CFD and rain droplet tracing, it remains impossible to predict the real transient rain pattern on a building enclosure. But, even if we succeeded in doing so, major questions remain. What precisely happens when rain droplets hit a surface? How much splashes away? How important is evaporation? Does a surface really react as predicted by the suction model, with complete suction first followed by runoff and moistening of the lower parts of the enclosure, or does each contact result in local suction and runoff? Why does runoff finger? All these questions are important, as wind driven rain is the main moisture load in many climates.

Spatial and time dependency

In almost all models actually in use, the inside and outside boundary values are considered as being time dependent only. This of course is not true. Temperatures also change with height. Surface film coefficients differ from spot to spot. Parts of the enclosure stay shadowed by parapets and overhangs, other get full solar gains during a large part of the day. Wind driven rain only hits the upper and corner parts, etc. All these variations affect the hygrothermal response of an enclosure. Even flat assemblies may react three-dimensionally in that way.

Building/building enclosure interaction

Most tools simplify the inside boundary conditions to a constant temperature, an inside vapour concentration excess and, if air transport is considered, a constant inside/outside air pressure difference. This, however, is not reality. Night setback and excessive solar gains subject the inside temperature to a daily periodicity. The local inside vapour pressure excess, although dampened by all sorption-active internal surfaces, oscillates with changes in vapour release, ventilation rate and air exchanges between rooms. Inside/outside air pressure differences not only depend on wind velocity, wind direction and inside/outside temperature difference but also on the overall layout of the building, the random distribution of air leaks in the enclosure and the way the ventilation system functions. Air mitigation especially couples the hygrothermal reaction of each part of an enclosure to the whole building response and stochastic behaviour of the users. All that turns heat, air, moisture performance of envelope parts into a risk analysis.

Gravity and pressure flow

Both were already mentioned under geometry and contact conditions. Gravity and pressure differences are a main cause of water flow through cracks, voids, open joints, etc. Activation demands a water film or a water head. Wind driven rain for example may cause a water film to develop on the exterior surface. That film may penetrate randomly through cracks in the rain screen. Voids in the head joints of a brick veneer wall get filled randomly that way.

4.5 Moisture tolerance

As they weep through to the cavity side, run off starts there. As a result, bricks turn moist from all sides, leading to a faster moisture build up and more moisture storage than calculated with the actual models, which cannot handle that phenomenon.

When the film sticks to horizontal mouldings, water heads develop that facilitate weeping through. No model predicts this from happening. The consequences in terms of leakage, however, could be problematic. Precisely that non-predictability of rain water ingress necessitates the simple control and design strategies explained above

Water heads also play a role below the freatic surface. Although no heat-air-moisture model is able to correctly assess that situation, special techniques for water-tightening cellars have been developed that are very efficient.

Moisture and durability

Despite professionals know moisture may cause severe durability problems, tools are lacking to judge if a situation is tolerable or not. Much for example is known about mould. However a judgement in a given situation is still difficult and uncertain. In general, durability depends on the limit state properties of a construction and the severity of the loads. The latter are always dynamic. The limit state properties may experience degradation over time, meaning that for the same load, safety decreases gradually. That decrease is not deterministic but stochastic, as the properties of a system are not constant but vary considerably. As a consequence, decay is only quantifiable as a risk and service life as a stochastic fact.

An example demonstrating model restrictions

In general

The university building under scrutiny had to house a very diverse program: underground parking, lecture theatres, library, smaller seminar rooms and individual office rooms. For that reason, the design team proposed a building volume, which narrowed from the basement to the top. The lecture theatres were situated just above the parking. The library was posted in between, while the seminar rooms and offices filled the higher floors. The result was a building with oblique façade walls (Figure 4.20). These were solved as cavity walls with masonry veneer, partially filled cavity with a PU-insulation and a reinforced concrete inside leaf with brick finish.

Figure 4.20. University building.

Complaints

The main complaint was the appearance of large moisture stains on the inside surface of the oblique cavity wall, while a second complaint concerned rain penetration along window sills (Figure 4.21).

Figure 4.21. Rain leakage on the inside of the oblique cavity wall, rain penetration along window sills.

Analysis

In a first step, the catch ratio pattern on the building envelope for wind driven rain coming from the main wind direction, which was south-west, was calculated. That proved the heaviness of rain exposure. Run-off was analysed on site, showing that the oblique facades functioned as very active drainage surfaces with a clear concentration of water at the edges, a thing that no model could predict (Figure 4.22). Because the veneer had been treated with a water repellent substance, the main run-off load was put on the contact surfaces between bricks and mortar, causing leakage to the cavity. Again, models failed to predict such an occurrence. There, the leaking water dripped on the insulation, ran-off, penetrated the joints between the insulation boards and wetted the concrete inside leaf, where shrinkage cracks directed the water to the brick-finish inside. That is also not predictable as no-one knows the locations of the joints and shrinkage cracks beforehand.

Rain penetration through the joints between windows and sills was caused by a lack of sill steps below the window frames (Figure 4.23).

Solution

In a first trial, one of the oblique veneer walls was replaced by a stepwise regressing veneer (Figure 4.24). That was not what could be called a success and the view was awful. The

4.5 Moisture tolerance

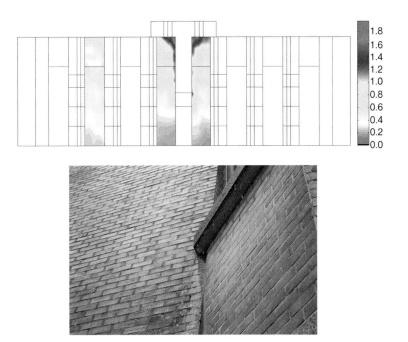

Figure 4.22. Catch ratio and run-off.

Figure 4.23. Lacking sill steps.

Figure 4.24. Solution, first trial.

solution also introduced quite some thermal bridging, lifting the U-value of the wall from 0.49 W/(m² · K) to 0.64 W/(m² · K). Trays at the bottom of the veneer wall at each step were forgotten, leaving room for further leakage An alternative proposed was exchanging the brick veneers for a watertight zinc cover on timber lathing. That permitted solving the sill problem.

4.6 Thermal bridges

4.6.1 Definition

The term 'thermal bridge' refers to spots in the envelope where heat transfer develops in two or three dimensions, resulting in higher heat flow and lower inside surface temperatures compared to the adjacent opaque envelope parts. The two aspects have to be considered: a heat flow increase and a surface temperature drop. The first was discussed when defining whole thermal transmittances; the second is evaluated through the temperature factor of (f_{h_i}):

$$f_{h_i} = \frac{\theta_{si,min} - \theta_e}{\theta_i - \theta_e} \tag{4.29}$$

The lower that value, the more problematic the thermal bridge in terms of dust deposit, mould risk, surface condensation and cracks developing.

4.6.2 Requirement

Looking to the lowest acceptable temperature factor on a thermal bridge, the value admitted in cool and cold climates should limit the mould germination risk to 5%, provided that the ventilation rate is quoted as normal and heating keeps the daily mean inside temperature above 12 °C. That minimum equals 0.7 in a cool climate. If maintained, double glazing will suffer from surface condensation rather than the thermal bridges from mould.

4.6.3 Consequences for the envelope

Two principles demand attention: (1) eliminate as much as possible structural thermal bridges by a consequent application of thermal breaks. That demands continuity of the insulation. Ideally it should be possible at the design stage to draw lines along the envelope in section and floor plan, without ever leaving the insulation layer. Due to structural requirements or buildability that is not always possible. So tricks are: long conduction paths (less heat loss), the inside surface larger than the outside surface (higher surface temperature); (2) Neutralize all geometrical thermal bridges via a consequent application of thermal breaks. In most cases this demands no tricks.

4.7 Contact coefficients

Contact coefficients intervene when feet comfort is at stake. Floor finishes gain in comfort with a low value. The value also has some importance for walls. Aluminium dwellings never gained market share because of the high contact coefficient of the material ($b = 23\,380$ J/(m² K s$^{0.5}$)). That turned contact with the walls into a nasty experience because they felt unpleasantly cold. The same happens with cast concrete walls. Inhabitants call such walls cold, not because they are not insulated and thus have low inside surface temperatures but because concrete has a high contact coefficient ($b = 2440$ J/(m² K s$^{0.5}$) and feels cold.

4.8 Hygrothermal stress and strain

That materials respond to humidity and temperature changes by deformation, stress and strain cannot be avoided. What should be excluded are cracks that disturb the performance or the aesthetics of finishing layers or whole envelope parts. To guarantee that, stresses should at no point exceed the tensile strength of a layer.

However translating that into general performance requirements remains difficult. Acceptability in the case of outside stucco for example differs according to its nature (water repellent or not), the substrate (capillary or not), the colour, etc. Also predicting stress and strain and valuing cracking risk is difficult. Such analysis combines transient heat and moisture modelling with stress/strain calculations due to hindered deformation, accounting for cracking conditions, fatigue, creep and relaxation.

Nonetheless the smaller the differences in temperature and relative humidity, the lower the stress and strain will be and the smaller the cracking risk. That rule involves two structural principles: (1) keep temperature and relative humidity fluctuations in the load bearing and massive parts of a building as small as possible; (2) for layers that are subjected to large hygrothermal fluctuations choose solutions that absorb related mechanical loading without cracking.

Applying principle (1) to an insulated massive envelope shows that outside insulation outperforms all other choices. All these induce larger temperature and relative humidity fluctuations in the massive parts of the envelope. Of course initial shrinkage during built-in moisture drying cannot be avoided. Principle (2) implies that extra attention must be given to all layers at the outside of the thermal insulation. The outside stucco of an EIFS-system should be deformable, get a reinforcing net and be unsaturated coloured. A veneer wall wins in quality when frost resisting and somewhat deformable. Scaled and plated elements must be fixed in a way deformations will not add, etc. Things get more complicated in massive constructions where the outside finish has the massive wall as a substrate. In such case, the massive part imposes the way the finish deforms. For that reason coupling between both should be as weak as admissible, which is why stucco is applied in two-layers, a first layer with low modulus of elasticity and high deformability and a second that is stiffer and harder.

4.9 Example of performance control: timber-framed walls

4.9.1 Assembly (from inside to outside)

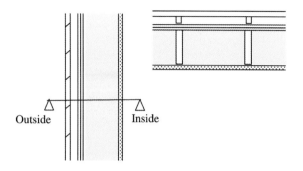

Layer	Thickness m	λ-value W/(m · K)	μ-value –	Air permeance ($K_a = a \, \Delta P_a^b$)	
				a kg/(m² · s · Pa^b)	$b - 1$ –
Gypsum board					
– Without leaks	0.012	0.2	7	$3.1 \cdot 10^{-5}$	–0.19
– With leaks				$3.8 \cdot 10^{-4}$	–0.39
Thermal insulation (glass wool)	X	0.04	1.2	$2.3 \cdot 10^{-3}$	–0.11
Sheathing (plywood)	0.022	0.14	20–30	$5.4 \cdot 10^{-4}$	–0.46
Building paper			In a first step not applied		
Cavity	0.04		0.3		
Timber siding	0.012	0.14	0	$4.1 \cdot 10^{-4}$	–0.32

4.9.2 Air-tightness

Assume the thermal insulation has a thickness of 10 cm. If the gypsum board is perfectly mounted, the wall will have an air permeance equalling: $3.03 \cdot 10^{-5} \, (\Delta P_a)^{-0.2}$ kg/(m² · Pa · s), i.e. hardly different from the inside gypsum board lining. If that lining is perforated by 1 leak ϕ20 mm per m², then the air permeance increases to $2.13 \cdot 10^{-4} \, (\Delta P_a)^{-0.38}$ kg/(m² · Pa · s), i.e. seven times the value without leaks at an air pressure difference of 1 Pa, but anyhow smaller than the air permeance of the leaky gypsum board thanks to the other layers. For the air flow rate, see Figure 4.25.

4.9 Example of performance control: timber-framed walls

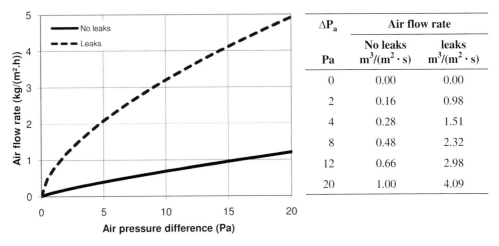

ΔP_a	Air flow rate	
	No leaks	leaks
Pa	m³/(m²·s)	m³/(m²·s)
0	0.00	0.00
2	0.16	0.98
4	0.28	1.51
8	0.48	2.32
12	0.66	2.98
20	1.00	4.09

Figure 4.25. Air flow rate across a timber framed wall.

4.9.3 Thermal transmittance

In case the assembly is airtight and thermal bridge free, thermal transmittance is given by:

$$U_0 = \frac{1}{0.043 + \dfrac{0.012}{0.14} + 0.17 + \dfrac{0.022}{0.14} + \dfrac{X}{0.04} + 0.06 + 0.125} = \frac{0.04}{0.026 + X}$$

See Figure 4.26 and added table, second column. Timber-framed walls have low clear wall U-values for limited wall thickness (20 cm insulation demands a wall thickness of 26 cm).

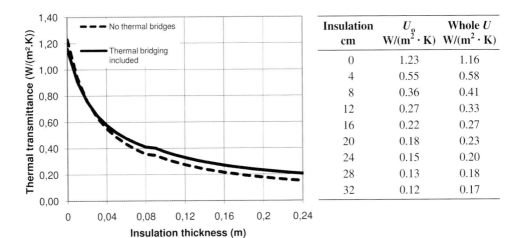

Insulation cm	U_0 W/(m²·K)	Whole U W/(m²·K)
0	1.23	1.16
4	0.55	0.58
8	0.36	0.41
12	0.27	0.33
16	0.22	0.27
20	0.18	0.23
24	0.15	0.20
28	0.13	0.18
32	0.12	0.17

Figure 4.26. Clear and whole wall thermal transmittance.

No thermal bridging however is fiction. Studs, bottom, top and connecting plates increase heat flow. For an insulation thickness above 8 cm, their linear thermal transmittances are:

Stud $\psi = 0.017$ W/(m · K)
Bottom plate $\psi = 0.01$ W/(m · K)
Top and connecting plates $\psi = 0.023$ W/(m · K)

For typical timber-framed walls with the studs at a distance centre to centre of 40 cm and a wall height of 2.6 m, each field contains 2.48 m of studs, 0.4 m of bottom plate and 0.4 m of top and connecting plate. Whole wall thermal transmittance, included thermal bridging, is shown in the third column of the added table. With thick insulation thermal bridging gains in importance, a sufficient motivation to look for less conducting alternatives for the timber studs (I-shaped engineered studs or doubled studs for example)

Perfect air-tightness is also fiction. Assume the wall is room-high (2.6 m) and the air leaks are uniformly distributed over the height. The room door is closed and wind velocity outside is zero. That way only thermal stack intervenes with as pressure difference over the wall:

$$\Delta P_a = \frac{1}{2}\left[0.043\,(z - H/2)\,(\theta_i - \theta_e)\right]$$

where H is room height and z the height ordinate. At a January mean temperature difference between in and outside of 17 °C the absolute pressure difference at a height of 0.65 and 1.95 m then becomes 0.48 Pa. with 0.05 m^3/h inflow at 0.65 m and 0.05 m^3/h outflow at 1.95 m for a leak-free lining and 0.41 m^3/h in and outflow with leaks. Table 4.3 lists the apparent thermal transmittances one should measure inside at both heights. Formula:

$$U_{app,si} = \frac{c_a\,g_a}{1 - \exp(c_a\,g_a\,R_T)}$$

At 0.65 m thermal transmittance apparently turns worse while at 1.95 m a lower value appears. The difference with the clear wall value also augments when leakage increases! Or, air leakage clearly turns thermal transmittance into a kind of fictitious quantity.

Table 4.3. Apparent thermal transmittance on the inside surface, air in and outflow.

Insulation	U_o	Gypsum board leak-free		Gypsum board leaky	
		$U_{0.65}$	$U_{1.95}$	$U_{0.65}$	$U_{1.95}$
cm	W/(m^2·K)	W/(m^2·K)	W/(m^2·K)	W/(m^2·K)	W/(m^2·K)
0	1.23	1.24	1.22	1.30	1.16
4	0.55	0.56	0.54	0.62	0.49
8	0.36	0.36	0.35	0.43	0.29
12	0.27	0.28	0.27	0.35	0.21
16	0.22	0.22	0.21	0.29	0.15
20	0.18	0.19	0.17	0.25	0.12
24	0.15	0.16	0.14	0.23	0.09

4.9.4 Transient response

Timber framed walls are too light to be thermally inert. Yet, temperature damping and the admittance will be low. In timber framed dwellings other means should be applied to avoid summer overheating in cool climates. These are: limited glass area, exterior solar shading at the sunny sides, heavy-weight floors and inner walls and night time ventilation in case heavy floors and inside walls are used.

It is nevertheless valuable to see how both properties change with increasing insulation thickness. The results are given in Figure 4.27. Apparently temperature damping increases more than linearly, though even with a 24 cm thick insulation layer a value of 15 is still not reached. Admittance decreases first, followed by some increase at higher thicknesses.

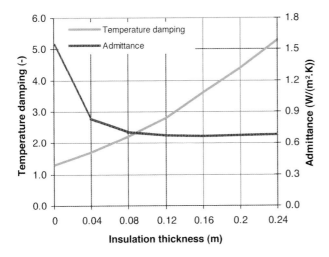

d_{ins}	Temperature damping		Admittance	
cm	Value −	Shift h	Value W/(m² · K)	Shift h
0	1.3	3:18	1.54	1:42
4	1.7	5:00	0.84	2:48
8	2.2	6:00	0.71	3:36
12	2.8	6:42	0.68	4:00
16	3.6	7:18	0.67	4:18
20	4.4	7:48	0.68	4:36
24	5.3	8:18	0.68	4:42

Figure 4.27. Timber framed wall: temperature damping and thermal admittance.

Infiltration degrades temperature damping, however without real impact on the admittance. Exfiltration instead upgrades temperature damping.

4.9.5 Moisture tolerance

4.9.5.1 Built-in moisture

This should not be a problem provided the timber for framing is not too moist and sheathing and inside lining have a moderate vapour resistance.

4.9.5.2 Rain

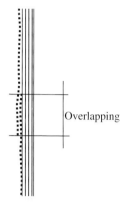

A redundant rain control engages three interfaces to act as drainage planes: the cladding's outer surface, the cladding's backside and the sheathing's cavity side. The last demands application of a building paper against the sheathing draped so that rain is conducted down into the cavity, where a tray should direct it outside. Application of a building paper adds air and wind-tightness to the wall.

One should take care when using a capillary brick veneer as outside clad. Solar radiation may drive buffered rain water as vapour to the inside, resulting in high relative humidity in and perhaps condensation against the sheathing, in the insulation and at the inside lining. Avoiding solar driven humidity demands a correct balance between the diffusion resistance of the building paper and the inside lining.

4.9.5.3 Rising damp

Rising damp is easily avoided by inserting a watertight layer below the lowest bottom plate.

4.9.5.4 Hygroscopic moisture and surface condensation

Should not be a problem with a properly insulated timber-framed wall, though the outside clad will suffer from larger hygroscopic moisture swings with increasing insulation thickness. More paint peeling could be one of the consequences.

4.9.5.5 Interstitial condensation

Wall airtight, Glaser (or dew point) calculation

A Glaser calculation assumes that no layer in an assembly is hygroscopic neither capillary. In such case, one may get droplets deposited, which is correctly called 'interstitial condensation'. Plywood, however, is highly hygroscopic and somewhat capillary, causing humidity deposited mainly to change the hygroscopic moisture content.

Glaser as applied uses the equivalent outside temperature for condensation and drying at Uccle, Belgium. Short wave absorptivity of the outside surface is 0.8. The wall looks north. The (humid) plywood sheathing has a diffusion resistance factor 10 and the vapour permeable building paper a diffusion thickness of 1 cm. The cavity between the outside clad and the building paper is vented, giving a vapour pressure there hardly different from outside.

For North-Western Europe the coldest month is January. Using the outside and inside boundary conditions for Uccle gives (residential building):

4.9 Example of performance control: timber-framed walls

Outside		Mean	Amplitude	January
Temperature		11.4	8.7	2.9
Vapour pressure		1042	430	621
Inside		**Mean**	**Amplitude**	**January**
Temperature		20	3	17.1
Vapour pressure	ICC1	1107	365	749
	ICC2	1312	220	1096
	ICC3	1442	220	1226

For the temperature, saturation pressure and ICC1 vapour pressure course in the wall, see Table 4.4. ICC1 does not give moisture deposit at the backside of the plywood, whose average relative humidity touches 100 (674 + 624) / (815 + 792) = 81%, giving a mean moisture ratio of 14 to 18 %kg/kg, i.e. below the mould limit of 20 %kg/kg.

Table 4.4

Layer	R $m^2 \cdot K/W$	ΣR $m^2 \cdot K/W$	Temp. °C	p_{sat} Pa	μ –	$\Sigma \mu d$ m	p Pa	$> p_{sat}$?
			17.1	1949			749	No
h_i	0.13	0.13	16.7	1909		0,007	748	No
Gypsum board	0.06	0.19	16.6	1890		0,091	729	No
Insulation (20 cm)	5.00	5.19	4.0	815	1.2	0,331	674	No
Plywood	0.16	5.35	3.6	792	10	0,551	624	No
Building paper	0.00	5.35	3.6	792	μd = 1 cm	0,561	622	No
Cavity	0.17	5.52	3.2	769		0,567	621	No
Clad	0.09	5.60	3.0	757				
h_e	0.04	5.64	2.9	752				

Table 4.5 summarizes the calculations for the Belgian climate classes ICC2 and ICC3. Some moisture deposit at the backside of the plywood is now a fact. Figure 4.28 shows the January mean saturation and vapour pressure in the wall for both climate classes together with the moisture deposited. Also the allowable amount for the case of moisture-proof plywood is represented. The threshold ICC2/ICC3 hardly gives deposit. At the threshold ICC3/ICC4, the deposit passes the allowable quantity at the end of february. Some extra vapour resistance at the inside is needed for a deposit below acceptable.

That timber-framed walls seem quite tolerant when it comes to interstitial condensation however is a premature conclusion. Assume for example the plywood has a vapour resistance factor of 30. Then, in ICC2, the maximum deposit at the backside of the plywood reaches 0.8 kg/m² which is far from acceptable! In ICC3, that maximum increases to a troubling 1.75 kg/m². A moisture tolerant timber-framed wall therefore demands an extra diffusion thickness μd inside of the thermal insulation layer of 1 m or more. That is much less than the diffusion thickness a perfectly mounted, non perforated polyethylene foil offers: > 40 m.

Table 4.5

Layer	Temp. °C	p_{sat} Pa	ICC2		ICC3	
			p Pa	p_{corr} Pa	p Pa	p_{corr} Pa
	17.1	1949	1096	1096	1226	1226
h_i	16.7	1909	1091	1090	1219	1218
Gypsum board	16.6	1890	1020	1019	1129	1113
Insulation (20 cm)	4.0	815	**819**	815	873	815
Plywood	3.6	792	634	634	638	634
Building paper	3.6	792	626	626	627	626
Cavity	3.2	769	621	621	621	621
Clad	3.0	757				
h_e	2.9	752				

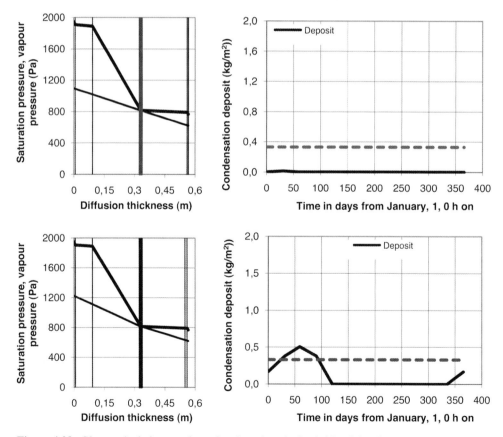

Figure 4.28. Glaser calculation, condensation deposit at the backside of the plywood.

4.9 Example of performance control: timber-framed walls

Wall not airtight, diffusion/convection calculation

Even a timber framed wall with perfectly mounted gypsum board shows some air leakage. If perforated, air permeance increases greatly, up to a mean of $2.13 \cdot 10^{-4} (\Delta P_a)^{-0.38}$ kg/ ($m^2 \cdot s \cdot Pa$). Diffusion/convection calculations are typically restricted to a cold week. Nonetheless, first, a whole year is considered, using mean values for wind pressure and stack. That way the same boundary conditions and allowance criteria as for Glaser hold.

The exposed timber-framed wall looks north. As the wind blows mainly south-west in North-Western Europe, north is an under-pressure orientation experiencing air outflow. The detached dwelling has a quadratic floor plan, giving as monthly mean wind and stack pressure at 1.95 m (Pa):

	J	F	M	A	M	J	J	A	S	O	N	D
Wind	−1.3	−1.3	−1.6	−1.3	−1.0	−0.9	−0.8	−0.9	−0.9	−1.0	−1.7	−1.5
Stack	−0.4	−0.4	−0.3	−0.3	−0.2	−0.2	−0.2	−0.2	−0.2	−0.3	−0.3	−0.4

Figure 4.29 shows what could be the consequence in ICC2 if this under-pressures last for a whole month, assuming the plywood has a vapour resistance factor of 10: a dramatic increase in moisture deposit, far beyond acceptable. Things turn more detrimental in ICC3, where the maximum touches 6.2 kg/m^2.

These results of course do not picture reality. Air leakage is not distributed uniformly. Where the leaks are is mostly unknown. Wind changes continuously in amplitude and direction, which makes steady state calculations unreliable, which is why a cold week was forwarded as period to be considered. The deposit during that cold week is given in Figure 4.30.

Note how the deposit varies with increasing diffusion resistance of plywood and building paper or when a vapour retarder is included, in both cases with constant air permeance (which of course is an assumption, not a fact). Figure 4.31 shows the result.

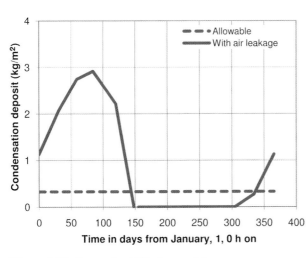

Figure 4.29. Convection/diffusion model, ICC2/ICC3.

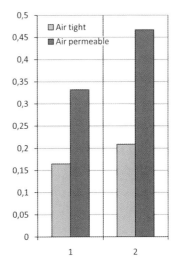

Figure 4.30. Cold week, deposit, 1 = ICC2/ICC3, 2 = ICC3/ICC4.

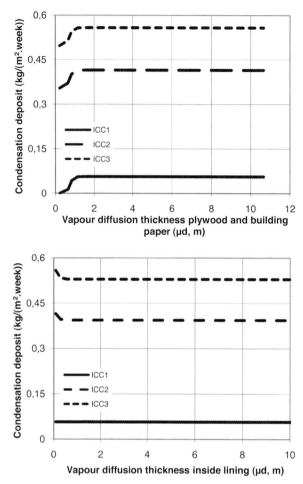

Figure 4.31. Timber-framed wall: condensation deposit by diffusion/convection, impact of vapour diffusion thickness of sheathing and building paper, effect of vapour retarding quality of the inside lining when a building paper with vapour diffusion thickness of 10 m is used.

Air outflow clearly fixes the moisture deposit. The vapour diffusion thickness (μd) of plywood and building paper hardly matters, as does the vapour retarder in case a building paper with high diffusion resistance is used. Even a combination 'inside lining with high vapour diffusion resistance/vapour permeable sheathing+building paper' does not exclude moisture deposit in ICC2 and ICC3!

Wall airtight, wet brick veneer

Under Uccle conditions, solar driven diffusion in a wall without vapour retarder inside but with vapour permeable building paper does not cause long-lasting condensation at the backside of the inside lining, though during warm, sunny days after a rainy period, some deposit may be possible. Yet, as Figure 4.32 shows for ICC2, the plywood sheathing suffers more because of the increased amounts of moisture deposit due to poorer drying to the outside in winter.

4.9 Example of performance control: timber-framed walls

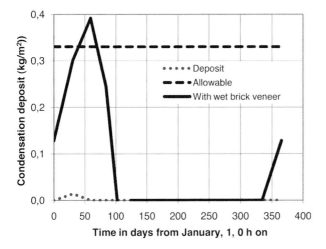

Figure 4.32. Wet brick veneer, moisture deposit against the plywood.

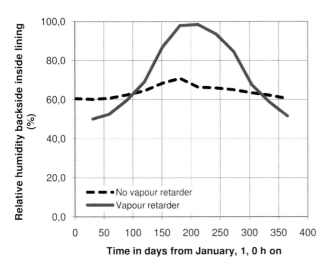

Figure 4.33. Relative humidity at the backside of the inside lining.

However, with a vapour retarder inside, things change. In fact, whereas the relative humidity at the backside of the inside line remains quite constant without vapour retarder, the value at the insulation and stud side increases to unacceptable levels with vapour retarder, see Figure 4.33. Or, with brick veneer, the building paper should have a vapour diffusion resistance in balance with the air and vapour retarder installed between the inside lining and thermal insulation.

Another remark concerns the air/vapour retarder. Perforations should be avoided, a requirement which is best guaranteed by leaving a cavity between the air/vapour retarder and the inside lining, wide enough for wiring, electrical outlet installation and piping.

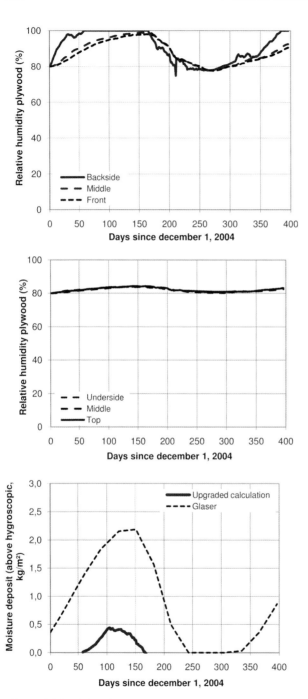

Figure 4.34. Timber framed wall, hygric inertia of the plywood considered, relative humidity in the plywood at the left without vapour retarder, at the right with vapour retarder, below comparing the condensation deposit with Glaser.

Better modelling

Neither Glaser nor the convection/diffusion method accounts for sorption by the sheathing. In fact, interstitial condensation may simply be an increase in sorption moisture. Take an airtight timber framed wall. Inertia is included considering the plywood as a chain of hygric capacities, linked by diffusion resistances with the first and last capacity coupled to both the inside and the outside. Written in terms of forward differences the equation per capacity is:

$$p_{pl}^{t+\Delta t} = p_{sat,2}^{t+\Delta t} \left[\frac{p_2^t}{p_{sat,2}^t} + \frac{\Delta t}{0.5 \, \rho \, \xi \left(\Delta d_{1,2} + \Delta d_{2,3}\right)_{pl}} \left(\frac{p_1^t - p_2^t}{Z_{1,2}} + \frac{p_3^t - p_2^t}{Z_{2,3}} \right) \right]$$

Calculations on a daily basis were done for an ICC3 situation. The wall considered is north-oriented and first had no vapour retarder and then had one with a diffusion thickness of 2.5 m. The building paper diffusion thickness numbers 1.8 m, while the initial sorption moisture in the plywood shows an 80% relative humidity. A lathed cladding with a non-vented cavity behind serves as finish. The daily thermal response was assumed steady state, which allowed direct calculation of the vapour saturation pressure in the plywood. For the calculation results, see Figure 4.34.

Without vapour retarder some condensate is still deposited at the backside of the plywood, though the amounts are a fraction of those calculated with Glaser. Of course, relative humidity in the plywood touches values high enough for mould. With a vapour retarder, the plywood sees its relative humidity stabilized around 82%.

4.10 References

[4.1] Glaser, H. (1958). *Wärmeleitung und Feuchtigkeitsdurchgang durch Kühlraumisolierungen.* Kältetechnik, 3/1958, pp. 86–91 (in German).

[4.2] Neufert (1964). *Bau Entwurfslehre.* Verlag Ullstein, Berlin (in German).

[4.3] A. J. Metric Handbook (1969). The Architectural Press, London.

[4.4] Mamillan, M., Boineau, A. (1974). *Etude de l'assèchement des murs soumis à des remomtées capillaires, paper 2.2.2.* CIB-RILEM 2d International Conference on Moisture Problems in Buildings, Rotterdam (in French).

[4.5] Interindustriële studiegroep voor de Bouwnijverheid IC-IB (1979). *Prestatiegids voor gebouwen, delen 1 tot 9.* WTCB (in Dutch).

[4.6] Nevander, L.E, Elmarsson, B. (1981). *Fukthandbok.* Svensk Byggtjänst (in Swedish).

[4.7] NRC-Canada (1982). *Exterior Walls, Understanding the Problems.* Proceedings of the Building Science Forum '82, 63 pp.

[4.8] WTCB-tijdschrift (1982). *De rekenmetode van Glaser,* 1/82, pp. 13–24 (in Dutch).

[4.9] NRC-Canada (1983). *Humidity, Condensation and Ventilation in Houses.* Proceedings of the Building Science Insight '83, 66 pp.

[4.10] Hens, H. (1984). *Buitenwandoplossingen voor de residentiële bouw, De spouwmuur.* Rapport R-D Energie (in Dutch).

[4.11] Andersen, N. E., Christensen, G., Nielsen, F. (1984). *Bygningers Fugtisolering.* SBI-Anvisning 139, Statens Byggeforskningsinstitut, Horsholm (in Danish).

[4.12] Künzel, H. (1985). *Regenschutz von Außenwänden durch mineralische Putze*. Der Stukkateur, Heft 5/85 (in German).

[4.13] PATO (1986). *Syllabus 'Leidt energiebesparing tot Vochtproblemen'*. Delft, 30 sept. – 1 Okt. (in Dutch).

[4.14] Verbruggen, A. (1988). *Investeringsanalyse*. Rationeel energiegebruik in kantoorgebouwen, Wilrijk (in Dutch).

[4.15] NRC-Canada (1989). *An Air Barrier for the Building Envelope*. Proceedings of the Building Insight '86, 24 pp.

[4.16] Hens, H. (1989). *Heat, Air, Moisture Transport in Building Components*. Seminar U.Lg, nov.

[4.17] American Hotel & Motel Association (1989). Mold & Mildew in Hotels and Motels, the Survey, 1989.

[4.18] Sagelsdorf, R. (1989). *Wasserdampfdiffusion, Grundlagen, Berechnungsverfahren, Diffusionsnachweis*. SIA-Dokumentation D 018, Zürich (in German).

[4.19] Bouwbesluit, Staatsblad 680 (1991). Staatsdrukkerij, Den Haag, Dec. (in Dutch).

[4.20] Uitvoeringsbesluiten Isolatiedecreet (1991).

[4.21] Nevander, L. E., Elmarsson, B. (1991). *Fuktdimensionering av träkonstruktioner, Riskanalys*. Byggforskningsrådet, rapport R38 (in Swedish).

[4.22] Hens, H. (1991). *Analysis of causes of dampness, influence of salt attack*. Seminar 'l'Umidita ascendente nelle murature: fenomenologia e sperimentazione', Bari.

[4.23] Anon. (1991). *Condensation and Energy, Guidelines and Practice*. Final Report IEA, EXCO ECBCS Annex 24, ACCO, Leuven.

[4.24] Winnepenninckx, E. (1992). *Het Vlaamse isolatie- en ventilatiedecreet, onuitgegeven eindwerk*. IHAM (in Dutch).

[4.25] Carmeliet, J. (1992). *Duurzaamheid van weefselgewapende pleisters voor buitenisolatie, een probabilistische benadering op basis van de niet locale continue schademechanica*. Doctoraal proefschrift, KU-Leuven (in Dutch).

[4.26] Künzel, H. (1994). *Verfahren zur ein und zweidimensionalen Berechnung des gekoppelten Wärme- und Feuchtetransport in Bauteilen mit einfachen Kennwerten*. Doktor Abhandlung, Iniversität Stuttgart, 104 p. + fig. (in German).

[4.27] Hens, H. (1996). *Heat, Air and Moisture Transport Through New and Retrofitted Insulated Envelope Parts, Modeling*. Final Report Task 1 IEA, EXCO ECBCS Annex 24, ACCO, Leuven.

[4.28] Anon. (1996). *Heat, Air and Moisture Transport Through New and Retrofitted Insulated Envelope Parts, Common Exercices*. Final Report Task 1 IEA, EXCO ECBCS Annex 24, ACCO, Leuven.

[4.29] Sanders, C. (1996). *Heat, Air and Moisture Transport Through New and Retrofitted Insulated Envelope Parts, Environmental Conditions*. Final Report Task 2 IEA, EXCO ECBCS Annex 24, ACCO, Leuven.

[4.30] Hagentoft, C. E. (1998). *Heat, Air and Moisture Transport Through New and Retrofitted Insulated Envelope Parts, Performances and Practice*. Final Report Task 5 IEA, EXCO ECBCS Annex 24, ACCO, Leuven.

[4.31] Straube, J. F. (1998). *Moisture Control and Enclosure Wall Systems*. PhD-Thesis, University of Waterloo.

[4.32] Janssens, A. (1997). Reliable *Control of Interstitial Condensation in Lightweight Roof Systems*. PhD, K. U. Leuven, 217 p.

4.10 References

[4.33] Hens, H. (1999). *Fungal Defacement in Buildings, a Performance Related Approach*. International Journal for Heating, Ventilation, Air Conditioning and Refrigeration Research, Vol. 5, No. 3, pp. 256–280.

[4.34] Hendriks, L., Hens, H. (2000). *Building Envelopes in a Holistic Perspective*. Final Report IEA, EXCO ECBCS Annex 32, ACCO, Leuven.

[4.35] ASHRAE (2001). *Handbook of Fundamentals*. Chapter 23, Atlanta.

[4.36] Sedlbauer, C. (2001). *Vorhersage von Schimmelpilzbildung auf und in Bauteilen*. Doktor-Ingenieurs Abhandlung, Universität Stuttgart (in German).

[4.37] Holm, A. (2001). *Ermittlung der Genauigkeit von instationären hygrothermischen Bauteilberechnungen mittels eines stochatischen Konzeptes*. Doktor-Ingenieur Abhandlung, Universität Stuttgart, 129 p. (in German).

[4.38] Majoor, G. J. M., de Jong, A. (2002). *Herziene normen gepubliceerd voor thermische isolatieberekening en EPN*. Bouwfysica, Vol. 15, nr 2 (in Dutch).

[4.39] Blocken, B., Hens, H., Carmeliet, J. (2002). *Methods for the Quantification of Driving Rain on Buildings*. ASHRAE Transactions, Vol. 108, part 2, p. 338–350.

[4.40] Van Mook, F. (2003). *Driving rain on building envelopes*. Doctoraal proefschrift TU/e (Bouwstenen 69).

[4.41] Carmeliet, J., Hens, H., Vermeir, G. (Eds.) (2003). *Research in Building Physics*. A. A. Balkema Publishers, Lisse/Abingdon/Exton/Tokyo.

[4.42] R. Zheng, J. Carmeliet, H. Hens, A. Janssens, W. Bogaerts (2004). *A Hot Box-Cold Box Investigation of the Corrosion Behavior of Highly Insulated Zinc Roofing Systems*. Journal of Thermal Envelope and Building Science, Vol. 28, nr. 1, July, pp. 27–44.

[4.43] Hens, H., S. Roels, W Desadeleer (2004). *Rain Leakage through veneer walls, built with concrete blocks*. Proceedings CIB-W40 meeting, Glasgow.

[4.44] Blocken, B. (2004). *Wind-driven rain on buildings, measurements, numerical modelling and applications*. PhD, K. U. Leuven, 323 p. + add.

[4.45] Houvenaghel, G., Hens, H., Horta, A. (2004). *The impact of airflow on the hygrothermal behavior of highly insulated pitched roof systems*. Proceedings of the Performance of the Exterior Envelopes of Whole Buildings IX Conference, Clear Water Beach, Florida.

[4.46] ASHRAE (2005). *Handbook of Fundamentals*. Chapter 23, Atlanta.

[4.47] Hens, H., Roels, S., Desadeleer, W. (2005). *Glued concrete block veneers with open head joints: rain leakage and hygrothermal performance*. Proceedings of the Nordic Building Physics Conference, Rejkjavik, June.

[4.48] Fazio, P., Hua, G., Rao, J., Desmarais, G. (Eds.) (2006). *Research in Building Physics and Building Engineering*. Taylor & Francis, London/Leiden/New York/Philadelphia/Singapore.

[4.49] Poupeleer, A. S. (2007). *Transport and crystallization of dissolved salts in cracked porous building materials*. PhD, K. U. Leuven, 299 p.

[4.50] Janssens, A., Hens, H. (2007). *Effects of wind on the transmission heat loss in duo-pitch insulated roofs: a field study*. Energy and Buildings 39, 1047–1054.

[4.51] Vinha, J. (2007). *Hygrothermal Performance of Timber-Framed External Walls in Finnish Climatic Conditions*. PhD, Tampereen Teknillinen Yliopisto, 338 p. + app.

[4.52] Hens, H. (2007). *Does heat, air, moisture modelling really helps in solving hygrothermal problems?* Keynote, Rakennusfysiikka, Tampere, 18 october.

[4.53] Peer, L. (2007). *Use of Thermal Breaks in Cladding Support Systems*. Proceedings of the Buildings X Conference, Clearwater Beach (CD-ROM).

[4.54] Williams, M. (2007). *Developing Innovative Drainage and Drying Solutions for the Building Enclosure*. Proceedings of the Buildings X Conference, Clearwater Beach (CD-ROM).

[4.55] Karagiosis, A., Desjarlais, A., Lstiburek, J. (2007). *Scientific Analysis of Vapour Retarder Recommendations for Wall Systems Constructed in North America*. Proceedings of the Buildings X Conference, Clearwater Beach (CD-ROM).

[4.56] Derome, D., Carmeliet, J., Karagiosis, A. (2007. *Cyclic temperature gradient driven moisture transport in walls with wetted masonry cladding*. Proceedings of the Buildings X Conference, Clearwater Beach (CD-ROM).

[4.57] Tariku, F., Cornick, S., Lacasse, M. (2007). *Simulation of Wind-Driven Rain Penetration Effects on the Performance of a Stucco-clad wall*. Proceedings of the Buildings X Conference, Clearwater Beach (CD-ROM).

[4.58] Fazio, P. (2008). *Developing a standardized test to evaluate the relative hygrothermal performance of different building envelope systems*. Proceedings of the Building Physics Symposium, Leuven, 29–31 October, pp. 63–66.

[4.59] Abuku Masaru (2009). *Moisture stress of wind-driven rain on building enclosures*. PhD, K. U. Leuven, 107 p.

[4.60] ASHRAE (2009). *Handbook of Fundamentals*. Chapter 23, Atlanta.

5 Heat-air-moisture material properties

5.1 Introduction

A whole set of material properties is needed to describe the heat-air-moisture characteristics of any building, insulating and finishing material. Density ρ and open porosity Ψ_o are two of them: the weight per unit volume of dry material and the volume per unit volume taken by pores that are accessible for water molecules. The others are listed in Table 5.1.

Table 5.1. Array of heat, air, moisture material properties.

	Heat	Air	Moisture
Storage	Specific heat capacity c Volumetric specific heat capacity ρ_c	Specific moisture ratio ξ Volumetric specific moisture ratio ρ_ξ	Specific air content
Transport	Thermal conductivity λ Thermal resistance R Radiation: – Absorptivity α – Emissivity e – Reflectivity ρ	Water vapour permeability δ Vapour resistance factor μ Diffusion thickness μ_d Moisture permeability k_m Thermal moisture diffusion coefficient K_θ	Air permeability k_a Air permeance K_a
Combined	Thermal diffusivity a Contact coefficient b	Moisture diffusivity D_w Water sorption coefficient A	
Consequences	Thermal expansion coefficient α	Hygric expansion ε	

Chapter 5 gives measured and practice values for a series of building, insulating and finishing materials. While the measured values are simple test results, practice values not only include a variance due to the application but also account for the spread in measured values. How to fix practice values is described in standards such as ISO 10456 for thermal conductivity.

5.2 Dry air and water

Dry air

Gas constant	287.055 J/(kg · K)
Atmospheric pressure	Normally 101.325 Pa, higher in cyclonic than in anti-cyclonic weather

Viscosity (η)	Pressure Bar	Temperature °C	Viscosity Pa · s
	1	0	0.0000174
	1	50	0.0000199
	1	100	0.0000222

Specific heat capacity (c_p)	1007 J/(kg · K)

Thermal conductivity (λ)	Pressure Bar	Temperature °C	λ W/(m · K)
	1	0	0.024
	1	100	0.031

Water

Density (ρ)	Pressure Bar	Temperature °C	ρ kg/m³
	1	0	999.9
	1	10	999.7
	1	20	998.2
	1	40	992.2
	1	60	983.2
	1	80	971.8
	1	100	958.4

Viscosity (η)	Pressure Bar	Temperature °C	η Pa · s
	1	0	$1.787 \cdot 10^{-3}$
	1	10	$1.307 \cdot 10^{-3}$
	1	20	$1.002 \cdot 10^{-3}$
	1	40	$0.653 \cdot 10^{-3}$
	1	60	$0.467 \cdot 10^{-3}$
	1	80	$0.354 \cdot 10^{-3}$
	1	100	$0.282 \cdot 10^{-3}$

Surface tension (σ)	$(75.9 - 0.17\,\theta) \cdot 10^{-3}$ N/m with θ temperature in °C
Specific heat capacity (c)	4187 J/(kg · K)
Heat of evaporation (l_b)	2 500 000 J/kg at 0 °C
Heat of solidification (l_s)	33 400 J/kg at 0 °C

Thermal conductivity	Pressure Bar	Temperature °C	λ W/(m · K)
	1	0	0.54
	1	60	0.67

Long-wave emissivity	0.95

5.3 Building and isolating materials

5.3.1 Thermal properties

5.3.1.1 Definitions

Thermal conductivity Declared value (λ_D)	Starts from the measured thermal conductivity of the dry material at a mean temperature of 10 °C. Testing must include a representative number of samples (20), allowing a statistical analysis of the data. The declared value coincides with the 90% percentile, i.e. the value assuring that 90% of the production will have a lower thermal conductivity.
Certified versus non-certified	A material is certified when the declared thermal conductivity is known. Conversely, a material is called non-certified when the declared thermal conductivity has not been measured by a certified laboratory.
Thermal conductivity Design value (λ_U)	The design value follows from the declared one by applying: $$\lambda_U = \lambda_D \exp\left[f_u(X_2 - X_1)\right] \quad \lambda_U = \lambda_D \exp\left[f_\psi(\Psi_2 - \Psi_1)\right] \quad (5.1)$$ where X_2, Ψ_2 is the moisture ratio in kg/kg or m³/m³ present and X_1, Ψ_1 the moisture ratio in kg/kg or m³/m³ as measured during testing for the declared thermal conductivity (normally zero). f_u and f_ψ are conversion factors given in the standards

5.3.1.2 List of design values, non-certified materials (ISO 10456)

Material group	Material	Density ρ kg/m³	Specific heat capacity c J/(kg · K)	Thermal conductivity, design value λ W/(m · K)
Metals	Aluminium alloys	2800	880	220
	Duralumin	2800	880	160
	Brass	8400	380	120
	Bronze	8700	380	65
	Copper	8900	380	380
	Iron	7900	450	75
	Iron, cast	7500	450	50
	Steel	7800	450	50
	Stainless steel	7900	460	17
	Lead	11300	130	35
	Zinc	7100	380	110

Material group	Material	Density ρ kg/m³	Specific heat capacity c J/(kg·K)	Thermal conductivity, design value λ W/(m·K)
Wood and wood-based materials	Soft wood	500	1600	0.13
	Hard wood	700	1600	0.18
	Plywood (from low to high density)	300	1600	0.09
		500	1600	0.13
		700	1600	0.17
		1000	1600	0.24
Wood and wood-based materials	Particle board – Soft	300	1700	0.10
	– Semi-hard	600	1700	0.14
	– Hard	900	1700	0.18
	Particle board, cement bounded	1200	1500	0.23
	OSB	680	1700	0.13
Wood and wood-based materials	Fibre board – Soft	400	1700	0.10
	– Semi-hard	600	1700	0.14
	– Hard	800	1700	0.18
Gypsum	Gypsum blocks	600	1000	0.18
		900	1000	0.30
		1200	1000	0.43
		1500	1000	0.56
	Gypsum board	900	1050	0.25
Mortars	Cement mortar, mixed on site	1800	1100	0.9
		1900	1100	1.0
Plasters	Gypsum plaster – Light	600	1000	0.18
	– Normal	1000	1000	0.40
	– Heavy	1300	1000	0.57
	Lime or gypsum plus sand	1600	1000	0.70
	Cement plus sand	1700	1000	1.0
Concrete	Medium to high density	1800	1000	1.15
		2000	1000	1.35
		2200	1000	1.65
		2400	1000	2.00
	Reinforced 1% steel	2300	1000	2.30
	2% steel	2400	1000	2.50

5.3 Building and isolating materials

Material group	Material	Density ρ kg/m^3	Specific heat capacity c J/(kg·K)	Thermal conductivity, design value λ W/(m·K)
Stone	Crystalline rock	2800	1000	3.50
	Sedimentary rock	2600	1000	2.30
	Sedimentary, light	1500	1000	0.85
	Lava	1600	1000	0.55
	Basalt	2700–3000	1000	3.50
	Gneiss	2400–2700	1000	3.50
	Granite	2500–3000	1000	2.80
	Marble	2800	1000	3.50
	Slate	2000–2800	1000	2.20
	Lime stone – Extra soft – Soft – Semi-hard – Hard – Extra hard	 1600 1800 2000 2200 2600	 1000 1000 1000 1000 1000	 0.85 1.10 1.40 1.70 2.30
	Sandstone	2600	1000	2.30
	Natural pumice	400	1000	0.12
	Artificial stone	1750	1000	1.30
Soils	Clay or silt	1200–1800	1670–2500	1.5
	Sand and gravel	1700–2200	910–1180	2.0
Water, ice, snow	Ice, –10 °C	920	2000	2.30
	Ice, 0 °C	900	2000	2.20
	Water, 10 °C	1000	4187	0.60
	Water, 40 °C	990	4187	0.63
	Snow – Fresh (< 30 mm) – Soft (30–70 mm) – Slightly compacted – Compacted	 100 200 300 500	 2000 2000 2000 2000	 0.05 0.12 0.23 0.60

Material group	Material	Density ρ kg/m³	Specific heat capacity c J/(kg·K)	Thermal conductivity, design value λ W/(m·K)
Plastics, solid	Acrylic	1050	1500	0.20
	Polycarbonates	1200	1200	0.20
	PFTE	2200	1000	0.25
	PVC	1390	900	0.17
	PMMA	1180	1500	0.18
	Polyacetate	1410	1400	0.30
	Polyamide (Nylon)	1150	1600	0.25
	Nylon, 25% glass fibre	1450	1600	0.30
	PE, High density	980	1800	0.50
	PE, low density	920	2100	0.33
	Polystyrene	1050	1300	0.16
	Polypropylene (PP)	910	1800	0.22
	PP, 25% glass fibre	1200	1800	0.25
	Polyurethane	1200	1800	0.25
	Epoxy resin	1200	1200	0.20
	Phenol resin	1300	1700	0.30
	Polyester resin	1400	1200	0.19
Rubbers	Natural	910	1100	0.13
	Neoprene	1240	2140	0.23
	Butyl	1200	1400	0.24
	Foam rubber	60–80	1500	0.06
	Hard rubber	1200	1400	0.17
	EPDM	1150	1000	0.25
	Polyisobutylene	920	1130	0.13
	Polysulfide	1700	1000	0.43
	Butadiene	980	1000	0.25
Glass	Quartz glass	2200	750	1.4
	Normal glass	2500	750	1.0
	Glass mosaic	2000	750	1.2

5.3 Building and isolating materials

Material group	Material	Density ρ kg/m³	Specific heat capacity c J/(kg · K)	Thermal conductivity, design value λ W/(m · K)
Gases	Air	1.23	1008	0.025
	Argon	1.70	519	0.017
	Carbon dioxide	1.95	820	0.014
	Sulphur hexafluoride	6.36	614	0.013
	Krypton	3.56	245	0.009
	Xenon	5.68	160	0.0054
Thermal breaks, sealants, weather stripping	Silica gel (desiccant)	720	1000	0.13
	Silicone foam	750	1000	0.12
	Silicone pure	1200	1000	0.35
	Silicone, filled	1450	1000	0.50
	Urethane/PUR (thermal break)	1300	1800	0.21
	PVC, flexible with 40% softener	1200	1000	0.14
	Elastomeric foam, flexible	60–80	1500	0.05
	Polyurethane foam	70	1500	0.05
	Polyethylene foam	70	2300	0.05
Roofing materials	Asphalt	2100–2300	1000	0.7
	Bitumen, pure	1050	1000	0.17
	Bitumen felt	1100	1000	0.23
	Clay tiles	2000	800	1.00
	Concrete tiles	2100	1000	1.50
Floor coverings	Rubber	1200	1400	0.17
	Plastic	1700	1400	0.25
	Linoleum	1200	1400	0.17
	Carpet/textile	200	1300	0.06
	Underlay, rubber	270	1400	0.10
	Underlay, felt	120	1300	0.05
	Underlay, wool	200	1300	0.06
	Underlay, cork	200	1500	0.05
	Ceramic tiles	2300	840	1.30
	Plastic tiles	1000	1000	0.20
	Cork tiles, light	200	1500	0.050
	Cork tiles, heavy	500	1500	0.065

5.3.1.3 Design values according to the standard NBN B 62-002 (edition 2001)

Definition

Design values for thermal conductivity according to NBN B62-002 (λ_U)	The standard differentiates between indoor and outdoor conditions. Indoor conditions prevail between the backside of the thermal insulation and inside. The design value there is fixed for a temperature of 23 °C at a relative humidity of 50%. Outdoors pertains to all layers outside of the thermal insulation. For capillary materials the design value coincides with 10 °C and a moisture ratio of 0.75 times the critical value (see Table 5.2 for masonry). For non-capillary materials the conditions are 10 °C and 80% relative humidity.

Table 5.2. Moisture ratio in masonry at indoor and outdoor conditions.

Masonry	Density ρ kg/m³	Moisture ratio 23 °C, 50% RH		Moisture ratio 0.75 (X_{cr} or Ψ_{cr})		Conversion factors	
		X_i kg/kg	Ψ_i m³/m³	X_e kg/kg	Ψ_e m³/m³	f_u kg/kg	f_Ψ m³/m³
Bricks, perforated	700–2100		0.007		0.075		10
Lime-sand stone	900–2200		0.012		0.090		10
Concrete blocks (massive, perforated)							
– Heavy	1600–2400		0.025		0.090		4
– Expanded clay	400–1700		0.020		0.090		4
– Lightweight	500–1800		0.030		0.090		4
Cellular concrete	300–1000	0.026			0.150		4

Design values

Metals

Material	Density ρ kg/m³	Specific heat capacity c J/(kg · K)	Thermal conductivity λ_{Ui} W/(m · K)	Thermal conductivity λ_{Ue} W/(m · K)
Lead	11340	130	35	35
Copper	$8300 \leq \rho \leq 8900$	390	384	384
Steel	7800	480–530	45	45
Aluminium, 99%	2700	880	203	203
Iron, cast	7500	530	56	56
Zinc	7000	390	113	113

5.3 Building and isolating materials

Stone

Material	Density ρ kg/m³	Specific heat capacity c J/(kg·K)	Thermal conductivity λ_{Ui} W/(m·K)	Thermal conductivity λ_{Ue} W/(m·K)
Heavy (granite, gneiss, basalt, porphyry)	$2750 \leq \rho \leq 3000$	1000	3.49	3.49
Limestone, hard	2700	1000	2.91	3.49
Marble	2750	1000	2.91	3.49
Sandstone				
– Hard	2550	1000	2.21	2.68
– Semi	2350	1000	1.74	2.09
– Soft	2200	1000	1.40	1.69

Glued masonry, joint width ≤ 3 mm (grey cells: not outdoors)

Bricks and perforated large format bricks

Density ρ kg/m³	Specific heat capacity c J/(kg·K)	Certified Thermal conductivity		Not certified Thermal conductivity	
		λ_{Ui} W/(m·K)	λ_{Ue} W/(m·K)	λ_{Ui} W/(m·K)	λ_{Ue} W/(m·K)
≤ 700	1000	0.20	0.39	0.22	0.43
$700 < \rho \leq 800$	1000	0.23	0.45	0.25	0.49
$800 < \rho \leq 900$	1000	0.26	0.51	0.28	0.56
$900 < \rho \leq 1000$	1000	0.29	0.57	0.32	0.63
$1000 < \rho \leq 1100$	1000	0.32	0.64	0.35	0.70
$1100 < \rho \leq 1200$	1000	0.35	0.70	0.39	0.77
$1200 < \rho \leq 1300$	1000	0.39	0.76	0.42	0.84
$1300 < \rho \leq 1400$	1000	0.43	0.85	0.47	0.93
$1400 < \rho \leq 1500$	1000	0.46	0.91	0.51	1.00
$1500 < \rho \leq 1600$	1000	0.50	0.99	0.55	1.09
$1600 < \rho \leq 1700$	1000	0.55	1.08	0.60	1.19
$1700 < \rho \leq 1800$	1000	0.59	1.16	0.65	1.28
$1900 < \rho \leq 1900$	1000	0.64	1.27	0.71	1.40
$1900 < \rho \leq 2000$	1000	0.69	1.35	0.76	1.49
$2000 < \rho \leq 2100$	1000	0.74	1.46	0.81	1.61

Lime-sand stone

Density ρ kg/m^3	Specific heat capacity c J/(kg·K)	Certified Thermal conductivity		Not certified Thermal conductivity	
		λ_{Ui} W/(m·K)	λ_{Ue} W/(m·K)	λ_{Ui} W/(m·K)	λ_{Ue} W/(m·K)
≤ 900	1000	0.33	0.71	0.36	0.78
900 < ρ ≤ 1000	1000	0.34	0.74	0.37	0.81
1000 < ρ ≤ 1100	1000	0.36	0.79	0.40	0.87
1100 < ρ ≤ 1200	1000	0.41	0.89	0.45	0.97
1200 < ρ ≤ 1300	1000	0.46	1.01	0.51	1.11
1300 < ρ ≤ 1400	1000	0.52	1.13	0.57	1.24
1400 < ρ ≤ 1500	1000	0.60	1.30	0.66	1.43
1500 < ρ ≤ 1600	1000	0.69	1.50	0.76	1.65
1600 < ρ ≤ 1700	1000	0.79	1.72	0.87	1.89
1700 < ρ ≤ 1800	1000	0.91	1.99	1.00	2.19
1800 < ρ ≤ 1900	1000	1.04	2.26	1.14	2.49
1900 < ρ ≤ 2000	1000	1.18	2.58	1.30	2.84
2000 < ρ ≤ 2100	1000	1.35	2.95	1.49	3.25
2100 < ρ ≤ 2200	1000	1.54	3.37	1.70	3.71

Normal concrete blocks

Density ρ kg/m^3	Specific heat capacity c J/(kg·K)	Certified Thermal conductivity		Not certified Thermal conductivity	
		λ_{Ui} W/(m·K)	λ_{Ue} W/(m·K)	λ_{Ui} W/(m·K)	λ_{Ue} W/(m·K)
≤ 1600	1000	0.97	1.26	1.07	1.39
1600 < ρ ≤ 1700	1000	1.03	1.33	1.13	1.47
1700 < ρ ≤ 1800	1000	1.12	1.45	1.23	1.59
1800 < ρ ≤ 1900	1000	1.20	1.56	1.33	1.72
1900 < ρ ≤ 2000	1000	1.32	1.71	1.45	1.88
2000 < ρ ≤ 2100	1000	1.44	1.86	1.58	2.05
2100 < ρ ≤ 2200	1000	1.57	2.04	1.73	2.24
2200 < ρ ≤ 2300	1000	1.72	2.24	1.90	2.46

5.3 Building and isolating materials

Expanded clay concrete blocks

Density ρ kg/m³	Specific heat capacity c J/(kg·K)	Certified Thermal conductivity		Not certified Thermal conductivity	
		λ_{Ui} W/(m·K)	λ_{Ue} W/(m·K)	λ_{Ui} W/(m·K)	λ_{Ue} W/(m·K)
$400 < \rho \leq 500$	1000	0.16		0.18	
$500 < \rho \leq 600$	1000	0.19	0.26	0.21	0.28
$600 < \rho \leq 700$	1000	0.23	0.30	0.25	0.33
$700 < \rho \leq 800$	1000	0.27	0.36	0.30	0.39
$800 < \rho \leq 900$	1000	0.30	0.40	0.33	0.44
$900 < \rho \leq 1000$	1000	0.35	0.46	0.38	0.50
$1000 < \rho \leq 1100$	1000	0.39	0.52	0.43	0.57
$1100 < \rho \leq 1200$	1000	0.44	0.59	0.49	0.65
$1200 < \rho \leq 1300$	1000	0.50	0.66	0.55	0.73
$1300 < \rho \leq 1400$	1000	0.55	0.73	0.61	0.80
$1400 < \rho \leq 1500$	1000	0.61	0.80	0.67	0.88
$1500 < \rho \leq 1600$	1000	0.68	0.90	0.75	0.99
$1600 < \rho \leq 1700$	1000	0.76	1.00	0.83	1.10

Concrete blocks with other lightweight aggregates

Density ρ kg/m³	Specific heat capacity c J/(kg·K)	Certified Thermal conductivity		Not certified Thermal conductivity	
		λ_{Ui} W/(m·K)	λ_{Ue} W/(m·K)	λ_{Ui} W/(m·K)	λ_{Ue} W/(m·K)
≤ 500	1000	0.27		0.30	
$500 < \rho \leq 600$	1000	0.30	0.39	0.33	0.43
$600 < \rho \leq 700$	1000	0.34	0.43	0.37	0.47
$700 < \rho \leq 800$	1000	0.37	0.47	0.41	0.52
$800 < \rho \leq 900$	1000	0.42	0.53	0.46	0.58
$900 < \rho \leq 1000$	1000	0.46	0.59	0.51	0.65
$1000 < \rho \leq 1100$	1000	0.52	0.66	0.57	0.73
$1100 < \rho \leq 1200$	1000	0.59	0.75	0.64	0.82
$1200 < \rho \leq 1300$	1000	0.65	0.83	0.72	0.91
$1300 < \rho \leq 1400$	1000	0.74	0.95	0.82	1.04
$1400 < \rho \leq 1500$	1000	0.83	1.06	0.92	1.17
$1500 < \rho \leq 1600$	1000	0.94	1.19	1.03	1.31
$1600 < \rho \leq 1800$	1000	1.22	1.55	1.34	1.70

Autoclaved cellular concrete blocks

Density ρ kg/m³	Specific heat capacity c J/(kg·K)	Certified Thermal conductivity		Not certified Thermal conductivity	
		λ_{Ui} W/(m·K)	λ_{Ue} W/(m·K)	λ_{Ui} W/(m·K)	λ_{Ue} W/(m·K)
≤ 300	1000	0.09		0.10	
300 < ρ ≤ 400	1000	0.12		0.13	
400 < ρ ≤ 500	1000	0.14		0.16	
500 < ρ ≤ 600	1000	0.18	0.29	0.20	0.32
600 < ρ ≤ 700	1000	0.20	0.33	0.22	0.36
700 < ρ ≤ 800	1000	0.23	0.38	0.26	0.42
800 < ρ ≤ 900	1000	0.27	0.44	0.29	0.48
900 < ρ ≤ 1000	1000	0.29	0.47	0.32	0.52

Fresh cellular concrete contains important amounts of moisture. The values given hold once dried.

Mortared masonry

Define the wall ratio taken by the blocks. Calculate the design value λ_U as:

$$\lambda_U = \lambda_{U,block}\, f_{block} + \lambda_{U,mortar}\left(1 - f_{block}\right) \qquad (5.2)$$

where $\lambda_{U,block}$ is the thermal conductivity of the blocks, f_{block} the wall ratio taken by the blocks and $\lambda_{U,mortar}$ the thermal conductivity of the mortar. If thermal conductivity of the blocks is unknown, then use the values given for glued masonry.

Normal concrete

Material	Density ρ kg/m³	Specific heat capacity c J/(kg·K)	Thermal conductivity λ_{Ui} W/(m·K)	Thermal conductivity λ_{Ue} W/(m·K)
Reinforced	2400	1000	1.7	2.2
Not reinforced	2200	1000	1.3	1.7

These values are low when compared to measured data and the values given in ISO 10456.

5.3 Building and isolating materials

Light-weight concrete used as slabs or screeds

Material	Density ρ kg/m³	Specific heat capacity c J/(kg·K)	Thermal conductivity λ_{Ui} W/(m·K)	Thermal conductivity λ_{Ue} W/(m·K)
Concrete with expanded clay, furnace slag, vermiculite, cork, perlite or polystyrene pearls as aggregates Cellular concrete	≤ 350	1000	0.12	
	300 < ρ ≤ 400	1000	0.14	
	400 < ρ ≤ 450	1000	0.15	
	450 < ρ ≤ 500	1000	0.16	
	500 < ρ ≤ 550	1000	0.17	
	550 < ρ ≤ 600	1000	0.18	
	600 < ρ ≤ 650	1000	0.20	0.31
	650 < ρ ≤ 700	1000	0.21	0.34
	700 < ρ ≤ 750	1000	0.22	0.36
	750 < ρ ≤ 800	1000	0.23	0.38
	800 < ρ ≤ 850	1000	0.24	0.40
	850 < ρ ≤ 900	1000	0.25	0.43
	900 < ρ ≤ 950	1000	0.27	0.45
	950 < ρ ≤ 1000	1000	0.29	0.47
	1000 < ρ ≤ 1100	1000	0.32	0.52
	1100 < ρ ≤ 1200	1000	0.37	0.58

Gypsum with or without light-weight aggregates

Material	Density ρ kg/m³	Specific heat capacity c J/(kg·K)	Thermal conductivity λ_{Ui} W/(m·K)	Thermal conductivity λ_{Ue} W/(m·K)
	≤ 800	1000	0.22	
	800 < ρ ≤ 1100	1000	0.35	
	> 1100	1000	0.52	

Plaster

Material	Density ρ kg/m³	Specific heat capacity c J/(kg·K)	Thermal conductivity λ_{Ui} W/(m·K)	Thermal conductivity λ_{Ue} W/(m·K)
Cement plaster	1900	1000	0.93	1.5
Lime plaster	1600	1000	0.70	1.2
Gypsum plaster[1]	1300	1000	0.52	

[1] Is a heavy-weight gypsum plaster. Most ready to mix gypsum plasters have a density not exceeding 1000 kg/m³ (see table for gypsum)

Wood and wood-based materials

Material	Density ρ kg/m³	Specific heat capacity c J/(kg·K)	Thermal conductivity λ_{Ui} W/(m·K)	Thermal conductivity λ_{Ue} W/(m·K)
Timber	≤ 600	1880	0.13	0.15
	> 600	1880	0.18	0.20
Plywood	≤ 400	1880	0.09	0.11
	400 < ρ ≤ 600	1880	0.13	0.15
	600 < ρ ≤ 850	1880	0.17	0.20
	> 850	1880	0.24	0.28
Particle board	≤ 450	1880	0.10	
	450 < ρ ≤ 750	1880	0.14	
	> 750	1880	0.18	
Fibre-cement board	1200	1470	0.23	
OSB	650	1880	0.13	
Fibreboard	≤ 375	1880	0.07	
	375 < ρ ≤ 500	1880	0.10	
	500 < ρ ≤ 700	1880	0.14	
	> 700	1880	0.18	

5.3 Building and isolating materials

Insulation materials

Insulation materials may under no circumstances become humid. Therefore, only λ_{Ui} is given.

Material	Density ρ kg/m³	Specific heat capacity c J/(kg · K)	Thermal conductivity λ_{Ui} W/(m · K)
Not certified			
Cork	90–160	1560	0.05
Glass and rockwool	10–200	1030	0.045
EPS	10–50	1450	0.045
Polyethylene (PE)	20–65	1450	0.045
Phenolfoam	20–50	1400	0.045
PUR, lined	28–55	1400	0.035
XPS	20–65	1450	0.040
Cellular glass	100–140	1000	0.055
Perlite board	200–300	900	0.060
Vermiculite	50–170	1080	0.065
Vermiculite board		900	0.090
Certified			
Glass and rockwool	10–200	1030	0.040
EPS	10–50	1450	0.040
Phenol foam	20–50	1400	0.025
PUR, lined	28–55	1400	0.028
XPS	20–65	1450	0.034
Cellular glass	100–140	1000	0.048
Perlite board	200–300	900	0.055

Miscellanea

Material	Density ρ kg/m³	Specific heat capacity c J/(kg·K)	Thermal conductivity λ_{Ui} W/(m·K)	Thermal conductivity λ_{Ue} W/(m·K)
Glass	2500	750	1.00	1.00
Clay tiles	1700	1000	0.81	1.00
Grès ttles	2000	1000	1.20	1.30
Rubber	1500	1400	0.17	0.17
Linoleum, PVC	1200	1400	0.19	
Fibre cement	1400 < ρ < 1900	1000	0.35	0.50
Asphalt	2100	1000	0.70	0.70
Bitumen	1100	1000	0.23	0.23

Perforated blocks, floor elements, gypsum board

Material	Thickness cm	Specific heat capacity c J/(kg·K)	Thermal resistance R_{Ui} W/(m·K)
Masonry from perforated blocks			
– Normal concrete (ρ > 1200 kg/m³)	14	1000	0.11
	19	1000	0.14
	29	1000	0.20
– Lightweight concrete (ρ < 1200 kg/m³)	14	1000	0.30
	19	1000	0.35
	29	1000	0.45
Clay floor elements			
– One cavity along span	8	1000	0.08
	12	1000	–0.11
– Two cavities along span	12	1000	–0.13
	16	1000	0.16
	20	1000	0.19
Concrete floor elements	12	1000	0.11
	16	1000	0.13
	20	1000	0.15
Gypsum board	< 1.4	1000	0.05
	≥ 1.4	1000	0.08

5.3 Building and isolating materials

5.3.1.4 Measured data

Building materials

Concrete

Density	kg/m³	2176, $\sigma = 40.5$		
		Mean for 39 samples		
Specific heat capacity (dry)	J/(kg · K)	840		
Thermal conductivity	W/(m · K)	$2.74 + 0.0032\, w$		
		w: moisture content in kg/m³		
		T (K)	300	6000
		a	0.88	0,6
		ρ	0.12	0.4
Absorptivity, reflectivity	–			
Thermal expansion coefficient	°C⁻¹	$12 \cdot 10^{-6}$		

Lightweight concrete

Density	kg/m³	$644 \leq \rho \leq 1187$	
Specific heat capacity (dry)	J/(kg · K)	840	
Thermal conductivity	W/(m · K)	$644 \leq \rho \leq 1187$ $\theta = 20$ °C, $w = 0$ kg/m³	$0.0414 \exp(0.00205\, \rho)$
		$1158 \leq \rho \leq 1187$ $\theta = 20$ °C, $w \leq 74$ kg/m³	$0.511 + 0.0026\, w$
		$1130 \leq \rho \leq 1138$ $\theta = 20$ °C, $w \leq 144$ kg/m³	$0.371 + 0.001\, w$
		$644 \leq \rho \leq 674$ $\theta = 20$ °C, $w \leq 39$ kg/m³	$0.161 + 0.0015\, w$

Autoclaved aerated concrete

Density	kg/m³	$455 \leq \rho \leq 800$	
Specific heat capacity (dry)	J/(kg · K)	840	
Thermal conductivity	W/(m · K)	$598 \leq \rho \leq 626$ $\theta = 10$ °C, $w \leq 425$ kg/m³	$0.176 + 0.0008\, w$
		$598 \leq \rho \leq 626$ $\theta = 20$ °C, $w \leq 425$ kg/m³	$0.177 + 0.001\, w$
		$455 \leq \rho \leq 492$ $\theta = 20$ °C, $w \leq 298$ kg/m³	$0.138 + 0.0009\, w$
As a function of density, temperature and moisture content		$\lambda = 0.172 - 1.67 \cdot 10^{-3}\, w - 9.34 \cdot 10^{-3}\, \theta$ $\quad - 2.97 \cdot 10^{-6}\, \rho + 3.77 \cdot 10^{-6}\, \rho\, w + 1.16 \cdot 10^{-4}\, w\, \theta$ $\quad + 1.6 \cdot 10^{-5}\, \rho\, \theta - 1.62 \cdot 10^{-7}\, \rho\, w\, \theta$	

Concrete with expanded polystyrene pearls as aggregate

Density	kg/m³	$259 \leq \rho \leq 792$			
Specific heat capacity (dry)	J/(kg · K)	Depends on the concentration of EPS-pearls: $\rho = 259$ kg/m³ $c = 1370$ J/(kg · K) $\rho = 792$ kg/m³ $c = 1018$ J/(kg · K)			
Thermal conductivity	W/(m · K)	$259 \leq \rho \leq 792$ $\theta = 10\ °C, w = 0$ kg/m³ $\theta = 20\ °C, w \leq 425$ kg/m³	$0.041 \exp(0.00232\rho)$ $A_1 + A_2\, w$		
			ρ (kg/m³)	A_1	$A_2 \cdot 10^{-4}$
			259–335	0.074	4.8
			357–382	0.111	4.6
			407–456	0.126	6.1
			641	0.151	7.9
			792	0.213	1.0
		$\rho = 422$ kg/m³ $0 \leq \theta \leq 30\ °C$	$B_1 + B_2\, \theta$		
			w (kg/m³)	B_1	$B_2 \cdot 10^{-4}$
			0	0.112	1.3
			94	0.171	12
			262	0.231	15

Lightweight and normal cement mortars

Density	kg/m³	$1055 \leq \rho \leq 1822$			
Specific heat capacity (dry)	J/(kg · K)	840			
Thermal conductivity	W/(m · K)	$1055 \leq \rho \leq 1822$ $\theta = 20\ °C, w = 0$ kg/m³ $\theta = 20\ °C, w \leq 330$ kg/m³	$0.088 \exp(0.00125\rho)$ $0.177 + 0.001\, w$ $0.138 + 0.0009\, w$		
			$A_1 + A_2\, w$		
			ρ (kg/m³)	A_1	$A_2 \cdot 10^{-3}$
			1072	0.346	1.2
			1512	0.526	3.1
			1800	0.854	4.5

5.3 Building and isolating materials

Lime-sand stone masonry

Density	kg/m³	$1170 \leq \rho \leq 1230$				
Specific heat capacity (dry)	J/(kg · K)	840				
Thermal resistance	m² · K/W	$\theta = 20\,°\text{C}$	$1/(A_1 + A_2\, u)$			
		$u \leq u_c$, u in %kg/kg	d (m)	ρ (kg/m³)	A_1	A_2
			0.14	1140	4.07	0.24

Large format perforated brick masonry

Density	kg/m³	$860 \leq \rho \leq 1760$				
Specific heat capacity (dry)	J/(kg · K)	840				
Thermal resistance	m² · K/W	$d = 14$ cm $860 \leq \rho \leq 1430$ $\theta = 20\,°\text{C}$, $w = 0$ kg/m³	$1/[0.98 \exp(0.001\,\rho)]$			
		$d = 19$ cm $830 \leq \rho \leq 1630$ $\theta = 20\,°\text{C}$, $w = 0$ kg/m³	$1/[0.59 \exp(0.0012\,\rho)]$			
		$\theta = 20\,°\text{C}$ $u \leq u_c$, u in %kg/kg	$1/(A_1 + A_2\, u)$			
			d (m)	ρ (kg/m³)	A_1	A_2
			0.09	1470	7.94	0.40
			0.14	863	2.09	0.09
				1100	3.35	0.18
				1120	2.72	0.28
				1180	3.37	0.31
				1200	3.13	0.32
				1240	4.17	0.38
				1360	2.96	0.34
				1430	4.17	0.42
			0.19	800	1.53	0.10
				830	1.51	0.08
				880	1.83	0.13
				1100	2.41	0.09
				1140	2.26	0.13
				1650	4.00	0.34

Concrete block masonry

Density	kg/m³	$860 \leq \rho \leq 1650$	
Specific heat capacity (dry)	J/(kg · K)	840	
Thermal resistance	m² · K/W	$d = 14$ cm $860 \leq \rho \leq 1650$ $\theta = 20\ °C, w = 0$ kg/m³	$1 / [0.73 \exp (0.0014\,\rho)]$
		$\theta = 20\ °C$ $u \leq u_c$, u in %kg/kg	$1 / (A_1 + A_2\,u)$

d (m)	ρ (kg/m³)	A_1	A_2
0.12	980	2.92	0.12
0.14	1080	3.12	0.24
0.14	1115	3.37	0.16
0.19	860	2.35	0.08

Autoclaved cellular concrete masonry

Density	kg/m³	$518 \leq \rho \leq 660$	
Specific heat capacity (dry)	J/(kg · K)	840	
Thermal resistance	m² · K/W	$\theta = 20\ °C$ $w \leq w_c$, w in kg/m³	$1 / (A_1 + A_2\,u)$

d (m)	ρ (kg/m³)	A_1	A_2
Mortar			
0.15	524	1.23	0.007
	660	1.44	0.007
Glue			
0.15	518	1.13	0.006
	634	1.20	0.007
Mortar			
0.18	550	1.25	0.009

Gypsum plaster

Density	kg/m³	975	
Specific heat capacity (dry)	J/(kg · K)	840	
Thermal conductivity	W/(m · K)	$569 \leq \rho \leq 981$ $\theta = 20\ °C, w$ in kg/m³	$0.263 + 0.001\,w$

5.3 Building and isolating materials

Outside rendering

Density	kg/m³	$878 \leq \rho \leq 1736$	
Specific heat capacity (dry)	J/(kg · K)	840	
Thermal expansion coefficient	K⁻¹	$\rho = 1736$	$10.9 \cdot 10^{-6}$

Wood

Density	kg/m³	400 (den) $\leq \rho \leq$ 690 (beuk)	
Specific heat capacity (dry)	J/(kg · K)	1880	
Thermal conductivity	W/(m · K)	den $\theta = 20$ °C, $w = 0$ kg/m³	0.11

Particle board

Density	kg/m³	$570 \leq \rho \leq 800$	
Specific heat capacity (dry)	J/(kg · K)	1880	
Thermal conductivity	W/(m · K)	$500 \leq \rho \leq 702$ $\theta = 20$ °C, $w = 0$ kg/m³	$0.098 + 0.0001 \, (\rho - 590)$
		$587 \leq \rho \leq 702$ $\theta = 20$ °C, w in kg/m³	$0.106 + 1.3 \cdot 10^{-4} \, w$ $+ 3.3 \cdot 10^{-7} \, w^2$

Plywood

Density	kg/m³	$445 \leq \rho \leq 799$	
Specific heat capacity (dry)	J/(kg · K)	1880	
Thermal conductivity	W/(m · K)	$445 \leq \rho \leq 692$ $\theta = 20$ °C, $w = 0$ kg/m³	$0.020 + 1.7 \cdot 10^{-4} \, \rho$
		$445 \leq \rho \leq 799$ $\theta = 20$ °C, w in kg/m³	$0.113 + 3.1 \cdot 10^{-4} \, w$

Fibre cement

Density	kg/m³	$823 \leq \rho \leq 2052$	
Specific heat capacity (dry)	J/(kg · K)	840	
Thermal conductivity	W/(m · K)	$823 \leq \rho \leq 866$ $\theta = 20$ °C	$0.14 + 5.8 \cdot 10^{-4} \, w$
		$\rho = 1495$ $\theta = 20$ °C	$0.113 + 3.1 \cdot 10^{-4} \, w$

Gypsum board

Weight	kg/m²	$6.5 \leq \rho \leq 13$	
Specific heat capacity (dry)	J/(kg · K)	840	
Equivalent thermal conductivity	W/(m · K)	$\theta = 20$ °C, $d = 9.5$ mm w in kg · m³	$0.07 + 2.1 \cdot 10^{-4} w$

Insulation materials

Cork

Density	kg/m³	111	
Specific heat capacity (dry)	J/(kg · K)	1880	
Thermal conductivity	W/(m · K)	$\theta = 20$ °C, $w = 0$ kg/m³	0.042

Cellular glass

Density	kg/m³	$113 \leq \rho \leq 140$	
Specific heat capacity (dry)	J/(kg · K)	840	
Thermal conductivity	W/(m · K)	$113 \leq \rho \leq 139$ $\theta = 20$ °C, $\psi = 0$ %m³/m³	$0.037 + 8.8 \cdot 10^{-5} \rho$
		$126 \leq \rho \leq 134$ $\theta = 20$ °C, ψ in %m³/m³	$0.047 + 8.2 \cdot 10^{-4} \psi$
		$\rho = 129$ $0 \leq \theta \leq 35$	$0.046 + 2.4 \cdot 10^{-4} \theta$

Glasswool

Density	kg/m³	$11.6 \leq \rho \leq 136$	
Specific heat capacity (dry)	J/(kg · K)	840	
Thermal conductivity	W/(m · K)	$\theta = 20$ °C, $\psi = 0$ %m³/m³	$0.0268 + 4.9 \cdot 10^{-5} \rho$ $+ 0.178 / \rho$

Mineral wool

Density	kg/m³	$32 \leq \rho \leq 191$	
Specific heat capacity (dry)	J/(kg · K)	840	
Thermal conductivity	W/(m · K)	$\theta = 20$ °C, $\psi = 0$ %m³/m³	$0.0317 + 2.6 \cdot 10^{-5} \rho$ $+ 0.206 / \rho$
		$\theta = 10$ °C, $\psi = 0$ %m³/m³	$0.026 + 5.5 \cdot 10^{-5} \rho$ $+ 0.331 / \rho$

5.3 Building and isolating materials

EPS (expanded polystyrene)

Density	kg/m³	$13 \leq \rho \leq 40$				
Specific heat capacity (dry)	J/(kg · K)	1470				
Thermal conductivity	W/(m · K)	$\psi = 0\ \%m^3/m^3$	$A_1 + A_2 \rho + A_3/\rho$			
			θ (°C)	A_1	$A_2 \cdot 10^{-4}$	A_3
			10	0.017	1.9	0.258
			20	0.021	1.2	0.235
		$\theta = 20\ °C,\ \psi$ in $\%m^3/m^3$	$B_1 + B_2 \psi$			
			ρ (kg/m³)	B_1	$B_2 \cdot 10^{-3}$	
			15	0.0390	2.0	
			20	0.0348	1.9	
			25	0.0326	2.7	
			30	0.0331	1.2	
		$\rho = 15$ kg/m³, d in m	$0.029 + 0.017\ d^{0,25}$			
		θ in °C	$0.0354 + 1.6 \cdot 10^{-4}\ \theta$			

XPS (extruded polystyrene)

Density	kg/m³	$25 \leq \rho \leq 55$	
Specific heat capacity (dry)	J/(kg · K)	1470	
Thermal conductivity	W/(m · K)	$\theta = 10\ °C,\ \psi = 0\ \%m^3/m^3$	$0.0174 + 1.6 \cdot 10^{-5} \rho$ $+ 0.263/\rho$
		$\theta = 20\ °C,\ \psi = 0\ \%m^3/m^3$	$0.0404 + 3.9 \cdot 10^{-4} \rho$ $+ 0.029/\rho$
		$\theta = 10\ °C, \rho = 35$ ψ in $0\ \%m^3/m^3$	$0.0240 + 1.6 \cdot 10^{-4} \psi$ $+ 5.8 \cdot 10^{-5} \psi^2$
		$\theta = 20\ °C, \rho = 35$ ψ in $0\ \%m^3/m^3$	$0.0251 + 5.2 \cdot 10^{-5} \psi$ $+ 7.0 \cdot 10^{-5} \psi^2$

PUR/PIR (polyurethane foam, polyisocyanurate foam)

Density	kg/m³	$20 \leq \rho \leq 40$	
Specific heat capacity (dry)	J/(kg · K)	1470	
Thermal conductivity	W/(m · K)	$\theta = 10\ °C,\ \psi = 0\ \%m^3/m^3$	$-0.112 + 1.9 \cdot 10^{-3} \rho$ $+ 2.36/\rho$
		$\theta = 20\ °C,\ \psi = 0\ \%m^3/m^3$	$-0.008 + 5.1 \cdot 10^{-4} \rho$ $+ 0.436/\rho$

Perlite board

Density	kg/m³	$135 \leq \rho \leq 215$				
Specific heat capacity (dry)	J/(kg · K)	1000				
Thermal conductivity	W/(m · K)	$\theta = 20\,°C, \psi = 0\ \%m^3/m^3$	$0.046 + 0.00014\,(\rho - 100)$			
		$\theta = 20\,°C, \psi$ in $\%m^3/m^3$	$B_1 + B_2\,\psi$			
			ρ (kg/m³)	θ (°C)	B_1	$B_2 \cdot 10^{-3}$
			142	10	0.052	1.5
			142	20	0.053	1.8
			171	10/20	0.059	2.9
			212	20	0.058	5.7
		$\rho = 140$ kg/m³	$D_1 + D_2\,\theta$			
			ψ (%m³/m³)		D_1	$D_2 \cdot 10^{-4}$
			0		0.047	1.2
			1.6		0.056	5.7

5.3.2 Air properties

5.3.2.1 Design values

There are no standard tables with design values.

5.3.2.2 Measured data

Masonry

Air permeance, per m^2 ($K_a = a \, \Delta P_a^{b-1}$)

Component	a kg/(s · m^2 · Pa^b)	b
Veneer walls		
Bricks of 19 × 9 × 4.5 cm^3, joints not pointed, badly filled	3.24 · 10^{-4}	0.69
Bricks of 19 × 9 × 4.5 cm^3, joints not pointed, well filled	1.08 · 10^{-4}	0.75
Bricks of 19 × 9 × 4.5 cm^3, joints pointed, 1	3.48 · 10^{-5}	0.80
Bricks of 19 × 9 × 6.5 cm^3, joints pointed, 2	3.60 · 10^{-5}	0.78
Bricks of 19 × 9 × 6.5 cm^3, joints not pointed, badly filled	7.20 · 10^{-4}	0.68
Bricks of 19 × 9 × 6.5 cm^3, joints not pointed, well filled	1.56 · 10^{-4}	0.71
Bricks of 19 × 9 × 6.5 cm^3, joints pointed, 1	3.60 · 10^{-5}	0.81
Bricks of 19 × 9 × 6.5 cm^3, joints pointed, 2	3.48 · 10^{-5}	0.82
Concrete blocks 19 × 9 × 9 cm^3 1955 kg/m^3, joints pointed	1.44 · 10^{-4}	0.88
Concrete blocks 19 × 9 × 9 cm^3, 1927 kg/m^3, joints pointed	1.92 · 10^{-4}	0.86
Concrete blocks 19 × 9 × 9 cm^3, 1881 kg/m^3, joints pointed	2.76 · 10^{-4}	0.82
Inside leafs		
Lightweight brickwork 19 × 14 × 14 cm^3, joints not pointed 1	2.64 · 10^{-3}	0.59
Lightweight brickwork 19 × 14 × 14 cm^3, joints not pointed 2	4.20 · 10^{-3}	0.57
Lightweight brickwork 19 × 14 × 14 cm^3, joints pointed 1	2.88 · 10^{-5}	0.72
Lightweight brickwork 19 × 14 × 14 cm^3, joints pointed 2	2.04 · 10^{-5}	0.81
Lightweight brickwork 19 × 14 × 14 cm^3, joints pointed 3	1.68 · 10^{-5}	0.82
Lightweight brickwork 19 × 14 × 14 cm^3, plastered at the inside 1	3.72 · 10^{-7}	0.96
Lightweight brickwork 19 × 14 × 14 cm^3, plastered at the inside 2	2.76 · 10^{-7}	0.97
Hollow concrete blocks 39 × 19 × 14 cm^3, 987 kg/m^3, joints not pointed	3.96 · 10^{-3}	0.58
Hollow concrete blocks, 39 × 19 × 14 cm^3 954 kg/m^3, joints not pointed	5.04 · 10^{-3}	0.56
Hollow concrete blocks 39 × 19 × 14 cm^3, 910 kg/m^3, joints not pointed	6.00 · 10^{-3}	0.55
Hollow concrete blocks 39 × 19 × 14, 987 kg/m^3, joints pointed	1.92 · 10^{-4}	0.91
Hollow concrete blocks 39 × 19 × 14, 954 kg/m^3, joints pointed	3.72 · 10^{-4}	0.79
Hollow concrete blocks 39 × 19 × 14, 910 kg/m^3, joints pointed	6.00 · 10^{-4}	0.70

Component	a $kg/(s \cdot m^2 \cdot Pa^b)$	b
Hollow concrete blocks 39 × 19 × 14, 987 kg/m³, plastered at the inside	$3.48 \cdot 10^{-7}$	0.96
Hollow concrete blocks 39 × 19 × 14, 954 kg/m³, plastered at the inside	$2.64 \cdot 10^{-7}$	0.95
Hollow concrete blocks 39 × 19 × 14, 910 kg/m³, plastered at the inside	$3.12 \cdot 10^{-7}$	0.97
Cellular concrete 60 × 24 × 14 cm³, 510 kg/m³, joints glued/not pointed, 1	$2.28 \cdot 10^{-3}$	0.61
Cellular concrete 60 × 24 × 14 cm³, 510 kg/m³, joints glued/not pointed, 2	$1.44 \cdot 10^{-3}$	0.62
Cellular concrete 60 × 24 × 14 cm³, 510 kg/m³, joints glued and pointed, 1	$9.60 \cdot 10^{-5}$	0.64
Cellular concrete 60 × 24 × 14 cm³, 510 kg/m³, joints glued and pointed, 2	$1.02 \cdot 10^{-4}$	0.63
Cellular concrete 60 × 24 × 14 cm³, 510 kg/m³, joints glued, plastered inside	$4.08 \cdot 10^{-7}$	0.97
Sand-lime stone, 29 × 14 × 14 cm³, 1140 kg/m³, joints not pointed	$3.24 \cdot 10^{-3}$	0.61
Sand-lime stone, 29 × 14 × 14 cm³, 1140 kg/m³, joints not pointed	$4.20 \cdot 10^{-3}$	0.57
Sand-lime stone, 29 × 14 × 14 cm³, 1140 kg/m³, joints pointed	$2.28 \cdot 10^{-5}$	0.75
Sand-lime stone, 29 × 14 × 14 cm³, 1140 kg/m³, joints pointed	$1.80 \cdot 10^{-5}$	0.80
Sand-lime stone, 29 × 14 × 14 cm³, 1140 kg/m³, plastered at the inside	$3.00 \cdot 10^{-7}$	0.95

Metal construction

Air permeance, per m² ($K_a = a \Delta P_a^{b-1}$)

Component	a $kg/(s \cdot m^2 \cdot Pa^b)$	b
No special care for air-tightness	$7.9 \cdot 10^{-5}$	0.97
Screw holes at the columns sealed	$6.7 \cdot 10^{-5}$	0.90
Screw holes sealed and joints between metal boxes taped	$1.6 \cdot 10^{-5}$	0.92

5.3 Building and isolating materials

Roof covers

Air permeance, per m^2 ($K_a = a\, \Delta P_a^{b-1}$)

Layer	a kg/(s · m² · Pa^b)	b
Ceramic tiles, single lock, 9.2 meter run of locked joints per m² ('Storm'-tiles)	0.0140	0.55
Ceramic tiles, double lock, 9.2 meter run of locked joints per m² ('Storm'-tiles)	0.0190	0.50
Ceramic tiles, double lock, 9.2 meter run of locked joints per m² ('Storm'-tiles), 1 ventilation tile per m²	0.0160	0.50
Ceramic tiles, double lock, 8.3 meter run of locked joints per m² (Flemish tiles)	0.0150	0.50
Concrete tiles, double lock and overlap, 6.2 meter run of locked joints and overlap per m²	0.0110	0.68
Schiefer slates, 3 slates thick	0.0140	0.62
Fiber cement slates, 3 slates thick	0.0081	0.54
Metal tiles, single lock, 3.5 meter run of locked joints per m²	0.0110	0.54
Fiber cement corrugated boards	0.0042	0.79
Metal roof with standing seam, outside the fixing zone (extends 15 cm at both sides of the fixation)	0.0054	0.64
Metal roof with standing seam, in the fixing zone (extends 15 cm at both sides of the fixation)	0.0014	0.70
Roof sandwich element, composed of (top to bottom): – corrugated 0.8 mm thick aluminum sheet as cover – 50 mm EPS with a top surface that follows the corrugations of the aluminum sheet – 0.17 mm thick aluminum vapor barrier Total surface: 4.93 m². Contains two joints, parallel to the slope, and one overlap, orthogonal to the slope. The parallel joints are formed by one corrugation overlapping at the cover side and a profiled insert inside	0.0017	0.79

Underlays

Air permeance, per m² ($K_a = a \, \Delta P_a^{b-1}$)

	a kg/(s·m²·Pa^b)	b
Fiber/cellulose cement plate, d = 3.2 mm, perfectly closed overlap	0.000420	0.66
Fiber/cellulose cement plate, d = 3.2 mm, overlap 1.4 mm open	0.000420	0.69
Fiber/cellulose cement plate, d = 3.2 mm, overlap 3.6 mm open	0.003200	0.60
Fiber/cellulose cement plate, d = 3.2 mm, overlap 13 mm open	0.010000	0.55
Spun-bonded PE, d = 0.42 mm, 140 g/m²		
– Sample 1	$1.5 \cdot 10^{-6}$	0.97
– Sample 2	$2.5 \cdot 10^{-6}$	1.00
– Sample 3	$2.8 \cdot 10^{-6}$	1.00
Spun-bonded PE, 11 perforations ⌀ 1.2 mm per m²	$4.6 \cdot 10^{-6}$	1.00
Spun-bonded PE, 11 perforations ⌀ 1.7 mm per m²	$7.6 \cdot 10^{-6}$	0.96
Spun-bonded PE, 11 perforations ⌀ 2.7 mm per m²	$1.7 \cdot 10^{-5}$	0.85
Spun-bonded PE, 11 perforations ⌀ 3.0 mm per m²	$2.2 \cdot 10^{-5}$	0.82
Spun-bonded PE, 11 perforations ⌀ 5.0 mm per m²	$9.2 \cdot 10^{-5}$	0.66
Spun-bonded PE, 11 perforations ⌀ 8.5 mm per m²	$3.2 \cdot 10^{-4}$	0.63
Spun-bonded PE, with overlap	0.000013	0.67
Spun-bonded PE, 11 staples per m²	$2.8 \cdot 10^{-6}$	1.00
Spun-bonded PE, 33 staples per m²	$3.2 \cdot 10^{-6}$	1.00
Spun-bonded PE, 55 staples per m²	$3.3 \cdot 10^{-6}$	1.00
Spun-bonded PE, 111 staples per m²	$3.3 \cdot 10^{-6}$	1.00
Spun-bonded PE, overlapping, overlaps continuous sealed		
– Sample 1	$2.8 \cdot 10^{-6}$	1.00
– Sample 2	$9.5 \cdot 10^{-6}$	0.94
– Sample 3	$3.1 \cdot 10^{-6}$	1.00
Bitumen impregnated polypropylene foil, d = 0.6 mm, 496 g/m²	$7.9 \cdot 10^{-7}$	0.43
Bitumen impregnated polypropylene foil, d = 0.6 mm, 496 g/m², with overlap	0.000065	0.37
Bituminous underlay foil		
– Sample 1	$5.3 \cdot 10^{-7}$	0.99
– Sample 2	$6.1 \cdot 10^{-7}$	0.97
Bituminous underlay foil, 11 perfor. ⌀ 1.2 mm per m²	$6.9 \cdot 10^{-6}$	0.74
Bituminous underlay foil, 11 perfor. ⌀ 1.7 mm per m²	$1.7 \cdot 10^{-5}$	0.67
Bituminous underlay foil, 11 perfor. ⌀ 2.7 mm per m²	$5.1 \cdot 10^{-5}$	0.60
Bituminous underlay foil, 11 perfor. ⌀ 3.0 mm per m²	$7.3 \cdot 10^{-5}$	0.59
Bituminous underlay foil, 11 perfor. ⌀ 5.0 mm per m²	$2.1 \cdot 10^{-4}$	0.58
Bituminous underlay foil, 11 perfor. ⌀ 8.5 mm per m²	$7.0 \cdot 10^{-4}$	0.54
Micro-perforated, glass-fiber reinforced plastic foil	0.000090	0.77
– with overlap of 20 cm, per m run	0.005000	0.20
– with overlap of 30 cm, per m run	0.002500	0.43
– with overlap of 50 cm, per m run	0.001200	0.43

5.3 Building and isolating materials

Insulating materials

Air permeability

Material	Air permeability (k_a) kg/(m · s · Pa)
Fiberglass (ρ: density in kg/m³)	$4.2 \cdot 10^{-3} \rho^{-1.24}$
Mineral fiber (ρ: density in kg/m³)	$1.8 \cdot 10^{-2} \rho^{-1.44}$
Fiberglass, 18 kg/m³, along the thickness	$7.2 \cdot 10^{-5}$
Fiberglass, 42 kg/m³, along the thickness	$5.2 \cdot 10^{-5}$
Fiberglass, 42 kg/m³, orthogonal to the thickness	$8.1 \cdot 10^{-5}$
Fiberglass, 78 kg/m³, along the thickness	$3.9 \cdot 10^{-5}$
Fiberglass, 78 kg/m³, orthogonal to the thickness	$5.4 \cdot 10^{-5}$
Fiberglass, 147 kg/m³, along the thickness	$2.2 \cdot 10^{-5}$
Fiberglass, 147 kg/m³, orthogonal to the thickness	$2.9 \cdot 10^{-5}$

Air permeance, per m² ($K_a = a\, \Delta P_a^{b-1}$)

Material	a kg/(s · m² · Pa^b)	b
EPS, d = 64.4 mm, ρ = 10.2 kg/m³	0.004700	0.50
EPS, d = 63.6 mm, ρ = 10.1 kg/m³	0.017000	0.823
EPS, d = 39.1 mm, ρ = 11.6 kg/m³	$7.3 \cdot 10^{-6}$	0.94
EPS, d = 39.1 mm, ρ = 11.5 kg/m³	$8.4 \cdot 10^{-6}$	0.93
EPS, d = 50.1 mm, ρ = 14.1 kg/m³	0.000045	0.94
EPS, d = 50.2 mm, ρ = 14.6 kg/m³	0.000053	0.90
EPS, d = 40.9 mm, ρ = 45.1 kg/m³	0.000030	0.95
EPS, d = 39.3 mm, ρ = 64.2 kg/m³	0.000058	0.90
EPS, d = 49.0 mm, ρ = 64.5 kg/m³	0.000028	0.92
EPS, d = 49.6 mm, ρ = 63.7 kg/m³	$5.5 \cdot 10^{-6}$	0.93
EPS, d = 80 mm, ρ = 19.8 kg/m³	0.000018	0.93
EPS, d = 80 mm, ρ = 20.2 kg/m³	0.000019	0.86
EPS, d = 100 mm, ρ = 20.6 kg/m³	0.000016	0.90
EPS, d = 100 mm, ρ = 21.2 kg/m³	0.000032	0.88
EPS, d = 120 mm, ρ = 20.5 kg/m³	0.000023	0.97
EPS, d = 120 mm, ρ = 19.7 kg/m³	0.000015	0.93
EPS, d = 140 mm, ρ = 19.9 kg/m³	$6.8 \cdot 10^{-6}$	0.91
EPS, d = 140 mm, ρ = 20.3 kg/m³	$6.7 \cdot 10^{-6}$	0.91
EPS, d = 80 mm, ρ = 28.3 kg/m³	$9.4 \cdot 10^{-6}$	0.89
EPS, d = 80 mm, ρ = 28.5 kg/m³	0.000013	0.83
EPS, d = 100 mm, ρ = 30.5 kg/m³	0.000015	0.91
EPS, d = 100 mm, ρ = 30.9 kg/m³	0.000013	0.87
EPS, d = 120 mm, ρ = 29.7 kg/m³	0.000017	0.89
EPS, d = 120 mm, ρ = 27.7 kg/m³	$8.9 \cdot 10^{-6}$	0.91
EPS, d = 140 mm, ρ = 29.0 kg/m³	$4.5 \cdot 10^{-6}$	0.81
EPS with aluminum foil backing, d = 80 mm, ρ = 28.7 kg/m³	$9.0 \cdot 10^{-6}$	0.74
EPS with aluminum foil backing, d = 100 mm, ρ = 29.7 kg/m³	$8.2 \cdot 10^{-7}$	0.97
EPS with aluminum foil backing, d = 120 mm, ρ = 30.0 kg/m³	$7.2 \cdot 10^{-6}$	0.81
EPS with aluminum foil backing, d = 140 mm, ρ = 29.9 kg/m³	$6.6 \cdot 10^{-6}$	0.80
Glass-fiber bat, d = 6 cm, bituminous paper backing, overlaps taped	0.000065 0.000092	0.71 0.70
Glass-fiber bat, d = 6 cm, bituminous paper backing, overlaps taped, staples alongside the tape	0.000075	0.74

5.3 Building and isolating materials

Layer	a kg/(s·m²·Pa^b)	b
Glass-fiber bat, $d = 6$ cm, bituminous paper backing, overlaps taped, 1 nail $\varnothing$ 3 mm per m² perforating the bat	0.000071 0.000097	0.75 0.64
Glass-fiber bat, $d = 6$ cm, bituminous paper backing, overlaps taped, bat ripped ($L = 40$ mm, $B = 40$ mm, 1 per m²)	0.000200	0.61
Glass-fiber bat, $d = 6$ cm, bituminous paper backing, overlaps held together with nailed lath	0.001000 0.000330	0.56 0.76
Glass-fiber bat, $d = 6$ cm, bituminous paper backing, overlaps not overlapping	0.003200 0.000890	1.00 0.85
Glass-fiber bat, $d = 6$ cm, aluminum paper backing, overlaps taped	0.000047	1.00
Glass-fiber bat, $d = 6$ cm, aluminum paper backing, overlaps taped, staples alongside the tape	0.000048	1.00
Glass-fiber bat, $d = 6$ cm, aluminum paper backing, overlaps taped, 1 nail $\varnothing$ 3 mm per m² perforating the bat	0.000050	1.00
Glass-fiber bat, $d = 6$ cm, aluminum paper backing, overlaps taped, bat ripped ($L = 40$ mm, $B = 40$ mm, 1 per m²)	0.000085	1.00
Mineral fiber, dense boards, 170 kg/m³, $d = 10$ cm	0.000230	0.89
XPS, $d = 5$ cm, mounted groove and tongue	0.000400	0.59
XPS, $d = 5$ cm, mounted with well closed joints	0.000470	0.54
XPS, $d = 5$ cm, mounted with open joints of 2 mm	0.002000	0.55
EPS, $d = 12$ m, 9.3 kg/m³, plain board	0.000077	0.91
Composite element particle board 4 mm/EPS 88 mm/particle board 4 mm with EPS/EPS joints – 0.5 mm wide – 7.5 mm wide – 15 mm wide	 0.000710 0.002300 0.010000	 0.56 0.60 0.53
Composite element particle board 4 mm/PUR 61 mm/particle board 4 mm with timber/timber joints – 0.3 mm wide – 3.5 mm wide – 8.0 mm wide – 16 mm wide	 0.000650 0.000740 0.001100 0.001500	 0.56 0.63 0.61 0.55
Composite element particle board 4 mm/PUR 88 mm/particle board 4 mm with PUR/PUR joints – 0.2 mm wide – 4.3 mm wide – 16 mm wide	 0.000150 0.000140 0.015000	 0.66 0.65 0.53
XPS board, $d = 60$ mm, $\rho = 39$ kg/m³ – Sample 1 – Sample 2 – Sample 3	 $5.7 \cdot 10^{-7}$ $4.4 \cdot 10^{-7}$ $6.7 \cdot 10^{-7}$	 1.00 1.00 1.00
XPS-board, $d = 60$ mm, $\rho = 39$ kg/m³ with gypsum board, glued to the XPS, as inside lining – Sample 1 – Sample 2 – Sample 3 – Sample 4	 $5.6 \cdot 10^{-7}$ $6.6 \cdot 10^{-7}$ $4.0 \cdot 10^{-7}$ $9.0 \cdot 10^{-7}$	 1.00 1.00 1.00 1.00

Air permeance, per m run ($K_a = a \, \Delta P_a^{b-1}$)

Layer	a kg/(s · m² · Pa^b)	b
Composite element plywood/30 mm air cavity/120 mm mineral fiber insulation/aluminum vapor retarder/plywood. The elements have timber/timber joints, with a wooden insert close to the outside surface and a special joint profile which should guarantee air tightness		
Element loaded with 65 kg/m²		
– Joint perfectly closed	0.000018	0.87
– Joint perfectly closed, joint profile removed	0.000240	0.64
– Joint 1 mm open	0.000086	0.69
– Joint 2 mm open	0.000480	0.68
– Joint 3 mm open	0.001300	0.57
Element unloaded		
– Joint 1 mm open	0.000240	0.66
– Joint 2 mm open	0.000320	0.75
– Joint 5 mm open	0.001700	0.57
XPS, d = 50 mm (manufacturer 1), tongue and groove (α)		
– Joint 0 mm open	0.000170	0.70
– Joint 5 mm open	0.000600	0.60
– Joint 10 mm open	0.001400	0.50
XPS, d = 50 mm (man. 2), tongue and groove (α)		
– Joint width 0 mm	0.000025	1.00
– Joint width 5 mm	0.000190	0.70
– Joint width 10 mm	0.000360	0.60
XPS, d = 50 mm (manufacturer 2, other batch), tongue and groove (α)		
– Joint width 0 mm	0.000042	1.00
– Joint width 5 mm	0.000140	0.70
– Joint width 10 mm	0.000350	0.60
XPS, d = 50 mm (manufacturer 3), tongue and groove (α)		
– Joint width 0 mm	$8.4 \cdot 10^{-6}$	1.00
– Joint width 5 mm	0.000120	0.70
– Joint width 10 mm	0.000960	0.60
EPs, d = 50 mm (manufacturer 1), tongue and groove (α)		
– Joint width 0 mm	0.000012	1.00
– Joint width 5 mm	0.000096	1.00
– Joint width 10 mm	0.000960	0.60

5.3 Building and isolating materials

Layer	a kg/(s · m² · Pa^b)	b
EPs, $d = 40$ mm (manufacturer 2), tongue and groove (α)		
– Joint width 0 mm	$4.8 \cdot 10^{-6}$	1.00
– Joint width 5 mm	0.000026	1.00
– Joint width 10 mm	0.000240	0.60
Two rebated (rebatted?) XPS boards on a rafter, joints between rafter and boards and between boards open		
– Sample 1	$6.7 \cdot 10^{-5}$	1.00
– Sample 2	$1.8 \cdot 10^{-3}$	0.75
Two rebattedXPS boards on a rafter, joints between rafter and boards sealed with a bituminous tape		
– Sample 1	$2.2 \cdot 10^{-5}$	0.98
– Sample 2	$1.1 \cdot 10^{-4}$	0.95
– Sample 3	$2.7 \cdot 10^{-5}$	0.80
– Sample 4	$4.1 \cdot 10^{-5}$	1.00
Two rebatted XPS boards on a rafter, joints between rafter and boards and at the top, between the two boards, sealed with a bituminous tape		
– Sample 1	$6.6 \cdot 10^{-6}$	0.96
– Sample 2	$6.8 \cdot 10^{-6}$	0.95
– Sample 3	$11.9 \cdot 10^{-6}$	0.95
– Sample 4	$4.1 \cdot 10^{-6}$	0.99
Two rebatted XPS boards on a rafter, joints between rafter and boards and at the top, between the two boards, sealed with silicone joint filler		
– Sample 1	$1.3 \cdot 10^{-6}$	0.97
– Sample 2	$8.0 \cdot 10^{-7}$	1.00
– Sample 3	$2.1 \cdot 10^{-6}$	0.91
– Sample 4	$2.5 \cdot 10^{-6}$	0.92

Vapour retarders and air barriers

Air permeance, per m² ($K_a = a\, \Delta P_a^{b-1}$)

Layer	a kg/(s·m²·Pa^b)	b
PE-foil, $d = 0.2$ mm		
– Sample 1	$5.5 \cdot 10^{-11}$	1.00
– Sample 2	$3.9 \cdot 10^{-7}$	1.00
– Sample 3	$4.8 \cdot 10^{-7}$	1.00
– Sample 4	$4.3 \cdot 10^{-7}$	1.00
PE-foil, $d = 0.2$ mm, 11 perforations $\varnothing$ 1.2 mm per m²	$1.1 \cdot 10^{-6}$	0.91
PE-foil, $d = 0.2$ mm, 11 perforations $\varnothing$ 1.7 mm per m²	$2.5 \cdot 10^{-6}$	0.82
PE-foil, $d = 0.2$ mm, 11 perforations $\varnothing$ 2.7 mm per m²	$9.2 \cdot 10^{-6}$	0.71
PE-foil, $d = 0.2$ mm, 11 perforations $\varnothing$ 3.0 mm per m²	$2.4 \cdot 10^{-5}$	0.65
PE-foil, $d = 0.2$ mm, 11 perforations $\varnothing$ 5.0 mm per m²	$1.1 \cdot 10^{-4}$	0.57
PE-foil, $d = 0.2$ mm, 11 perforations $\varnothing$ 8.5 mm per m²	$4.7 \cdot 10^{-4}$	0.53
PE-foil, $d = 0.2$ mm, with overlap, overlap sealed		
– Sample 1 (bad)	$6.0 \cdot 10^{-6}$	0.77
– Sample 2	$4.9 \cdot 10^{-7}$	1.00
– Sample 3	$4.4 \cdot 10^{-7}$	1.00
Plastic foil SD2		
– Sample 1	$4.6 \cdot 10^{-7}$	1.00
– Sample 2	$5.1 \cdot 10^{-7}$	1.00
– Sample 3	$4.7 \cdot 10^{-7}$	1.00
Plastic foil SD2, 11 perforations $\varnothing$ 1.2 mm per m²	$1.9 \cdot 10^{-6}$	0.86
Plastic foil SD2, 11 perforations $\varnothing$ 1.7 mm per m²	$6.6 \cdot 10^{-6}$	0.73
Plastic foil SD2, 11 perforations $\varnothing$ 2.7 mm per m²	$2.4 \cdot 10^{-5}$	0.63
Plastic foil SD2, 11 perforations $\varnothing$ 3.0 mm per m²	$4.6 \cdot 10^{-5}$	0.60
Plastic foil SD2, 11 perforations $\varnothing$ 5.0 mm per m²	$1.2 \cdot 10^{-4}$	0.59
Plastic foil SD2, 11 perforations $\varnothing$ 8.5 mm per m²	$3.5 \cdot 10^{-4}$	0.60
Plastic foil SD2, 22 staples per m²	$6.5 \cdot 10^{-7}$	1.00
Plastic foil SD2, 44 staples per m²	$7.6 \cdot 10^{-7}$	0.96
Plastic foil SD2, 111 staples per m²	$1.1 \cdot 10^{-6}$	0.94
Plastic foil SD2		
– Sample 1	$7.3 \cdot 10^{-7}$	0.97
– Sample 2	$7.1 \cdot 10^{-7}$	1.00
– Sample 3 (not well done)	$5.3 \cdot 10^{-6}$	0.80

Internal linings

Air permeance, per m² ($K_a = a \, \Delta P_a^{b-1}$)

Layer	a kg/(s · m² · Pa^b)	b
Lathed ceiling, d = 1 cm, groove and tongue	0.000270	0.63
	0.000410	0.68
Lathed ceiling, d = 1 cm, groove and tongue, leak ⌀ 20 mm per m²	0.000760	0.63
Gypsum board, no joint	$1.5 \cdot 10^{-7}$	0.89
Gypsum board, no joint, painted	$1.2 \cdot 10^{-7}$	0.82
Gypsum board drywall, joint reinforced and plastered	0.000031	0.81
Gypsum board drywall, joint reinforced and plastered, leak ⌀ 20 mm per m²	0.000380	0.61
Gypsum board drywall, joint not plastered	0.000330	0.73
Gypsum board drywall, joint not plastered, leak ⌀ 20 mm per m²	0.000630	0.73
Gypsum board drywall with aluminum paper backing, joint reinforced and plastered	0.000013	1.00
Gypsum board drywall with aluminum paper backing, joint reinforced and plastered, leak ⌀ 20 mm per m²	0.000470	0.53
Gypsum board drywall with aluminum paper backing, joint not plastered, leak ⌀ 20 mm per m²	0.000560	0.59

5.3.3 Moisture properties

5.3.3.1 Design values (ISO 10456)

The table in ISO 10456 gives vapour resistance factors.

Building materials

Material group	Material	Density ρ kg/m³	Vapour resistance factor μ –	
			Wet cup	Dry cup
Metals	Aluminium alloys	2800	∞	∞
	Duralumin	2800		
	Brass	8400		
	Bronze	8700		
	Copper	8900		
	Iron	7900		
	Iron, cast	7500		
	Steel	7800		
	Stainless steel	7900		
	Lead	11300		
	Zinc	7100		

Material group	Material	Density ρ kg/m³	Vapour resistance factor μ	
			Wet cup	Dry cup
Wood and wood-based materials	Soft wood	500	50	20
	Hard wood	700	200	50
	Plywood (from low to high density)	300	150	50
		500	200	70
		700	220	90
		1000	250	110
	Particle board – Soft	300	50	10
	– Semi-hard	600	50	15
	– Hard	900	50	20
	OSB	680	50	30
	Fibre board – Soft	400	10	5
	– Semi-hard	600	20	12
	– Hard	800	30	20
Gypsum	Gypsum blocks	600	10	4
		900	10	4
		1200	10	4
		1500	10	4
	Gypsum board	900	10	4
Mortars	Normal mortar, mixed at site	1800	10	6
Plasters	Gypsum plaster – Light	600	10	6
	– Normal	1000	10	6
	– Heavy	1300	10	6
	Lime or gypsum, sand	1600	10	6
	Cement, sand	1700	10	6
Concrete	Medium to high density	1800	100	60
		2000	100	60
		2200	120	70
		2400	130	80
	Reinforced – 1% steel	2300	130	80
	– 2% steel	2400	130	80

5.3 Building and isolating materials

Material group	Material	Density ρ kg/m³	Vapour resistance factor μ	
			Wet cup	Dry cup
Stone	Crystalline rock	2800	10000	10000
	Sedimentary rock	2600	250	200
	Sedimentary, light	1500	30	20
	Lava	1600	20	15
	Basalt	2700–3000	10000	10000
	Gneiss	2400–2700	10000	10000
	Granite	2500–3000	10000	10000
	Marble	2800	10000	10000
	Slate	2000–2800	1000	800
	Lime stone			
	– Extra soft	1600	30	20
	– Soft	1800	40	25
	– Semi-hard	2000	50	40
	– Hard	2200	200	150
	– Extra hard	2600	250	200
	Sandstone	2600	40	302.30
	Natural pumice	400	8	6
	Artificial stone	1750	50	40
Soils	Clay or silt	1200–1800	50	50
	Sand and gravel	1700–2200	50	50
Water, ice, snow	Ice, –10 °C	920	No data	No data
	Ice, 0 °C	900		
	Water, 10 °C	1000		
	Water, 40 °C	990		
	Snow			
	– Fresh (< 30 mm)	100		
	– Soft (30–70 mm)	200		
	– Slightly compacted (70–100 mm)	300		
	– Compacted (> 200 mm)	500		

Material group	Material	Density ρ kg/m³	Vapour resistance factor μ –	
			Wet cup	Dry cup
Plastics, solid	Acrylic	1050	10000	10000
	Polycarbonates	1200	5000	5000
	PFTE	2200	10000	10000
	PVC	1390	50000	50000
	PMMA	1180	50000	50000
	Polyacetate	1410	100000	100000
	Polyamide (Nylon)	1150	50000	50000
	Nylon, 25% glass fibre	1450	50000	50000
	PE, high density	980	100000	100000
	PE, low density	920	100000	100000
	Polystyrene	1050	100000	100000
	Polypropylene (PP)	910	10000	10000
Plastics, solid	PP, 25% glass fibre	1200	10000	10000
	Polyurethane	1200	6000	6000
	Epoxy resin	1200	10000	10000
	Phenol resin	1300	100000	100000
	Polyester resin	1400	10000	10000
Rubbers	Natural	910	10000	10000
	Neoprene	1240	10000	10000
	Butyl	1200	200000	200000
	Foam rubber	60–80	7000	7000
	Hard rubber	1200	∞	∞
	EPDM	1150	6000	6000
	Polyisobutylene	920	10000	10000
	Polysulfide	1700	10000	10000
	Butadiene	980	100000	100000
Glass	Quartz glass	2200	∞	∞
	Normal glass	2500	∞	∞
	Glass mosaic	2000	∞	∞
Gases	Air	1.23	1	1
	Argon	1.70	1	1
	Carbon dioxide	1.95	1	1
	Sulphur hexafluoride	6.36	1	1
	Krypton	3.56	1	1
	Xenon	5.68	1	1

5.3 Building and isolating materials

Material group	Material	Density ρ kg/m³	Vapour resistance factor μ — Wet cup	Vapour resistance factor μ — Dry cup
Thermal breaks, sealants, weather stripping	Silica gel (desiccant)	720	∞	∞
	Silicone foam	750	10000	10000
	Silicone pure	1200	5000	5000
	Silicone, filled	1450	5000	5000
	Urethane/PUR (thermal break)	1300	60	60
	PVC, flexible with 40% softener	1200	100000	100000
	Elastomeric foam, flexible	60–80	10000	10000
	Polyurethane foam	70	60	60
	Polyethylene foam	70	100	100
Roofing materials	Asphalt	2100–2300	50000	50000
	Bitumen, pure	1050	50000	50000
	Bitumen felt	1100	50000	50000
	Clay tiles	2000	40	30
	Concrete tiles	2100	100	60
Floor coverings	Rubber	1200	10000	10000
	Plastic	1700	10000	10000
	Linoleum	1200	1000	800
	Carpet/textile	200	5	5
	Underlay, rubber	270	10000	10000
	Underlay, felt	120	20	15
	Underlay, wool	200	20	15
	Underlay, cork	200	20	15
	Ceramic tiles	2300	∞	∞
	Plastic tiles	1000	10000	10000
	Cork tiles, light	200	20	15
	Cork tiles, heavy	500	40	20

Insulation materials

Material	Density ρ kg/m³	Vapour resistance factor μ —
Cork	90–160	10
Glas- and mineral wool	10–200	1
EPS	10–50	30–70
XPS	25–65	100–300
PUR	28–55	50–100

5.3.3.2 Measured data
Building materials
Natural stone

Baumberger

Density	kg/m³	$\rho = 1980$					
Hygroscopic curve	kg/m³	$\phi = 0.1$	0.3	0.5	0.65	0.8	0.9
– Sorption		8.5	17.6		27.5	35.6	43.1
Capillary moisture content	kg/m³	210					
Saturation moisture content	kg/m³	230					
Diffusion resistance factor	–	$\phi = 0.265$	0.535		0.715		0.85
		20	17		14		8.8
Capillary water absorption coefficient	kg/(m² · s^{0.5})	0.044					

Obernkirchner

Density	kg/m³	$\rho = 2150$					
Hygroscopic curve	kg/m³	$\phi = 0.1$	0.3	0.5	0.65	0.8	0.9
– Sorption		0.6	1.3		2.6	3.4	4.3
Capillary moisture content	kg/m³	110					
Saturation moisture content	kg/m³	140					
Diffusion resistance factor	–	$\phi = 0.265$	0.535		0.715		0.85
		32	30		28		18
Capillary water absorption coefficient	kg/(m² · s^{0.5})	0.046					

Rüthener

Density	kg/m³	$\rho = 1950$					
Hygroscopic curve	kg/m³	$\phi = 0.1$	0.3	0.5	0.65	0.8	0.9
– Sorption		1.8	4.5		8.0	12.4	16.9
Capillary moisture content	kg/m³	200					
Saturation moisture content	kg/m³	240					
Diffusion resistance factor	–	$\phi = 0.265$	0.535		0.715		0.85
		17	16		13		9.4
Capillary water absorption coefficient	kg/(m² · s^{0.5})	0.30					

5.3 Building and isolating materials

Sander

Density	kg/m³	$\rho = 2120$					
Hygroscopic curve	kg/m³	$\phi = 0.1$	0.3	0.5	0.65	0.8	0.9
– Sorption			4.4	10.2	15.2		22.6
Capillary moisture content	kg/m³	130					
Saturation moisture content	kg/m³	170					
Diffusion resistance factor	–	$\phi = 0.265$		0.535	0.715		0.85
		33		30	22		13
Capillary water absorption coefficient	kg/(m²·s$^{0.5}$)	0.02					

Savonnières

Density	kg/m³	$\rho = 1661$ ($\sigma = 34$, 120 monsters)
Hygroscopic curve	kg/m³	
– Sorption		
Capillary moisture content	kg/m³	160, ($\sigma = 21.5$, 120 monsters)
Saturation moisture content	kg/m³	382 ($\sigma = 14.2$, 120 monsters)
Diffusion resistance factor	–	
Capillary water absorption coefficient	kg/(m²·s$^{0.5}$)	0.085 (//, $\sigma = 0.062$), 0.054 ($\perp$, $\sigma = 0.038$)

Concrete

Density	kg/m³	$\rho \leq 2176$
Hygroscopic curve	kg/m³	
– Sorption	$0.2 \leq \phi \leq 0.98$	$147.5 \left(1 - \dfrac{\ln \phi}{0.0453}\right)^{-\frac{1}{1.67}}$
– Desorption	$0.2 \leq \phi \leq 0.98$	$147.5 \left(1 - \dfrac{\ln \phi}{0.570}\right)^{-\frac{1}{0.64}}$
Critical moisture content	kg/m³	100–110
Capillary moisture content	kg/m³	110
Saturation moisture content	kg/m³	153
Diffusion resistance factor	–	
	$0.2 \leq \phi \leq 0.98$	$\dfrac{1}{6.8 \cdot 10^{-3} + 8.21 \cdot 10^{-5} \exp(5.66\, \phi)}$
Capillary water absorption coefficient	kg/(m² · s$^{0.5}$)	0.018
Moisture diffusivity (±)	m²/s, w in kg/m³	$1.8 \cdot 10^{-1} \exp(0.053\, w)$

Light weight concrete

Density	kg/m³	$644 \leq \rho \leq 1442$
Hygroscopic curve	kg/m³	
– Sorption	$938 \leq \rho \leq 1442$ $0.2 \leq \phi \leq 0.98$	$110 \left(1 - \dfrac{\ln \phi}{0.0277}\right)^{-\frac{1}{2.14}}$
– Desorption	$0.2 \leq \phi \leq 0.98$	$110 \left(1 - \dfrac{\ln \phi}{0.0221}\right)^{-\frac{1}{2.91}}$
Critical moisture content	kg/m³, $\rho = 935$	140
Capillary moisture content	kg/m³, $872 \leq \rho \leq 980$	97–190
Saturation moisture content	kg/m³, $\rho = 973$	584
Diffusion resistance factor	–	
	$0.2 \leq \phi \leq 0.98$, $\rho = 975$	$\dfrac{1}{6.76 \cdot 10^{-2} + 1.21 \cdot 10^{-3} \exp(3.94\, \phi)}$
Capillary waterabsorption coefficient	kg/(m² · s$^{0.5}$) $\rho = 975$ $\rho = 1410$	0.08 0.029
Moisture diffusivity (±)	m²/s, w in kg/m³ $\rho = 975$	$1.3 \cdot 10^{-9} \exp(0.035\, w)$

5.3 Building and isolating materials

Cellular concrete

Density	kg/m^3	$455 \leq \rho \leq 800$
Hygroscopic curve	kg/m^3	
– Sorption	$465 \leq \rho \leq 621$ $0.2 \leq \phi \leq 0.98$	$300 \left(1 - \dfrac{\ln \phi}{0.0011}\right)^{-\frac{1}{1.99}}$
– Desorption	$0.2 \leq \phi \leq 0.98$	$300 \left(1 - \dfrac{\ln \phi}{0.0038}\right)^{-\frac{1}{1.32}}$
Critical moisture content	kg/m^3	180
Capillary moisture content	kg/m^3	$109 + 0.383 \rho$
Saturation moisture content	kg/m^3	$972 - 0.350 \rho$
Diffusion resistance factor	–	
	$0.2 \leq \phi \leq 0.98$ $458 \leq \rho \leq 770$	$\dfrac{1}{1.16 \cdot 10^{-1} + 6.28 \cdot 10^{-3} \exp(4.19 \phi)}$
Capillary water absorption coefficient	kg/(m$^2 \cdot$ s$^{0.5}$)	0.02–0.08
Moisture diffusivity (±)	m^2/s, w in kg/m^3 $\rho = 511$	$9.2 \cdot 10^{-1} \exp(0.0215\, w)$

Cellular concrete 2

Density	kg/m^3	$\rho = 600$					
Hygroscopic curve	kg/m^3	$\phi = 0.1$	0.3	0.5	0.65	0.8	0.9
– Sorption				7.3	12.5	17	38
Capillary moisture content	kg/m^3	290					
Saturation moisture content	kg/m^3	720					
Diffusion resistance factor	–	$\phi = 0.265$		0.535		0.715	0.85
		7.6				6.7	
Capillary water absorption coefficient	kg/(m$^2 \cdot$ s$^{0.5}$)	0.09					

Polystyrene concrete

Density	kg/m³	$259 \leq \rho \leq 792$
Hygroscopic curve	kg/m³	
– Sorption	$\rho \leq 422$ $0.2 \leq \phi \leq 0.98$	$235\left(1 - \dfrac{\ln\phi}{0.0097}\right)^{-\frac{1}{1.55}}$
– Desorption		
Saturation moisture content	kg/m³, $\rho \leq 422$	489
Diffusion resistance factor	– $0.2 \leq \phi \leq 0.98$ $357 \leq \rho \leq 425$	$\dfrac{1}{8.18 \cdot 10^{-2} + 3.16 \cdot 10^{-3} \exp(2.88\phi)}$
Capillary water absorption coefficient	kg/(m²·s$^{0.5}$) $360 \leq \rho \leq 457$	0.026
Moisture diffusivity (±)	m²/s, w in kg/m³ $\rho = 422$	$4.6 \cdot 10^{-10} \exp(0.064\,w)$
Hygric strain	m/m $\rho = 422$	$0.0024 \cdot 10^{-4}\,\phi^2$

Mortel

Density	kg/m³	$1050 \leq \rho \leq 1940$
Hygroscopic curve	kg/m³	
– Sorption	$\rho = 1940$ $0.2 \leq \phi \leq 0.98$	$283\left(1 - \dfrac{\ln\phi}{0.029}\right)^{-\frac{1}{1.39}}$
– Desorption		$300\left(1 - \dfrac{\ln\phi}{0.061}\right)^{-\frac{1}{1.77}}$
Capillary moisture content	kg/m³	283
Diffusion resistance factor	– $0.2 \leq \phi \leq 0.98$	$\dfrac{1}{7.69 \cdot 10^{-2} + 2.43 \cdot 10^{-3} \exp(3.61\phi)}$
Capillary water absorption coefficient	kg/(m²·s$^{0.5}$)	0.042–0.80
Moisture diffusivity (±)	m²/s, w in kg/m³	$C_1 \exp(C_2\,w)$

ρ (kg/m³)	C_1	C_2
1072	$2.0 \cdot 10^{-9}$	0.022
1500	$2.7 \cdot 10^{-9}$	0.020
1807	$1.4 \cdot 10^{-9}$	0.027

5.3 Building and isolating materials

Brick

Density	kg/m³	$1505 \leq \rho \leq 2047$
Hygroscopic curve	kg/m³	
– Sorption	$0.2 \leq \phi \leq 0.98$	$200 \left(1 - \dfrac{\ln \phi}{1.46 \cdot 10^{-4}}\right)^{-\frac{1}{1.59}}$
Critical moisture content	kg/m³	100
Capillary moisture content	kg/m³ $1505 \leq \rho \leq 2047$	$730.3 - 0.287\rho$
Saturation moisture content	kg/m³, ρ idem	$1033 - 0.404\rho$
Diffusion resistance factor	–	
	$0.2 \leq \phi \leq 0.98$ $1505 \leq \rho \leq 1860$	$\dfrac{1}{5.6 \cdot 10^{-2} + 4.67 \cdot 10^{-3} \exp(2.79\phi)}$
Capillary water absorption coefficient	kg/(m² · s$^{0.5}$)	
– Clay	$1505 \leq \rho \leq 2000$	$0.653 - 0.00030\rho$
– Loam	$1628 \leq \rho \leq 1868$	$1.954 - 0.00087\rho$
Moisture diffusivity (±)	m²/s, w in kg/m³	$C_1 \exp(C_2 w)$

ρ (kg/m³)	C_1	C_2
1529	$2.1 \cdot 10^{-9}$	0.032
1619	$1.9 \cdot 10^{-8}$	0.022
1918	$7.4 \cdot 10^{-9}$	0.032

Brick 2

Density	kg/m³	$\rho = 1700$					
Hygroscopic curve	kg/m³	$\phi = 0.1$	0.3	0.5	0.65	0.8	0.9
– Sorption				7.5	8.4	18	34
Capillary moisture content	kg/m³	270					
Saturation moisture content	kg/m³	380					
Diffusion resistance factor	–	$\phi = 0.265$		0.535	0.715		0.85
		9.5		8.8	8.0		6.9
Capillary water absorption coefficient	kg/(m² · s$^{0.5}$)	0.25					

Lime sand stone

Density	kg/m³	$1685 \leq \rho \leq 1807$
Hygroscopic curve	kg/m³	
– Sorption	$0.2 \leq \phi \leq 0.98$ $1685 \leq \rho \leq 1726$	$210 \left(1 - \dfrac{\ln \phi}{3.56 \cdot 10^{-3}}\right)^{-\frac{1}{2.39}}$
– Desorption		$330 \left(1 - \dfrac{\ln \phi}{6.58 \cdot 10^{-3}}\right)^{-\frac{1}{1.81}}$
Critical moisture content	kg/m³ $\rho \leq 1807$	120
Capillary moisture content	kg/m³ $1711 \leq \rho \leq 1777$	233
Diffusion resistance factor	–	
	$0.2 \leq \phi \leq 0.98$ $1505 \leq \rho \leq 1860$	$\dfrac{1}{4.3 \cdot 10^{-2} + 4.56 \cdot 10^{-5} \exp(9.86 \phi)}$
Capillary water absorption coefficient	kg/(m²·s$^{0.5}$) $\rho \leq 1807$	0.042
Moisture diffusivity (±)	m²/s, w in kg/m³ $\rho \leq 1807$	$2.2 \cdot 10^{-10} \exp(0.027 w)$

Lime sand stone 2

Density	kg/m³	$\rho = 1900$					
Hygroscopic curve	kg/m³	$\phi = 0.1$	0.3	0.5	0.65	0.8	0.9
– Sorption				17	18	24.9	40.2
Capillary moisture content	kg/m³	250					
Saturation moisture content	kg/m³	290					
Diffusion resistance factor	–	$\phi = 0.265$		0.535		0.715	0.85
		28		24		18	13
Capillary water absorption coefficient	kg/(m²·s$^{0.5}$)	0.045					

5.3 Building and isolating materials

Gypsum plaster

Density	kg/m³	975
Hygroscopic curve	kg/m³	
– Sorption	$0.2 \leq \phi \leq 0.98$	$310 \left(1 - \dfrac{\ln \phi}{3.21 \cdot 10^{-2}}\right)^{-\frac{1}{1.59}}$
– Desorption		$\dfrac{\phi}{-0.004\, \phi^2 + 0.01\, \phi - 0.0004}$
Capillary moisture content	kg/m³	310
Diffusion resistance factor	–	
	$0.2 \leq \phi \leq 0.98$	$\dfrac{1}{0.133 + 7.45 \cdot 10^{-4} \exp(5.1\, \phi)}$
Capillary water absorption coefficient	kg/(m² · s^{0.5})	0.155
Moisture diffusivity (±)	m²/s, w in kg/m³	$1.7 \cdot 10^{-9} \exp(0.0206\, w)$

Gypsum plaster 2

Density	kg/m³	$\rho = 850$					
Hygroscopic curve	kg/m³	$\phi = 0.1$	0.3	0.5	0.65	0.8	0.9
– Sorption					3.6	5.2	6.3	11
Capillary moisture content	kg/m³	400					
Saturation moisture content	kg/m³	650					
Diffusion resistance factor	–	$\phi = 0.265$	0.535		0.715		0.85
		8.3			7.3		
Capillary water absorption coefficient	kg/(m² · s^{0.5})	0.29					

EIFS stucco

Density	kg/m^3	$878 \leq \rho \leq 1736$		
Capillary moisture content	kg/m^3, $878 \leq \rho \leq 1736$	185		
Diffusion resistance factor	–	$1.8 \exp(0.0016\rho)$		
	$880 \leq \rho \leq 1709$ $\phi = 86\%$	ρ kg/m^3	ϕ_m %	μ –
		1680	86	43
			95	15
Capillary water absorption coefficient	kg/(m$^2 \cdot$ s$^{0.5}$)	$0.0039 \leq A \leq 0.029$ Gemiddeld 0.0128		
Moisture diffusivity (±)	m^2/s$^{0.5}$	$C_1 \exp(C_2 w)$		
		ρ kg/m^3	C_1	C_2
		878	$4.4 \cdot 10^{-12}$	0.027
		1341	$2.1 \cdot 10^{-10}$	0.019

Timber

Density	kg/m^3	$400 \leq \rho \leq 690$		
Hygroscopic curve	kg/m^3			
– Sorption	$0.2 \leq \phi \leq 0.98$	$100 \left(1 - \dfrac{\ln \phi}{0.642}\right)^{-\frac{1}{0.64}}$		
– Desorption		$120 \left(1 - \dfrac{\ln \phi}{0.248}\right)^{-\frac{1}{1.22}}$		
Diffusion resistance factor	–	ρ kg/m^3	ϕ_m %	μ –
		Oak, 640	30	110
			87	21
		Beech, 435	25	180
			75	20
		Spruce, 390	53	57
			86	13
Capillary water absorption coefficient	kg/(m$^2 \cdot$ s$^{0.5}$) $\rho = 390$ (spruce)	$0.004 \perp$ fibres $0.016\ //$ fibres		

5.3 Building and isolating materials

Particle board

Density	kg/m^3	$570 \leq \rho \leq 800$			
Hygroscopic curve	%kg/kg				
– Sorption	$0.2 \leq \phi \leq 0.98$	$35\left(1 - \dfrac{\ln \phi}{0.0328}\right)^{-\frac{1}{1.89}}$			
– Desorption		$35\left(1 - \dfrac{\ln \phi}{0.081}\right)^{-\frac{1}{1.63}}$			
Critical moisture content	%kg/kg	85			
Capillary moisture content	%kg/kg	90			
Saturation moisture content	%kg/kg	99			
Diffusion resistance factor	–	$A_1 \exp(A_2 \rho)$			
		ϕ %	A_1	A_2	r^2
		25	0.543	0.00757	0.70
		86	0.508	0.00676	0.75
Capillary water absorption coefficient	kg/(m$^2 \cdot$ s$^{0.5}$) UF, … glue used	(UF) 0.0035 (UMF) 0.0035 (FF) 0.022			
Moisture diffusivity (±)	m^2/s w in kg/m^3	(UF) $2.3 \cdot 10^{-13} \exp(0.01\,w)$ (UMF) $2.3 \cdot 10^{-13} \exp(0.01\,w)$ (FF) $4.5 \cdot 10^{-12} \exp(0.01\,w)$			

Plywood

Density	kg/m³	$445 \leq \rho \leq 799$		
Hygroscopic curve	%kg/kg			
– Sorption	$0.2 \leq \phi \leq 0.98$	$75\left(1 - \dfrac{\ln \phi}{6.14 \cdot 10^{-3}}\right)^{-\frac{1}{1.91}}$		
– Desorption		$75\left(1 - \dfrac{\ln \phi}{9.63 \cdot 10^{-3}}\right)^{-\frac{1}{1.93}}$		
Critical moisture content	%kg/kg	75		
Capillary moisture content	%kg/kg	75		
Saturation moisture content	%kg/kg	75		
Diffusion resistance factor	–	$15.3 \exp[(0.0045(\rho - 437))]$		
	$437 \leq \rho \leq 591$ $\phi = 86\%$	ρ kg/m³	ϕ_m %	μ –
		548–580	54	97
		437–591	86	24
Capillary water absorption coefficient	kg/(m²·s$^{0.5}$)	0.003		
Moisture diffusivity (±)	m²/s$^{0.5}$	$3.2 \cdot 10^{-13} \exp(0.015 \, w)$		

Woodwool cement boards

Density	kg/m³	$314 \leq \rho \leq 767$
Hygroscopic curve	%kg/kg	
– Sorption	$0.2 \leq \phi \leq 0.98$ $\rho = 767$	$15\left(1 - \dfrac{\ln \phi}{0.172}\right)^{-\frac{1}{0.84}}$
Capillary moisture content	kg/m³, $\rho = 360$	180
Saturation moisture content	kg/m³, $\rho = 360$	240
Diffusion resistance factor	–	≈ 4
Capillary water absorption coefficient	kg/(m²·s$^{0.5}$)	0.007
Moisture diffusivity (±)	m²/s$^{0.5}$, w in kg/m³	$6.2 \cdot 10^{-12} \exp(0.027 \, w)$

5.3 Building and isolating materials

Fibre cement boards

Density	kg/m^3	$823 \leq \rho \leq 2052$
Hygroscopic curve	kg/m^3	
– Sorption	$0.2 \leq \phi \leq 0.98$	
	$\rho = 990$	$300\left(1 - \dfrac{\ln\phi}{0.0077}\right)^{-\frac{1}{1.93}}$
	$\rho = 1495$	$358\left(1 - \dfrac{\ln\phi}{0.0415}\right)^{-\frac{1}{1.36}}$
– Desorption		
	$\rho = 990$	$350\left(1 - \dfrac{\ln\phi}{0.1076}\right)^{-\frac{1}{1.22}}$
	$\rho = 1495$	$358\left(1 - \dfrac{\ln\phi}{0.197}\right)^{-\frac{1}{0.8}}$
Critical moisture content	kg/m^3, $\rho = 840$	350
Capillary moisture content	kg/m^3, $\rho = 1495$	358
Saturation moisture content	kg/m^3, $\rho = 1495$	430
Diffusion resistance factor	–, $0.2 \leq \phi \leq 0.98$	
	$\rho = 840$	$\dfrac{1}{0.0565 + 5.58 \cdot 10^{-5} \exp(7.85\,\phi)}$
	$\rho = 1495$	$\dfrac{1}{0.00642 + 1.4 \cdot 10^{-4} \exp(4.92\,\phi)}$
Capillary water absorption coefficient	kg/(m$^2 \cdot$ s$^{0.5}$)	0.024
Moisture diffusivity (±)	m^2/s$^{0.5}$, w in kg/m^3 $840 \leq \rho \leq 1495$	$3.4 \cdot 10^{-1} \exp(0.018\,w)$

Gipskarton

Weight per m^2	kg/m^2	$6.5 \leq \rho \leq 13$
Hygroscopic curve	kg/m^3	
– Sorption	$0.2 \leq \phi \leq 0.98$	$150\left(1 - \dfrac{\ln\phi}{2.99 \cdot 10^{-4}}\right)^{-\frac{1}{4.81}}$
– Desorption		$150\left(1 - \dfrac{\ln\phi}{0.026}\right)^{-\frac{1}{7.86}}$
Diffusion resistance factor	– $d = 9.5$, $d = 12.5$ mm	$\dfrac{1}{0.0712 + 2.81 \cdot 10^{-3} \exp(4.1\,\phi)}$

Masonry walls

Bricks

Density	kg/m³		$830 \leq \rho \leq 1760$		
Equivalent diffusion thickness	m	d m	wall	ϕ_m %	$\mu\,d$ m
		0.09	Bricks 6.5 × 9 × 19 cm	54 / 86	1.20 / 0.51
		0.09	Idem, painted with acrylic paint	54	2.20
		0.09	idem, water repellant treated	86	0.65
		0.09	Glazed bricks	86	4.00
		0.19	Bricks 6.5 × 9 × 19 cm	59 / 81	1.60 / 0.61
		0.14	Hollow blocks 14 × 19 × 29 cm	58 / 84	1.30 / 0.84
		0.19	Bricks 6.5 × 9 × 19 cm	88	0.53

Lime sand stone

Density	kg/m³		$860 \leq \rho \leq 1650$		
Equivalent diffusion thickness	m	d m	wall	ϕ_m %	$\mu\,d$ m
		0.14	$\rho = 1140$ kg/m³ 14 × 14 × 29 cm	25	3.70

Concrete blocks

Density	kg/m³		$860 \leq \rho \leq 1650$		
Equivalent diffusion thickness	m	d m	wall	ϕ_m %	$\mu\,d$ m
		0.14	Blocks $\rho = 960$ kg/m³ 14 × 14 × 29 cm	60 / 86	1.30 / 0.54
		0.14	Blocks $\rho = 1450$ kg/m³ 14 × 14 × 29 cm	61 / 64 / 83 / 90	0.61 / 0.58 / 0.61 / 0.28

5.3 Building and isolating materials

Insulation materials

Cork

Density	kg/m³	$\rho \leq 111$
Hygroscopic curve	%kg/kg	
– Sorption	$0.2 \leq \phi \leq 0.98$	$60\left(1 - \dfrac{\ln \phi}{1.02 \cdot 10^{-5}}\right)^{-\frac{1}{3.64}}$
Critical moisture content	kg/m³	60
Diffusion resistance factor		≈ 22

Cellular glass

Density	kg/m³	$114 \leq \rho \leq 140$
Diffusion resistance factor	–	5000 à 70000

Glass wool

Density	kg/m³	$12 \leq \rho \leq 133$
Diffusion resistance factor	–, $19 \leq \rho \leq 102$	1.2

Mineral wool

Density	kg/m³	$32 \leq \rho \leq 191$
Diffusion resistance factor	–, $148 \leq \rho \leq 172$	1.5

Expanded polystyrene (EPS)

Density	kg/m³	$13 \leq \rho \leq 40$
Diffusion resistance factor	–, $\phi = 0.86$, $15 \leq \rho \leq 40$	$4.9 + 1.97\,\rho$

Extruded polystyrene (XPS)

Density	kg/m³	$25 \leq \rho \leq 55$
Diffusion resistance factor	–, $\phi = 0.86$, $25 \leq \rho \leq 53$	$48.6 + 3.35\,\rho$

PUR/PIR

Density	kg/m³	$20 \leq \rho \leq 40$
Diffusion resistance factor	–, $\phi = 0.8$, $20 \leq \rho \leq 40$	$1.67 \exp(0.008\,\rho)$

Expanded perlite

Density	kg/m³	$135 \leq \rho \leq 215$
Hygroscopic curve	kg/m³	
– Sorption	$0.2 \leq \phi \leq 0.98$ $187 \leq \rho \leq 213$	$150 \left(1 - \dfrac{\ln \phi}{1.15 \cdot 10^{-4}}\right)^{-\frac{1}{2.63}}$
Critical moisture content	kg/m³	150–200
Capillary moisture content	kg/m³	550
Diffusion resistance factor	–, $0 \leq \phi \leq 1$	$\dfrac{1}{0.0143 + 6.54 \cdot 10^{-8} \exp(16.5 \phi)}$

Miscellaneous

Wall paper

Weight per m², thickness	kg/m², mm	Type	kg/m²	d (mm)
		1. textile	0.291	0.425
		2. vinyl	0.216	0.325
		3. textile	0.333	0.700
		4. vinyl	0.212	0.450
		5. paper	0.168	0.280
		6. paper	0.151	0.280

Hygroscopic curve	%kg/kg						
	Paper			ϕ (–)			
		1	2	3	4	5	6
– Sorption	0.33	3.2	1.4	2.9	1.2	1.6	1.8
	0.52	5.5	2.8	5.5	2.3	2.9	4.0
	0.75	7.9	5.0	6.4	3.3	4.6	5.5
	0.86	11.2	8.2	9.8	4.8	7.2	6.8
	0.97	21.4	40.0	24.9	13.3	15.8	16.9
– Desorption	0.33	5.3	4.6	4.0	7.0	2.7	3.9
	0.52	8.5	5.6	6.6	8.3	4.2	6.0
	0.75	11.8	8.7	9.9	9.8	6.9	9.4
	0.86	16.1	10.7	14.0	10.0	9.7	12.3
	0.97	42.8	52.2	38.3	17.9	23.8	36.2

Diffusion thickness	m						
	Paper			ϕ (–)			
		1	2	3	4	5	6
	0.42	0.280	2.14	0.155	0.09	0.035	0.025
	0.75	0.006	0.18	0.019	0.025	0.012	0.008

5.3 Building and isolating materials

Paint

Diffusion thickness	m			
			ϕ (–)	
		0.425	0.750	0.86
	Gypsum board			
	primer + two layers latex 1 paint			0.17
	primer + two layers latex 2 paint	4.50		1.10
	primer + two layers acrylic paint			0.46
	primer + two layers synthetic paint	3.20		1.00
	primer + two layers oil paint			0.76
	Cellular concrete			
	primer+two layers acrylic paint		0.43	
	Structured paint		1.10	

Carpet

Weight per m^2	kg/m^2	2.18	
Hygroscopic curve	%kg/kg		
– Sorption	$0.2 \leq \phi \leq 0.98$	ϕ (–)	X (%kg/kg)
		0.33	8.0
		0.52	9.9
		0.75	13.4
		0.86	17.2
		0.97	29.7
		0.98	29.8

Timber groove and tongue lathing

Weight per m²	kg/m²	4.0 ($d = 10$ mm)		
Hygroscopic curve	%kg/kg			
– Sorption	$0.2 \leq \phi \leq 0.98$	$50\left(1 - \dfrac{\ln\phi}{0.0213}\right)^{-\frac{1}{1.96}}$		
– Desorption		$50\left(1 - \dfrac{\ln\phi}{0.0397}\right)^{-\frac{1}{1.86}}$		
		ϕ (–)	Sorption X (%kg/kg)	desorption X (%kg/kg)
		0.33	5.8	7.1
		0.52	9.4	12.3
		0.75	13.7	15.4
		0.86	17.6	20.8
		0.97	31.4	37.3
		0.98	34.5	
Diffusion thickness	m, $\phi = 0.55$	0.86 ($\sigma = 0.12$)		

5.3 Building and isolating materials

Vapour retarders

			φ (-) =		μ (-), · 10³		
Diffusion resistance factor		d (mm)	0.28	0.52	0.70	0.75	0.86
	PE	0.1–0.2	321			289	271
Diffusion thickness	m		φ (-) =		μ d (m)		
	Vapour retrade		d (mm)	0.52	0.70	0.75	0.86
	Bituminous paperr		0.1				1.80
							2.80
	Idem		0.2	0.70			
	Idem		1.4			1.70	
						6.90	
	Idem		0.4		3.90		
					8.10		
	Aluminum paper		0.1				2.00
							2.80
	Idem		0.2				0.17
							0.33
	Idem		0.24		17.80		
					77.30		
	Idem		–				6.80
							17.80
	Glass fibre reainforced alupaper		0.4		3.80		
					4.70		
	Glass fibre reainforced PVC-foil		0.4				12.00
							29.00
	Stapled PE-foil		0.15			7.70	

Newspaper

Weight per m²	kg/m²	0.041		
Hygroscopic curve	%kg/kg $0.2 \leq \phi \leq 0.98$		φ (−)	Sorption X (%kg/kg)
			0.33	5.3
			0.52	9.0
			0.75	12.9
			0.86	16.9
			0.97	32.1
			0.98	40.9

Journal

Weight per m^2	kg/m^2	0.047	
Hygroscopic curve	%kg/kg $0.2 \leq \phi \leq 0.98$	ϕ (−)	Sorption X (%kg/kg)
		0.33	2.9
		0.52	4.7
		0.75	6.5
		0.86	8.4
		0.97	16.6
		0.98	19.0

5.3 Building and isolating materials

5.3.4 Radiant properties

Material, surface	Emissivity Long wave –	Absorptivity Short wave –
Snow		0.15
White paint	0.85	0.25
Black paint	0.97	
Oil paint	0.94	
White washed surface		0.30
Light colours, polished aluminum		0.3–0.5
Yellow brick	0.93	0.55
Red brick	0.93	0.75
Concrete, light coloured floor cover		0.6–0.7
Grass and leaves		0.75
Dark coloured floor cover		0.8–0.9
Carpet		0.8–0.9
Moist bottom		0.90
Dark gray slates		0.90
Bituminous felt	0.92	0.93
Gold, silver en copper polished	0.02	
Copper, oxydized	0.78	
Aluminum, polished	0.05	
Aluminium, oxydized	0.30	
Steel, hot casted	0.77	
Steel, oxydized	0.61	
Steel with silver finish	0.26	
Steel, polished	0.27	
Lead, oxydized	0.28	
Glass	0.92	
Porselan	0.92	
Plaster	0.93	
Timber, unpainted	0.90	
Marmor, polished	0.55	
Paper	0.93	
Water	0.95	
Ice	0.97	

5.4 References

[5.1] Stichting Bouwresearch (1974). *Eigenschappen van bouw- en isolatiematerialen* (Properties of building and insulating materials), 2^e, gewijzigde druk (in Dutch).

[5.2] Hens, H. (1975). *Theoretische en experimentale studie van het hygrothermisch gedrag van bouw- en isolatiematerialen bij inwendige condensatie en droging met toepassing op de platte daken* (Theoretical and experimental study of the hygrothermal response of building and insulating materials during interstitial condensation and drying with application on low-sloped roofs), PhD K. U. Leuven, 311 p. (in Dutch).

[5.3] Hens, H. (1984). *Cataloog van hygrothermische eigenschappen van bouw- en isolatiematerialen* (Catalogue of hygrothermal properties of building and insulating materials), rapport E/VI/2, R-D Energy, 100 p. + appendices (in Dutch).

[5.4] Trechsel/Bomberg (Eds.) (1989). *Water Vapor Transmission Through Building Materials and Systems*, ASTM, STP 1039.

[5.5] Hens, H. (1991). *Catalogue of Material Properties*, Final Report, IEA-EXCO on Energy Conservation in Buildings and Community Systems, Annex 14 'Condensation and Energy'.

[5.6] Krus, M. (1995). *Feuchtetransport- und Speicherkoeffizienten poröser mineralischer Baustoffe. Theoretische Grundlagen und Neue Meßtechniken*. Doctor Abhandlung, Universität Stuttgart (in German).

[5.7] Kumaran, K. (1996). *Material Properties.* Final Report, IEA, EXCO on Energy Conservation in Buildings and Community Systems, Annex 24 'Heat, Air and Moisture Transfer in Insulated Envelope Parts, Task 3.

[5.8] CEN/TC 89 (1998). B*uilding materials and products – Hygrothermal properties – Tabulated design values*, working draft.

[5.9] Roels, S. (2000). *Modelling unsaturated moisture transport in heterogeneous limestone.* Doctoral Thesis, K. U. Leuven.

[5.10] NBN B62-002/A1 (2001). *Berekening van de warmtedoorgangscoëfficiënten van wanden van gebouwen*, 2^e uitgave (in Dutch).

[5.11] Hagentoft, C. E. (2001). *Introduction to Building Physics*. Studentlitteratur, Lund.

[5.12] Kumaran, K. (2002). *A thermal and Moisture Property Database for Common Building and Insulating Materials*. Final report from ASHRAE Research Project 1018-RP, 231 p.

[5.13] prCEN/TR 14613 (2003). *Thermal performance of building materials and components – principles for the determination of thermal properties of moist materials and components*. Final draft.

[5.14] Roels, S., Carmeliet, J., Hens, H. (2003). *Moisture transfer properties and material characterization.* Final Report EU HAMSTAD project.

[5.15] ISO 10456 (2007). *Building materials and products-Hygrothermal properties – Tabulated design values and procedures for determining declared and design thermal values*, 24 p.

[5.16] ASHRAE (2009). *Handbook of Fundamentals*.

Appendix

Solar radiation for Uccle, Belgium, 50° 51′ North, 4° 21′ East

Table 1 gives the beam insolation on a surface perpendicular to the solar rays under clear sky conditionns for the 15th of each month. Table 2 lists the diffuse insolation on a horizontal surface under clear sky conditions for the 15th of each month. Table 3 compiles beam plus diffuse solar gains under clear sky conditions for surfaces with different orientation and slope. Table 4 gives the totals for each month. In Table 5, the reflected component is included.

Table 1. Beam insolation under clear sky conditions on a surface upright the solar rays on the 15th of each month (W/m^2).

Hour	J	F	M	A	M	J	J	A	S	O	N	D
4.0	0	0	0	0	0	103	45	0	0	0	0	0
4.5	0	0	0	0	133	223	165	0	0	0	0	0
5.0	0	0	0	29	248	321	270	119	0	0	0	0
5.5	0	0	0	169	355	412	370	246	48	0	0	0
6.0	0	0	0	293	445	489	454	356	192	0	0	0
6.5	0	0	138	399	518	552	524	446	314	112	0	0
7.0	0	0	274	484	578	603	580	519	414	251	0	0
7.5	0	163	383	551	625	645	626	577	494	359	153	0
8.0	102	279	469	604	664	680	664	623	555	445	268	97
8.5	218	378	536	646	695	708	694	660	603	510	360	219
9.0	315	455	587	678	719	730	718	689	640	559	430	310
9.5	389	513	626	704	739	749	738	712	668	596	482	379
10.0	445	557	655	723	754	763	754	730	689	623	519	430
10.5	484	587	676	736	765	774	765	743	704	640	544	464
11.0	509	608	689	745	772	781	773	751	713	651	558	485
11.5	523	620	697	749	776	785	778	756	717	654	561	494
12.0	525	623	698	749	775	784	778	756	716	650	556	491
12.5	513	612	693	740	767	776	770	748	706	632	532	468
13.0	483	592	678	728	756	768	761	739	692	609	500	435
13.5	450	569	654	710	739	752	748	723	670	581	460	392
14.0	403	536	629	690	722	736	733	705	646	543	405	333
14.5	338	490	595	664	700	717	714	683	613	492	331	254
15.0	252	428	549	629	673	693	690	655	571	425	236	156
15.5	142	347	490	586	639	664	661	619	217	336	118	0
16.0	0	243	413	531	596	628	625	573	448	229	0	0
16.5	0	80	315	461	544	584	580	517	359	102	0	0
17.0	0	0	196	373	478	530	526	448	250	0	0	0
17.5	0	0	55	265	398	464	459	358	120	0	0	0
18.0	0	0	0	142	302	385	378	252	0	0	0	0
18.5	0	0	0	0	191	291	282	133	0	0	0	0
19.0	0	0	0	0	70	187	175	0	0	0	0	0
19.5	0	0	0	0	0	75	60	0	0	0	0	0

Table 2. Diffuse insolation under clear sky conditions on a horizontal surface on the 15th of each month (W/m^2).

Hour	J	F	M	A	M	J	J	A	S	O	N	D
4.0	0	0	0	0	0	20	9	0	0	0	0	0
4.5	0	0	0	0	25	44	32	0	0	0	0	0
5.0	0	0	0	5	48	64	53	22	0	0	0	0
5.5	0	0	0	31	70	84	74	47	8	0	0	0
6.0	0	0	0	55	90	102	93	69	34	0	0	0
6.5	0	0	24	76	107	118	110	89	57	19	0	0
7.0	0	0	48	95	123	133	125	106	78	43	0	0
7.5	0	27	69	111	137	146	139	121	95	63	24	0
8.0	16	47	87	125	149	158	151	134	109	79	43	15
8.5	34	61	101	137	160	168	162	145	122	93	59	34
9.0	51	79	114	147	169	177	171	155	132	104	72	49
9.5	64	91	124	156	177	185	180	164	141	113	82	61
10.0	74	100	132	163	183	192	187	171	147	120	90	70
10.5	81	107	138	168	188	197	192	176	153	124	95	76
11.0	86	112	142	172	192	200	196	180	156	127	98	80
11.5	89	115	144	173	193	202	198	182	157	128	98	82
12.0	89	116	145	173	193	202	198	182	157	127	97	81
12.5	87	113	141	170	190	199	195	180	154	122	93	77
13.0	81	109	139	166	185	195	191	176	149	117	86	71
13.5	75	104	132	160	179	188	186	170	142	110	78	64
14.0	66	96	125	153	172	181	179	163	135	101	68	53
14.5	55	87	116	144	163	173	171	155	125	89	54	40
15.0	40	74	105	133	153	164	162	145	114	76	38	24
15.5	22	59	92	121	142	153	151	134	101	59	19	2
16.0	0	40	75	107	129	141	139	120	85	39	0	0
16.5	0	13	56	90	114	128	126	106	66	17	0	0
17.0	0	0	34	71	98	113	111	89	45	0	0	0
17.5	0	0	9	49	80	97	95	70	21	0	0	0
18.0	0	0	0	26	59	78	76	48	0	0	0	0
18.5	0	0	0	0	37	58	58	25	0	0	0	0
19.0	0	0	0	0	13	37	34	0	0	0	0	0
19.5	0	0	0	0	0	14	11	0	0	0	0	0

Table 3. Insolation under permanent clear sky conditions on the 15th of each month (direct and diffuse, W/m²).

Horizontal surface

Hour	J	F	M	A	M	J	J	A	S	O	N	D
4.0	0	0	0	0	0	26	8	0	0	0	0	0
4.5	0	0	0	0	36	68	46	0	0	0	0	0
5.0	0	0	0	4	84	122	95	33	0	0	0	0
5.5	0	0	0	50	146	188	157	83	10	0	0	0
6.0	0	0	0	107	219	262	228	147	58	0	0	0
6.5	0	0	37	178	297	338	305	220	117	28	0	0
7.0	0	0	93	256	375	414	381	299	188	79	0	0
7.5	0	37	159	335	451	487	456	376	263	138	34	0
8.0	16	87	230	410	522	559	527	449	337	202	77	16
8.5	54	143	302	479	591	629	597	517	405	265	124	52
9.0	96	201	367	542	656	694	663	582	465	322	171	90
9.5	138	256	424	599	714	748	722	640	518	371	214	128
10.0	176	304	472	647	758	785	766	691	562	410	250	161
10.5	208	342	510	685	787	799	791	730	596	438	276	187
11.0	232	370	537	711	802	798	799	755	618	454	292	205
11.5	245	386	552	724	806	794	799	767	629	460	296	212
12.0	247	391	555	723	805	794	798	768	626	453	290	210
12.5	239	384	545	708	797	799	799	758	610	435	272	197
13.0	220	365	524	680	777	798	794	734	582	406	244	175
13.5	191	335	491	640	742	781	772	698	543	366	207	145
14.0	155	295	448	590	692	741	732	649	485	316	163	109
14.5	115	245	395	533	631	685	676	591	439	258	115	71
15.0	72	190	333	469	564	618	611	527	375	195	69	34
15.5	32	131	264	399	493	548	542	459	304	131	26	0
16.0	0	76	192	323	420	476	470	387	229	73	0	0
16.5	0	28	122	244	344	402	397	311	155	24	0	0
17.0	0	0	62	167	265	327	321	232	89	0	0	0
17.5	0	0	12	98	188	250	244	157	35	0	0	0
18.0	0	0	0	42	119	177	171	92	0	0	0	0
18.5	0	0	0	0	63	113	107	40	0	0	0	0
19.0	0	0	0	0	19	61	55	0	0	0	0	0
19.5	0	0	0	0	0	21	15	0	0	0	0	0

Vertical surface (s = 90°) north (a_s = 180°)

Hour	J	F	M	A	M	J	J	A	S	O	N	D
4.0	0	0	0	0	0	83	49	0	0	0	0	0
5.0	0	0	0	22	119	169	144	61	0	0	0	0
6.0	0	0	0	53	120	163	152	89	25	0	0	0
7.0	0	0	29	52	66	82	77	59	43	25	0	0
8.0	6	25	43	57	68	73	71	62	51	39	23	7
9.0	26	38	49	60	68	72	71	64	55	46	35	25
10.0	36	44	53	62	67	71	69	65	58	50	41	34
11.0	40	48	56	62	67	70	69	65	59	52	44	38
12.0	41	49	56	62	67	70	69	65	59	52	44	39
13.0	39	48	55	62	67	70	69	65	59	50	41	36
14.0	34	44	53	61	68	71	70	65	56	46	35	28
15.0	22	37	48	59	68	73	71	64	52	39	22	13
16.0	0	24	40	55	67	74	71	61	46	24	0	0
17.0	0	0	23	47	101	134	113	58	32	0	0	0
18.0	0	0	0	48	131	179	161	90	0	0	0	0
19.0	0	0	0	0	60	128	113	24	0	0	0	0

Tilted surface, slope 60° (s = 60°) north (a_s = 180°)

Hour	J	F	M	A	M	J	J	A	S	O	N	D
4.0	0	0	0	0	0	82	45	0	0	0	0	0
5.0	0	0	0	21	138	198	164	65	0	0	0	0
6.0	0	0	0	88	198	256	231	138	37	0	0	0
7.0	0	0	39	90	196	254	235	145	59	34	0	0
8.0	8	34	58	81	165	227	208	116	71	53	31	9
9.0	34	50	67	86	129	194	174	94	77	62	46	33
10.0	46	58	73	88	101	162	142	95	81	67	54	44
11.0	51	63	75	89	101	141	121	95	82	69	57	49
12.0	53	64	76	89	101	137	116	95	83	69	57	49
13.0	50	62	75	89	101	151	128	95	81	67	53	46
14.0	43	58	72	87	118	180	154	95	78	62	45	36
15.0	29	49	65	84	155	213	188	94	73	52	28	16
16.0	0	31	55	78	189	244	221	131	63	32	0	0
17.0	0	0	31	96	203	261	239	149	43	0	0	0
18.0	0	0	0	57	163	231	213	115	0	0	0	0
19.0	0	0	0	0	59	135	120	21	0	0	0	0

Tilted surface, slope 30° ($s = 30°$) north ($a_s = 180°$)

Hour	J	F	M	A	M	J	J	A	S	O	N	D
4.0	0	0	0	0	0	61	30	0	0	0	0	0
5.0	0	0	0	14	124	180	145	55	0	0	0	0
6.0	0	0	0	108	233	291	258	158	51	0	0	0
7.0	0	0	47	190	319	375	346	246	117	40	0	0
8.0	9	40	91	253	382	439	411	312	172	64	36	10
9.0	39	61	130	300	435	495	466	362	214	78	55	38
10.0	54	72	159	336	472	527	504	401	244	89	64	51
11.0	61	78	176	356	484	521	509	423	261	100	68	57
12.0	62	80	181	360	485	516	506	427	263	99	68	58
13.0	59	77	173	346	478	528	511	416	250	88	63	53
14.0	51	71	152	317	448	513	493	386	224	77	53	42
15.0	33	59	120	275	399	464	445	342	185	63	33	19
16.0	0	37	80	219	340	404	381	286	133	38	0	0
17.0	0	0	36	144	262	330	314	211	70	0	0	0
18.0	0	0	0	55	158	230	215	115	0	0	0	0
19.0	0	0	0	0	44	110	98	12	0	0	0	0

Vertical surface ($s = 90°$) east ($a_s = 90°$)

Hour	J	F	M	A	M	J	J	A	S	O	N	D
4.0	0	0	0	0	0	117	70	0	0	0	0	0
5.0	0	0	0	82	228	353	300	156	0	0	0	0
6.0	0	0	0	366	532	562	529	428	243	0	0	0
7.0	0	0	335	582	652	654	645	608	492	287	0	0
8.0	116	293	517	634	648	646	648	640	582	444	245	111
9.0	247	401	527	566	566	568	577	572	529	441	313	221
10.0	257	361	418	425	417	417	435	437	393	327	256	217
11.0	168	228	248	236	222	225	245	251	211	160	128	127
12.0	54	68	78	83	86	88	89	87	78	66	54	48
13.0	44	55	64	69	72	74	75	72	65	55	44	39
14.0	35	46	54	60	64	66	66	63	56	47	35	29
15.0	22	37	47	54	58	60	60	57	49	38	22	13
16.0	0	23	38	46	52	54	54	50	41	22	0	0
17.0	0	0	21	36	44	47	47	42	27	0	0	0
18.0	0	0	0	16	31	38	37	27	0	0	0	0
19.0	0	0	0	0	8	21	20	0	0	0	0	0

Appendix

Tilted surface, slope 60° ($s = 60°$) east ($a_s = 90°$)

Hour	J	F	M	A	M	J	J	A	S	O	N	D
4.0	0	0	0	0	0	112	64	0	0	0	0	0
5.0	0	0	0	73	285	358	300	149	0	0	0	0
6.0	0	0	0	364	560	606	562	435	234	0	0	0
7.0	0	0	329	621	739	760	736	664	510	281	0	0
8.0	106	291	552	740	807	823	809	764	660	475	244	103
9.0	255	437	626	745	801	820	813	769	676	530	346	229
10.0	300	451	582	672	718	731	738	703	603	472	334	258
11.0	248	366	462	534	565	566	585	569	468	344	240	199
12.0	138	225	296	353	386	397	412	391	292	182	107	93
13.0	60	78	114	157	199	232	241	200	107	77	61	53
14.0	46	63	76	86	93	98	98	92	79	63	47	38
15.0	29	49	62	73	80	84	84	78	66	49	28	16
16.0	0	30	48	61	68	73	73	66	53	29	0	0
17.0	0	0	26	46	56	62	61	54	34	0	0	0
18.0	0	0	0	20	40	48	48	35	0	0	0	0
19.0	0	0	0	0	10	27	25	0	0	0	0	0

Tilted surface, slope 30° ($s = 30°$) east ($a_s = 90°$)

Hour	J	F	M	A	M	J	J	A	S	O	N	D
4.0	0	0	0	0	0	78	41	0	0	0	0	0
5.0	0	0	0	44	209	273	224	103	0	0	0	0
6.0	0	0	0	268	444	485	450	331	166	0	0	0
7.0	0	0	240	500	636	670	638	549	397	204	0	0
8.0	69	214	446	657	760	790	764	693	569	385	182	68
9.0	198	363	566	735	832	865	844	771	651	485	294	181
10.0	269	429	600	752	843	865	858	795	663	501	330	236
11.0	270	416	567	708	778	776	788	753	616	451	299	226
12.0	214	345	479	607	673	673	685	656	517	354	219	166
13.0	124	236	352	466	546	577	580	522	380	228	115	82
14.0	56	111	207	306	387	439	437	363	229	95	56	45
15.0	34	59	78	150	227	279	277	205	85	59	32	19
16.0	0	34	57	74	88	137	134	84	63	33	0	0
17.0	0	0	29	53	68	76	75	64	38	0	0	0
18.0	0	0	0	23	46	57	56	39	0	0	0	0
19.0	0	0	0	0	12	31	29	0	0	0	0	0

Vertical surface (s = 90°) south ($a_s = 0°$)

Hour	J	F	M	A	M	J	J	A	S	O	N	D
4.0	0	0	0	0	0	11	4	0	0	0	0	0
5.0	0	0	0	2	28	34	30	14	0	0	0	0
6.0	0	0	0	37	50	52	49	42	33	0	0	0
7.0	0	0	111	139	102	72	70	107	157	137	0	0
8.0	91	195	289	291	238	191	192	246	322	331	217	100
9.0	290	402	466	436	370	316	320	384	473	510	418	293
10.0	471	572	602	555	475	411	423	500	590	635	569	462
11.0	585	675	686	627	527	448	472	570	658	698	644	557
12.0	615	707	709	640	532	451	479	585	667	696	640	568
13.0	561	668	669	592	496	434	460	547	614	630	558	493
14.0	426	558	571	491	405	363	388	455	509	502	401	339
15.0	233	382	422	355	278	245	270	327	366	320	198	145
16.0	0	174	239	202	141	117	139	186	204	128	0	0
17.0	0	0	73	64	156	160	161	57	59	0	0	0
18.0	0	0	0	18	35	42	42	31	0	0	0	0
19.0	0	0	0	0	8	22	20	0	0	0	0	0

Tilted surface, slope 60° (s = 60°) south ($a_s = 0°$)

Hour	J	F	M	A	M	J	J	A	S	O	N	D
4.0	0	0	0	0	0	15	5	0	0	0	0	0
5.0	0	0	0	3	37	45	40	19	0	0	0	0
6.0	0	0	0	63	84	75	68	62	50	0	0	0
7.0	0	0	134	232	255	238	223	223	217	151	0	0
8.0	85	205	353	440	445	421	406	418	432	376	220	93
9.0	292	439	573	631	627	598	586	604	626	590	438	293
10.0	487	636	743	786	770	726	728	759	775	741	607	471
11.0	612	757	847	881	836	765	787	851	863	817	692	575
12.0	646	795	876	898	843	765	793	871	874	815	688	587
13.0	586	749	826	835	798	753	773	821	806	735	594	505
14.0	438	619	703	702	675	662	680	698	772	580	419	341
15.0	232	416	518	524	501	498	517	527	489	364	200	139
16.0	0	183	291	320	311	315	333	335	274	140	0	0
17.0	0	0	87	124	129	139	154	146	86	0	0	0
18.0	0	0	0	24	47	57	57	42	0	0	0	0
19.0	0	0	0	0	11	29	27	0	0	0	0	0

Tilted surface, slope 30° (s = 30°) south ($a_s = 0°$)

Hour	J	F	M	A	M	J	J	A	S	O	N	D
4.0	0	0	0	0	0	17	5	0	0	0	0	0
5.0	0	0	0	3	44	54	47	22	0	0	0	0
6.0	0	0	0	92	165	183	158	113	59	0	0	0
7.0	0	0	126	274	354	365	338	293	227	129	0	0
8.0	57	165	331	482	549	555	529	492	436	328	168	62
9.0	220	364	536	669	731	735	711	675	622	520	347	217
10.0	378	536	694	819	873	862	852	828	764	657	489	360
11.0	481	644	791	910	936	893	906	918	847	727	562	445
12.0	510	678	818	927	942	890	909	937	857	725	558	454
13.0	459	636	771	866	900	886	895	889	793	651	478	387
14.0	338	521	657	737	779	800	805	768	666	511	331	256
15.0	172	344	484	565	605	634	641	600	491	317	152	98
16.0	0	146	273	363	412	446	453	408	284	119	0	0
17.0	0	0	83	161	218	258	263	210	96	0	0	0
18.0	0	0	0	27	65	96	99	58	0	0	0	0
19.0	0	0	0	0	13	34	32	0	0	0	0	0

Vertical vlak (s = 90°) west ($a_s = -90°$)

Hour	J	F	M	A	M	J	J	A	S	O	N	D
4.0	0	0	0	0	0	11	4	0	0	0	0	0
5.0	0	0	0	2	26	32	28	13	0	0	0	0
6.0	0	0	0	29	41	44	42	35	20	0	0	0
7.0	0	0	27	43	50	52	50	46	38	24	0	0
8.0	6	25	40	51	56	58	57	53	47	38	23	7
9.0	26	38	49	57	62	63	62	59	54	47	36	26
10.0	37	47	56	65	69	70	69	67	63	56	45	37
11.0	47	56	67	76	81	82	80	77	74	67	55	46
12.0	75	77	106	148	162	145	130	134	157	169	141	98
13.0	100	244	297	349	362	336	323	334	347	335	265	201
14.0	266	370	455	514	529	512	500	502	499	445	312	230
15.0	222	397	539	615	631	621	613	610	579	440	231	147
16.0	0	274	485	623	661	661	657	641	535	274	0	0
17.0	0	0	262	476	583	616	614	550	321	0	0	0
18.0	0	0	0	198	368	453	447	315	0	0	0	0
19.0	0	0	0	0	109	211	200	57	0	0	0	0

Tilted surface, slope 60° ($s = 60°$) west ($a_s = -90°$)

Hour	J	F	M	A	M	J	J	A	S	O	N	D
4.0	0	0	0	0	0	14	5	0	0	0	0	0
5.0	0	0	0	2	33	41	36	17	0	0	0	0
6.0	0	0	0	37	52	57	54	44	26	0	0	0
7.0	0	0	34	55	65	68	66	59	49	31	0	0
8.0	8	32	53	67	76	79	77	71	62	50	30	9
9.0	35	50	66	80	89	92	89	83	75	64	48	34
10.0	50	64	80	96	144	158	133	99	91	78	62	49
11.0	64	82	163	270	331	329	310	281	241	191	120	65
12.0	72	241	343	461	513	492	480	470	423	352	251	175
13.0	270	377	499	621	649	666	652	633	571	477	339	250
14.0	298	455	602	722	785	794	779	740	663	530	342	246
15.0	222	429	620	753	812	830	819	776	676	468	229	141
16.0	0	269	506	689	769	796	790	736	567	268	0	0
17.0	0	0	252	486	626	685	680	582	316	0	0	0
18.0	0	0	0	188	370	471	463	311	0	0	0	0
19.0	0	0	0	0	102	208	196	49	0	0	0	0

Tilted surface, slope 30° ($s = 30°$) west ($a_s = -90°$)

Hour	J	F	M	A	M	J	J	A	S	O	N	D
4.0	0	0	0	0	0	16	5	0	0	0	0	0
5.0	0	0	0	3	38	47	41	19	0	0	0	0
6.0	0	0	0	42	62	68	64	51	29	0	0	0
7.0	0	0	38	66	81	87	83	72	57	35	0	0
8.0	9	37	63	91	183	216	188	123	78	60	35	10
9.0	41	61	101	238	338	371	340	270	188	101	58	40
10.0	61	123	246	399	501	524	499	432	339	235	125	61
11.0	151	248	389	551	639	635	625	583	482	360	228	147
12.0	234	355	507	672	749	730	726	702	594	455	304	215
13.0	276	421	581	741	830	835	824	779	656	502	330	239
14.0	257	426	598	749	843	877	863	793	660	482	286	201
15.0	167	351	543	698	786	828	817	744	599	377	169	99
16.0	0	196	397	577	679	726	720	641	453	193	0	0
17.0	0	0	178	371	508	577	571	464	230	0	0	0
18.0	0	0	0	130	278	369	361	229	0	0	0	0
19.0	0	0	0	0	69	152	142	28	0	0	0	0

Table 4. Beam and diffuse solar gains under permanent clear sky conditions (MJ/(m^2 · month)).

Surface	J	F	M	A	M	J	J	A	S	O	N	D
Hor.	136	232	424	610	787	823	825	705	497	324	168	111
North												
$s = 30°$	42	59	150	352	564	643	627	461	236	82	48	36
$s = 60°$	35	48	77	120	236	330	301	171	90	63	41	31
$s = 90°$	28	37	56	81	134	170	161	105	64	47	31	24
E + W												
$s = 30°$	136	226	406	572	731	761	764	659	471	313	167	113
$s = 60°$	129	207	359	484	604	623	627	552	408	281	156	109
$s = 90°$	102	159	268	345	419	428	768	388	297	213	122	88
South												
$s = 30°$	292	410	618	742	844	829	849	801	660	521	334	252
$s = 60°$	377	489	662	695	708	656	686	709	663	591	419	333
$s = 90°$	365	442	538	479	415	352	380	450	500	511	396	327
SE + SW												
$s = 30°$	242	353	554	695	817	817	832	765	605	457	282	209
$s = 60°$	292	392	564	641	700	673	695	679	586	487	329	258
$s = 90°$	268	336	444	456	459	422	442	464	440	401	295	242
NE + NW												
$s = 30°$	52	106	245	422	611	674	663	521	317	167	68	42
$s = 60°$	40	69	177	270	405	456	446	340	201	106	49	34
$s = 90°$	29	49	104	175	259	294	287	220	133	73	36	26

Table 5. Beam, diffuse and reflected solar gains under permanent clear sky conditions (MJ/(m^2 · month), albedo 0.2).

Surface	J	F	M	A	M	J	J	A	S	O	N	D
Hor.	136	232	424	610	787	823	825	705	497	324	168	111
North												
$s = 30°$	43	62	156	360	574	654	638	470	242	86	50	37
$s = 60°$	42	60	98	150	276	371	342	206	115	79	49	36
$s = 90°$	41	60	99	142	213	252	243	176	114	80	48	35
E + W												
$s = 30°$	138	229	412	580	741	772	775	668	478	317	169	115
$s = 60°$	135	218	380	515	644	664	669	587	433	297	164	115
$s = 90°$	116	182	310	406	498	510	515	459	347	245	139	99
South												
$s = 30°$	294	414	624	750	854	840	860	811	667	526	337	254
$s = 60°$	384	500	683	725	747	697	728	744	688	608	427	338
$s = 90°$	379	465	580	540	494	434	463	521	550	544	413	338
SE + SW												
$s = 30°$	244	356	560	703	828	828	843	774	612	461	284	210
$s = 60°$	299	404	585	671	740	714	736	714	611	503	338	263
$s = 90°$	283	360	487	517	537	504	525	534	490	433	312	253
NE + NW												
$s = 30°$	54	110	251	430	622	685	674	531	323	172	71	43
$s = 60°$	46	81	176	300	444	497	487	375	226	122	57	39
$s = 90°$	43	73	146	236	338	376	369	291	182	106	53	37